Institute for Mathematics and
its Applications
IMA

e Institute for Mathematics and its Applications was established by
t from the National Science Foundation to the University of Minnesota in
The IMA seeks to encourage the development and study of fresh mathemat-
ncepts and questions of concern to the other sciences by bringing together
maticians and scientists from diverse fields in an atmosphere that will stim-
discussion and collaboration.

e IMA Volumes are intended to involve the broader scientific community in
rocess.

Avner Friedman, Director
Willard Miller, Jr., Associate Director

* * * * * * * * * *

IMA PROGRAMS

* * * * * * * * * *

SPRINGER LECTURE NOTES FROM THE IMA:

D1067299

The IMA Volu
in Mathemat
and Its Applica

Volume 29

Series Editors
Avner Friedman   Willard

Th
a gram
1982.
ical co
mathe
ulate

T
this p

1982-
1983-

1984-
1985-
1986
1987
1988
1989
1990

The

Ori

Nev

M

James Glimm   Andrew J. Majda

Editors

Multidimensional Hyperbolic Problems and Computations

With 86 Illustrations

Springer-Verlag

New York Berlin Heidelberg London
Paris Tokyo Hong Kong Barcelona

James Glimm
Department of Applied
  Mathematics and Statistics
SUNY at Stony Brook
Stony Brook, NY 11794-3600

Andrew J. Majda
Department of Mathematics
  and Program in Applied and
  Computational Mathematics
Princeton University
Princeton, NJ 08544

Series Editors
Avner Friedman
Willard Miller, Jr.
Institute for Mathematics and its Applications
University of Minnesota
Minneapolis, Minnesota 55455
USA

Mathematics Subject Classification: 35, 69, 76, 80.

Printed on acid-free paper.

Camera-ready copy prepared by the IMA.
Printed and bound by Edwards Brothers, Inc., Ann Arbor, Michigan.
Printed in the United States of America.

9 8 7 6 5 4 3 2 1

ISBN 0-387-97485-7 Springer-Verlag New York Berlin Heidelberg
ISBN 3-540-97485-7 Springer-Verlag Berlin Heidelberg New York

Dedication

Ron DiPerna (1947 – 1989)

Ron DiPerna was a uniquely talented mathematician. He was a leading researcher of his generation in the mathematical theory of systems of hyperbolic conservation laws, incompressible flow, and the kinetic theory of gases. Ron DiPerna died tragically in January 1989 after a courageous struggle with cancer. His work and its impact was known to virtually all of the several hundred participants in this meeting. For many of us, he was a warm and loyal friend with a sharp wit and keen sense of humor. He left us too soon and at the height of his creative power.

The IMA Volumes
in Mathematics and its Applications

Current Volumes:

Forthcoming Volumes:

FOREWORD

This IMA Volume in Mathematics and its Applications

MULTIDIMENSIONAL HYPERBOLIC PROBLEMS AND COMPUTATIONS

is based on the proceedings of a workshop which was an integral part of the 1988-89 IMA program on NONLINEAR WAVES. We are grateful to the Scientific Committee: James Glimm, Daniel Joseph, Barbara Keyfitz, Andrew Majda, Alan Newell, Peter Olver, David Sattinger and David Schaeffer for planning and implementing an exciting and stimulating year-long program. We especially thank the Workshop Organizers, Andrew Majda and James Glimm, for bringing together many of the major figures in a variety of research fields connected with multidimensional hyperbolic problems.

Avner Friedman

Willard Miller

PREFACE

A primary goal of the IMA workshop on Multidimensional Hyperbolic Problems and Computations from April 3–14, 1989 was to emphasize the interdisciplinary nature of contemporary research in this field involving the combination of ideas from the theory of nonlinear partial differential equations, asymptotic methods, numerical computation, and experiments. The twenty-six papers in this volume span a wide cross-section of this research including some papers on the kinetic theory of gases and vortex sheets for incompressible flow in addition to many papers on systems of hyperbolic conservation laws. This volume includes several papers on asymptotic methods such as nonlinear geometric optics, a number of articles applying numerical algorithms such as higher order Godunov methods and front-tracking to physical problems along with comparison to experimental data, and also several interesting papers on the rigorous mathematical theory of shock waves. In addition, there are at least two papers in this volume devoted to open problems with this interdisciplinary emphasis.

The organizers would like to thank the staff of the IMA for their help with the details of the meeting and also for the preparation of this volume. We are especially grateful to Avner Friedman and Willard Miller for their help with the organization of the special day of the meeting in memory of Ron Diperna, Tuesday, April 4, on short notice.

James Glimm

Andrew J. Majda

CONTENTS

MACROSCOPIC LIMITS OF KINETIC EQUATIONS

CLAUDE BARDOS, FRANÇOIS GOLSE* AND DAVID LEVERMORE†

Abstract. The connection between kinetic theory and the macroscopic equations of fluid dynamics is described. In particular, our results concerning the incompressible Navier-Stokes equation are compared with the classical derivation of Hilbert and Chapman-Enskog. Some indications of the validity of these limits are given. More specifically, the connection between the DiPerna-Lions renormalized solution for the Boltzmann equation and the Leray-Hopf solution for the Navier-Stokes equation is considered.

I. Introduction. This paper is devoted to the connection between kinetic theory and macroscopic fluid dynamics. Formal limits are systematically derived and, in some cases, rigorous results are given concerning the validity of these limits. To do that several scalings are introduced for standard kinetic equations of the form

$$(1) \qquad \partial_t F_\epsilon + v \cdot \nabla_x F_\epsilon = \frac{1}{\epsilon} C(F_\epsilon).$$

Here $F_\epsilon(t, x, v)$ is a nonnegative function representing the density of particles with position x and velocity v in the single particle phase space $\mathbf{R}_x^3 \times \mathbf{R}_v^3$ at time t. The interaction of particles through collisions is modelled by the operator $C(F)$; this operator acts only on the variable v and is generally nonlinear. In section V the classical Boltzmann form of the operator will be considered.

The connection between kinetic and macroscopic fluid dynamics results from two types of properties of the collision operator:

(i) conservation properties and an entropy relation that implies that the equilibria are Maxwellian distributions for the zeroth order limit;

(ii) the derivative of $C(F)$ satisfies a formal Fredholm alternative with a kernel related to the conservation properties of (i).

The macroscopic limit is obtained when the fluid becomes dense enough that particles undergo many collisions over the scales of interest. This situation is described by the introduction of a small parameter ϵ, called the Knudsen number, that represents the ratio of the mean free path of particles between collisions to some characteristic length of the flow (e.g. the size of an obstacle). Properties (i) are sufficient to derive the compressible Euler equations from equation (1); they arise as the leading order dynamics from a formal expansion of F in ϵ (the Chapman-Enskog or Hilbert expansion described briefly in section III). Properties (ii) are used to obtain the Navier-Stokes equations; they depend on a more detailed knowledge of the collision operator. The compressible Navier-Stokes equations arise as corrections to those of Euler at the next order in the Chapman-Enskog expansion.

In a compressible fluid one also introduces the Mach Number Ma which is the ratio of the bulk velocity to the sound speed and the Reynolds number Re which is

*Département de Mathématiques, Université Paris VII , 75251 Paris Cédex 05, France
†Department of Mathematics, University of Arizona, Tucson, Arizona 85721, USA

a dimensionless reciprocal viscosity of the fluid. These numbers (cf.[LL] and [BGL]) are related by the formula

$$(2) \qquad \epsilon = \frac{Ma}{Re} \, .$$

Our main contribution concerns the incompressible limit; due to relation (2) it is the only case where one obtains, when ϵ goes to zero, an equation with a finite Reynolds number. This is the only regime where global weak solutions of fluid dynamic equations are known to exist. Related results have been obtained simultaneously by A. De Masi, R. Esposito and J.L. Lebowitz [DMEL]. Our considerations on the relation between the renormalized solution of the Boltzmann equation and the eray [L] solution of the Navier-Stokes equations rely on the pioneering work of DiPerna and Lions [DiPL], giving one more example of the importance of Ron DiPerna's influence in our community.

II. The Compressible Euler Limit. In this section the integral of any scalar or vector valued function $f(v)$ with respect to the variable v will be denote by $\langle f \rangle$;

$$(3) \qquad \langle f \rangle = \int f(v) \, dv \, .$$

The operator C is assumed to satisfy the conservation properties

$$(4) \qquad \langle C(F) \rangle = 0 \, , \quad \langle v C(F) \rangle = 0 \, , \quad \langle |v|^2 C(F) \rangle = 0 \, .$$

These relations represent the physical laws of mass, momentum and energy conservation during collisions and imply the local conservation laws

$$\partial_t \langle F \rangle + \nabla_x \cdot \langle v F \rangle = 0 \, ,$$

$$(5) \qquad \partial_t \langle v F \rangle + \nabla_x \cdot \langle v \otimes v \, F \rangle = 0 \, ,$$

$$\partial_t \langle \tfrac{1}{2} |v|^2 F \rangle + \nabla_x \cdot \langle v \tfrac{1}{2} |v|^2 F \rangle = 0 \, .$$

Additionally, $C(F)$ is assumed to have the property that the quantity $\langle C(F) \log F \rangle$ is nonpositive. This is the entropy dissipation rate and implies the local entropy inequality

$$(6) \qquad \partial_t \langle F \log F \rangle + \nabla_x \cdot \langle v F \log F \rangle = \langle C(F) \log F \rangle \le 0 \, .$$

Finally, the equilibria of $C(F)$ are assumed to be characterized by the vanishing of the entropy dissipation rate and given by the class of Maxwellian distributions, i.e. those of the form

$$(7) \qquad F = \frac{\rho}{(2\pi\theta)^{3/2}} \exp\left(-\frac{1}{2} \frac{|v - u|^2}{\theta} \right) \, .$$

More precisely, for every nonnegative measurable function F the following properties are equivalent:

(8)
$$\text{i)} \quad C(F) = 0\,,$$
$$\text{ii)} \quad \langle C(F) \log F \rangle = 0\,,$$
$$\text{iii)} \quad F \text{ is a Maxwellian with the form (7).}$$

These assumptions about $C(F)$ merely abstract some of the consequences of Boltzmann's celebrated H-theorem.

The parameters ρ, u and θ introduced in the right side of (7) are related to the fluid dynamic moments giving the mass, momentum and energy densities:

$$\langle F \rangle = \rho\,, \qquad \langle vF \rangle = \rho u\,, \qquad \langle \tfrac{1}{2}|v|^2 F \rangle = \rho(\tfrac{1}{2}|u|^2 + \tfrac{3}{2}\theta)\,.$$

They are called respectively the (mass) density, velocity and temperature of the fluid. In the compressible Euler limit, these variables are shown to satisfy the system of compressible Euler equations (11 below).

The main obstruction to proving the validity of this fluid dynamical limit is the fact that solutions of the compressible Euler equations generally become singular after a finite time (cf. Sideris [S]). Therefore any global (in time) convergence proof cannot rely on uniform regularity estimates. The only reasonable assumptions would be that the limiting distribution exists and that the relevant moments converge pointwise. With this hypothesis, it is shown that the above assumptions regarding $C(F)$ imply that the fluid dynamic moments of solutions converge to a solution of the Euler equations that satisfies the macroscopic entropy inequality.

THEOREM I. *Given a collision operator C with properties (i), let $F_\epsilon(t, x, v)$ be a sequence of nonnegative solutions of the equation*

(9)
$$\partial_t F_\epsilon + v \cdot \nabla_x F_\epsilon = \frac{1}{\epsilon} C(F_\epsilon)\,,$$

such that, as ϵ goes to zero, F_ϵ converges almost everywhere to a nonnegative function F. Moreover, assume that the moments

$$\langle F_\epsilon \rangle\,, \qquad \langle vF_\epsilon \rangle\,, \qquad \langle v \otimes v\, F_\epsilon \rangle\,, \qquad \langle v|v|^2 F_\epsilon \rangle\,,$$

converge in the sense of distributions to the corresponding moments

$$\langle F \rangle\,, \qquad \langle vF \rangle\,, \qquad \langle v \otimes v\, F \rangle\,, \qquad \langle v|v|^2 F \rangle\,;$$

the entropy densities and fluxes converge in the sense of distributions according to

$$\lim_{\epsilon \to 0} \langle F_\epsilon \log F_\epsilon \rangle = \langle F \log F \rangle\,, \quad \lim_{\epsilon \to 0} \langle vF_\epsilon \log F_\epsilon \rangle = \langle vF \log F \rangle\,;$$

while the entropy dissipation rates satisfy

$$\limsup_{\epsilon \to 0} \langle C(F_\epsilon) \log F_\epsilon \rangle \le \langle C(F) \log F \rangle\,.$$

Then the limit $F(t, x, v)$ is a Maxwellian distribution,

$$(10) \qquad F(t, x, v) = \frac{\rho(t, x)}{(2\pi\theta(t, x))^{3/2}} \exp\left(-\frac{1}{2}\frac{|v - u(t, x)|^2}{\theta(t, x)}\right),$$

where the functions ρ, u and θ solve the compressible Euler equations

$$\partial_t \rho + \nabla_x \cdot (\rho u) = 0,$$

$$(11) \qquad \partial_t(\rho u) + \nabla_x \cdot (\rho u \otimes u) + \nabla_x(\rho\theta) = 0,$$

$$\partial_t(\rho(\tfrac{1}{2}|u|^2 + \tfrac{3}{2}\theta)) + \nabla_x \cdot (\rho u(\tfrac{1}{2}|u|^2 + \tfrac{5}{2}\theta)) = 0.$$

and satisfy the entropy inequality

$$(12) \qquad \partial_t\left(\rho \log\left(\frac{\rho^{2/3}}{\theta}\right)\right) + \nabla_x \cdot \left(\rho u \log\left(\frac{\rho^{2/3}}{\theta}\right)\right) \leq 0.$$

Remark 1. The above theorem shows that any type of equation of the form (9) leads to the compressible Euler equations with a pressure p given by the ideal gas law $p = \rho\theta$ and an internal energy of $\frac{3}{2}\rho\theta$ (corresponding to a $\frac{5}{3}$-law perfect gas). This is a consequence of the fact that the kinetic equation considered here describes a monoatomic fluid in a three dimensional domain. Other equations of state may be obtained by introducing additional degrees of freedom that take into account the rotational and vibrational modes of the particles.

The proof of this theorem, as well as those of subsequent ones, can be found in our paper [BGL2].

III. The Compressible Navier-Stokes Limit. As has been noticed above, the form of the limiting Euler equation is independent of the choice of the collision operator C within the class of operators satisfying the conservation and the entropy properties. The choice of the collision operator appears at the macroscopic level only in the construction of the Navier-Stokes limit. The compressible Navier-Stokes equations are obtained by the classical Chapman-Enskog expansion. To compare this approach with the situation leading to the incompressible Navier-Stokes equation, a short description of this approach is given below.

Given (ρ, u, θ), denote the corresponding Maxwellian distribution by

$$(13) \qquad M_{(\rho, u, \theta)} = \frac{\rho}{(2\pi\theta)^{3/2}} \exp\left(-\frac{1}{2}\frac{|v - u|^2}{\theta}\right).$$

The subscript (ρ, u, θ) will often be omitted when it is convenient. Introduce the Hilbert space L_M^2 defined by the scalar product

$$(14) \qquad (f|g)_M = \langle fg \rangle_M = \int f(v)g(v)\, M(v)\, dv.$$

Denote by L and Q the first two Fréchet derivatives of the operator $G \mapsto M^{-1}C(MG)$ at $G = 1$:

$$(15) \qquad L(g) = \frac{1}{M}DC(M) \cdot (Mg), \qquad Q(g,g) = \frac{1}{M}D^2C(M):(Mg \vee Mg).$$

Taylor's formula then gives

$$(16) \qquad \frac{1}{M}C(M(1+\epsilon g)) = \epsilon L(g) + \epsilon^2 \tfrac{1}{2}Q(g,g) + O(\epsilon^3).$$

The linear operator L is assumed to be self-adjoint and to satisfy a Fredholm alternative in the space L^2_M with a five dimensional kernel spanned by the functions $\{1, v_1, v_2, v_3, |v|^2\}$. The fact that L must be nonpositive definite follows directly from examining the second variation of the entropy dissipation rate at M.

Denote by V, $A = \{A_i\}$ and $B = \{B_{ij}\}$ the following vectors and tensors:

$$V = \frac{v - u}{\sqrt{\theta}}, \qquad A(V) = (\tfrac{1}{2}|V|^2 - \tfrac{5}{2})V, \qquad B(V) = V \otimes V - \tfrac{1}{3}|V|^2 I.$$

By symmetry, the functions A_i and B_{ij} are orthogonal to the kernel of L; therefore the equations

$$(17) \qquad L(A') = A, \qquad L(B') = B,$$

have unique solutions in $\mathrm{Ker}(L)^\perp$. Assume (this would be a consequence of rotational invariance for the collision operator) that these solutions are given by the formulas

$$(18) \qquad A'(V) = -\alpha(\rho, \theta, |V|)\, A(V), \qquad B'(V) = -\beta(\rho, \theta, |V|)\, B(V),$$

where α and β are positive functions depending on ρ, θ and $|V|$. If $C(F)$ homogeneous of degree two (for example, quadratic) then a simple scaling shows that the ρ dependence of α and β is just proportionality to ρ^{-1}.

A function $H_\epsilon(t, x, v)$ is said to be an approximate solution of order p to the kinetic equation (1) if

$$(19) \qquad \partial_t H_\epsilon + v \cdot \nabla_x H_\epsilon = \frac{1}{\epsilon}C(H_\epsilon) + O(\epsilon^p),$$

where $O(\epsilon^p)$ denotes a term bounded by ϵ^p in some convenient norm. An approximate solution of order two can be constructed in the form

$$(20) \qquad H_\epsilon = M_\epsilon(1 + \epsilon g_\epsilon + \epsilon^2 w_\epsilon),$$

where $(\rho_\epsilon, u_\epsilon, \theta_\epsilon)$ solve the compressible Navier-Stokes equations with dissipation of the order ϵ (denoted $CNSE_\epsilon$):

$$\partial_t \rho_\epsilon + \nabla_x \cdot (\rho_\epsilon u_\epsilon) = 0,$$

(21) $$\rho_\epsilon(\partial_t + u_\epsilon \cdot \nabla_x)u_\epsilon + \nabla_x(\rho_\epsilon \theta_\epsilon) = \epsilon \nabla_x \cdot [\mu_\epsilon \sigma(u_\epsilon)],$$

$$\tfrac{3}{2}\rho_\epsilon(\partial_t + u_\epsilon \cdot \nabla_x)\theta_\epsilon + \rho_\epsilon \theta_\epsilon \nabla_x \cdot u_\epsilon = \epsilon \tfrac{1}{2}\mu_\epsilon \sigma(u_\epsilon) : \sigma(u_\epsilon) + \epsilon \nabla_x \cdot [\kappa_\epsilon \nabla_x \theta_\epsilon].$$

In these equations $\sigma(u)$ denotes the strain-rate tensor given by

$$\sigma_{ij}(u) = (u^i_{x_j} + u^j_{x_i}) - \tfrac{2}{3}\nabla_x \cdot u\, \delta_{ij},$$

while the viscosity $\mu_\epsilon = \mu(\rho_\epsilon, \theta_\epsilon)$ and the thermal diffusivity $\kappa_\epsilon = \kappa(\rho_\epsilon, \theta_\epsilon)$ are defined by the relations

(22)

$$\mu(\rho, \theta) = \tfrac{1}{10}\theta\langle\beta(\rho, \theta, |V|)|B(V)|^2\rangle_M = \tfrac{2}{15}\rho\theta\frac{1}{\sqrt{2\pi}}\int_0^\infty \beta(\rho, \theta, r)r^6 e^{-\frac{1}{2}r^2}\,dr,$$

$$\kappa(\rho, \theta) = \tfrac{1}{3}\theta\langle\alpha(\rho, \theta, |V|)|A(V)|^2\rangle_M = \tfrac{1}{6}\rho\theta\frac{1}{\sqrt{2\pi}}\int_0^\infty \alpha(\rho, \theta, r)(r^2 - 5)^2 r^4 e^{-\frac{1}{2}r^2}\,dr.$$

Notice that in the case where $C(F)$ is homogeneous of degree two, the left sides become independent of ρ; this is why classical expressions for the viscosity and thermal diffusivity depend only on θ.

The Chapman-Enskog derivation can be formulated according to the following.

THEOREM II. *Assume that* $(\rho_\epsilon, u_\epsilon, \theta_\epsilon)$ *solve the* $CNSE_\epsilon$ *with the viscosity* $\mu(\rho, \theta)$ *and thermal diffusivity* $\kappa(\rho, \theta)$ *given by* (22). *Then there exist* g_ϵ *and* w_ϵ *in* $\mathrm{Ker}(L)^\perp$ *such that* H_ϵ, *given by* (20), *is an approximate solution of order two to equation* (1). *Moreover,* g_ϵ *is given by the formula*

(23) $$g_\epsilon = -\tfrac{1}{2}\beta(\rho_\epsilon, \theta_\epsilon, |V|)B(V):\sigma(u_\epsilon) - \alpha(\rho_\epsilon, \theta_\epsilon, |V|)\frac{A(V)\cdot\nabla_x\theta_\epsilon}{\sqrt{\theta_\epsilon}}.$$

Remark 2. Let F_ϵ be a solution of the kinetic equation that coincides with a local Maxwellian at $t = 0$. Let $(\rho_\epsilon, u_\epsilon, \theta_\epsilon)$ be the solution of the $CNSE_\epsilon$ with initial data equal to the corresponding moments of $F_\epsilon(0, x, v)$. Then the expression given by

$$H_\epsilon = \frac{1}{(2\pi\theta_\epsilon(t, x))^{3/2}} \exp\left(-\frac{1}{2}\frac{|v - u_\epsilon(t, x)|^2}{\theta_\epsilon(t, x)}\right)(\rho_\epsilon(t, x) + \epsilon g_\epsilon + \epsilon^2 w_\epsilon)$$

is an approximation of order two of F_ϵ. Since $M_\epsilon g_\epsilon$ is orthogonal to the functions $1, v, |v|^2$, the quantities $\rho_\epsilon, \rho_\epsilon u_\epsilon$ and $\rho_\epsilon(\tfrac{1}{2}|u_\epsilon|^2 + \tfrac{3}{2}\theta_\epsilon)$ provide approximations of order two to the corresponding moments of F_ϵ. In fact, this observation was used to do the Chapman-Enskog derivation by the so called projection method (cf. Caflisch [C]).

IV. The Incompressible Navier-Stokes Limit. The purpose of this section is to construct a connection between the kinetic equation and the incompressible Navier-Stokes equations. As in the previous section, this will describe the range of parameters for which the incompressible Navier-Stokes equations provide a good approximation to the solution of the Boltzmann equation. However in this case the connection is drawn between the Boltzmann equation and macroscopic fluid dynamic equations with a finite Reynolds number. It is clear from formula (2), $\epsilon = Ma/Re$, that in order to obtain a fluid dynamic regime (corresponding to a vanishing Knudsen number) with a finite Reynolds number, the Mach number must vanish (cf. [LL] or [BGL1]).

In order to realize distributions with a small Mach number it is natural to consider them as perturbations about a given absolute Maxwellian (constant in space and time). By the proper choice of Galilean frame and dimensional units this absolute Maxwellian can be taken to have velocity equal to 0, and density and temperature equal to 1; it will be denoted by M. The initial data $F_\epsilon(0, x, v)$ is assumed to be close to M where the order of the distance will be measured with the Knudsen number. Furthermore, if the flow is to be incompressible, the kinetic energy of the flow in the acoustic modes must be smaller than that in the rotational modes. Since the acoustic modes vary on a faster timescale than rotational modes, they may be suppressed by assuming that the initial data is consistent with motion on a slow timescale; this scale separation will also be measured with the Knudsen number.

This scaling is quantified by the introduction of a small parameter ϵ such that the timescale considered is of order ϵ^{-1}, the Knudsen number is of order ϵ^q, and the distance to the absolute Maxwellian M is of order ϵ^r with q and r being greater or equal to one. Thus, solutions F_ϵ to the equation

$$(24) \qquad \epsilon \partial_t F_\epsilon + v \cdot \nabla_x F_\epsilon = \frac{1}{\epsilon^q} C(F_\epsilon),$$

are sought in the form

$$(25) \qquad F_\epsilon = M(1 + \epsilon^r g_\epsilon).$$

The basic case $r = q = 1$ is the unique scaling compatible with the usual incompressible Navier-Stokes equations.

The notation introduced in the previous section regarding the collision operator and its Fréchet derivatives is conserved but here the Maxwellian M is absolute so that L and Q no longer depend on the fluid variables.

THEOREM III. *Let $F_\epsilon(t, x, v)$ be a sequence of nonnegative solutions to the scaled kinetic equation (24) such that, when it is written according to formula (25), the sequence g_ϵ converges in the sense of distributions and almost everywhere to a function g as ϵ goes to zero. Furthermore, assume the moments*

$$\langle g_\epsilon \rangle_M, \qquad \langle v g_\epsilon \rangle_M, \qquad \langle v \otimes v g_\epsilon \rangle_M, \qquad \langle v |v|^2 g_\epsilon \rangle_M,$$

$$\langle L^{-1}(A(v)) g_\epsilon \rangle_M, \quad \langle L^{-1}(A(v)) \otimes v g_\epsilon \rangle_M, \quad \langle L^{-1}(A(v)) Q(g_\epsilon, g_\epsilon) \rangle_M,$$

$$\langle L^{-1}(B(v)) g_\epsilon \rangle_M, \quad \langle L^{-1}(B(v)) \otimes v g_\epsilon \rangle_M, \quad \langle L^{-1}(B(v)) Q(g_\epsilon, g_\epsilon) \rangle_M$$

converge in $D'\,(\mathbf{R}_t^+ \times \mathbf{R}_x^3)$ to the corresponding moments

$$\langle g \rangle_M, \qquad \langle v g \rangle_M, \qquad \langle v \otimes v g \rangle_M, \qquad \langle v |v|^2 g \rangle_M,$$

$$\langle L^{-1}(A(v))\,g\rangle_M\,,\quad \langle L^{-1}(A(v))\otimes vg\rangle_M\,,\quad \langle L^{-1}(A(v))\,Q(g,g)\rangle_M\,,$$

$$\langle L^{-1}(B(v))\,g\rangle_M\,,\quad \langle L^{-1}(B(v))\otimes vg\rangle_M\,,\quad \langle L^{-1}(B(v))\,Q(g,g)\rangle_M\,.$$

Then the limiting g has the form

$$(26)\qquad\qquad g=\rho+v\!\cdot\!u+(\tfrac{1}{2}|v|^2-\tfrac{3}{2})\theta\,,$$

where the velocity u is divergence free and the density and temperature fluctuations, ρ and θ, satisfy the Boussinesq relation

$$(27)\qquad\qquad \nabla_x\!\cdot\!u=0\,,\qquad \nabla_x(\rho+\theta)=0\,.$$

Moreover, the functions ρ, u and θ are weak solutions of the equations

$$(28)\quad \partial_t u+u\!\cdot\!\nabla_x u+\nabla_x p=\mu_*\Delta u\,,\quad \partial_t\theta+u\!\cdot\!\nabla_x\theta=\kappa_*\Delta\theta\,,\quad \text{if } r=1,\,q=1;$$

$$(29)\qquad\qquad \partial_t u+\nabla_x p=\mu_*\Delta u\,,\quad \partial_t\theta=\kappa_*\Delta\theta\,,\quad \text{if } r>1,\,q=1;$$

$$(30)\quad \partial_t u+u\!\cdot\!\nabla_x u+\nabla_x p=0\,,\quad \partial_t\theta+u\!\cdot\!\nabla_x\theta=0\,,\quad \text{if } r=1,\,q>1;$$

$$(31)\qquad\qquad \partial_t u+\nabla_x p=0\,,\quad \partial_t\theta=0\,,\quad \text{if } r>1,\,q>1\,.$$

In these equations the expressions μ_* and κ_* denote the function values $\mu(1,1)$ and $\kappa(1,1)$ obtained from (22) in the previous section.

Remark 3. The equation (31) is completely trivial; it corresponds to a situation where the initial fluctuations and the Knudsen number are too small to produce any evolution over the timescale selected. However, this limit would be nontrivial if it corresponded to a timescale on which an external potential force acts on the system (Bardos, Golse, Levermore [BGL1]).

Remark 4. The second equations of (27), (28) and (29) that describe the evolution of the temperature do not contain a viscous heating term $\tfrac{1}{2}\mu_*\sigma(u)\!:\!\sigma(u)$ such as appears in the $CNSE_\epsilon$:

$$\partial_t\rho_\epsilon+\nabla_x\!\cdot\!(\rho_\epsilon u_\epsilon)=0\,,$$

$$(32)\qquad\qquad \rho_\epsilon\big(\partial_t+u_\epsilon\!\cdot\!\nabla_x\big)u_\epsilon+\nabla_x(\rho_\epsilon\theta_\epsilon)=\epsilon\nabla_x\!\cdot\![\mu_\epsilon\sigma(u_\epsilon)]\,,$$

$$\tfrac{3}{2}\rho_\epsilon\big(\partial_t+u_\epsilon\!\cdot\!\nabla_x\big)\theta_\epsilon+\rho_\epsilon\theta_\epsilon\nabla_x\!\cdot\!u_\epsilon=\epsilon\tfrac{1}{2}\mu_\epsilon\sigma(u_\epsilon)\!:\!\sigma(u_\epsilon)+\epsilon\nabla_x\!\cdot\![\kappa_\epsilon\nabla_x\theta_\epsilon]\,.$$

This is consistent with the scaling used here when it is applied directly to the $CNSE_\epsilon$ to derive the incompressible Navier-Stokes equations. More precisely with the change of variable

$$(33)\qquad \rho_\epsilon=\rho_0+\epsilon\tilde\rho(t,\tfrac{x}{\epsilon})\,,\quad u_\epsilon=\epsilon\tilde u(t,\tfrac{x}{\epsilon})\,,\quad \theta_\epsilon=\theta_0+\epsilon\tilde\theta(t,\tfrac{x}{\epsilon})\,,$$

the system (27), (28) is obtained for $\tilde{\rho}(t,\tilde{x})$, $\tilde{u}(t,\tilde{x})$ and $\tilde{\theta}(t,\tilde{x})$ as ϵ vanishes. In this derivation every term of the last equation of (62) is of the order ϵ^2 except $\frac{1}{2}\mu_\epsilon\sigma(u_\epsilon) : \sigma(u_\epsilon)$, which is of order three. The viscous heating term would have appeared in the limiting temperature equation had the scaling in (33) been chosen with the density and temperature fluctuations of order ϵ^2 [BLP].

Remark 5. In the case where $q = r = 1$ a system is obtained that has some structure in common with a diffusion approximation. A formal expansion for g_ϵ, the solution of the equation

$$(34) \qquad \epsilon\partial_t g_\epsilon + v\cdot\nabla_x g_\epsilon = \frac{1}{\epsilon}L(g_\epsilon) + \frac{1}{2}Q(g_\epsilon,g_\epsilon) + O(\epsilon)\,,$$

can be constructed in the form

$$(35) \qquad g_\epsilon = g^{(1)} + \epsilon g^{(2)} + \epsilon^2 g^{(3)} + \cdots\,.$$

This approach is related to the method of the previous section and to the work of De Masi, Esposito and Lebowitz.

V. Remarks Concerning the Proof of the Fluid Dynamical Limit. In this section the collision operator is given by the classical Boltzmann formula,

$$(36) \qquad C_B(F) = \iint_{\mathbf{R}^3 \times S^2} (F(v_1')F(v') - F(v_1)F(v))\,b(v_1 - v,\omega)\,d\omega\,dv_1\,,$$

where ω ranges over the unit sphere, v_1 over the three dimensional velocity space and $b(v_1 - v,\omega)$ is a smooth function; v' and v_1' are given in term of v, v_1 and ω by the classical relations of conservation of mass, momentum and energy (cf. [CC]) .

Any proof concerning the fluid dynamical limit for a kinetic model will, as a by-product, give an existence proof for the corresponding macroscopic equation. However, up to now no new result has been obtained by this type of method. Uniform regularity estimates would likely be needed in order to obtain the limit of the nonlinear term. These estimates, if they exist, must be sharp because it is known (and is proved by Sideris [S] for a very general situation) that the solutions of the compressible nonlinear Euler equations become singular after a finite time.

In agreement with these observations and in the absence of boundary layers (full space or periodic domain), the following theorems are proved:

i) Existence and uniqueness of the solution to the $CNSE_\epsilon$ for a finite time that depends on the size of the initial data, provided the initial data is smooth enough (say in H^s with $s > 3/2$). This time of existence is independent of ϵ and when ϵ goes to zero the solution converges to a solution of the compressible Euler equations.

ii) Global (in time) existence of a smooth solution (cf. [KMN]) to the $CNSE_\epsilon$ provided the initial data is small enough with respect to ϵ.

These two points have their counterparts at the level of the Boltzmann equation:

i) Existence and uniqueness (under stringent smallness assumptions) during a finite time independent of the Knudsen number, as proved by Nishida [N] (cf. also

Caflisch [C]). When the Knudsen number goes to zero this solution converges to a local thermodynamic equilibrium solution governed by the compressible Euler equations.

ii) Global existence for the solution to the Boltzmann equation provided the initial data is small enough with respect to the Knudsen number.

Concerning a proof of existence, the situation for the incompressible Euler equations in three space variables is similar; their solution (defined during a finite time) is the limit of a sequence of corresponding incompressible Navier-Stokes solutions with viscosities of the order of ϵ that remains uniformly smooth over a time interval that is independent of ϵ.

However, there are two other types of results concerning weak solutions. First, the global existence of weak solutions to the incompressible Navier-Stokes equations has been proved by Leray [L]. Second, using a method with many similarities to Leray's, R. DiPerna and P.-L. Lions [DiPL] have proved the global existence of a weak solution to a class of normalized Boltzmann equations, their so-called renormalized solution. This solution exists without assumptions concerning the size of the initial data with respect to the Knudsen number. Such a result also holds for the equation

$$(37) \qquad \epsilon \partial_t F_\epsilon + v \cdot \nabla_x F_\epsilon = \frac{1}{\epsilon^q} C_B(F_\epsilon)$$

over a periodic spatial domain $\mathbf{T}^3$.

The situation concerning the convergence to fluid dynamical limits (with ϵ going to zero) of solutions of the Boltzmann equation (37) with initial data of the form

$$(38) \qquad F_\epsilon = M(1 + \epsilon^r g_\epsilon)$$

continues to reflect this similarity. Following Nishida [N], it can be shown that for smooth initial data (indeed very smooth) the solution of (37) is smooth for a time on the order of ϵ^{1-r}. For $r = 1$ this time turns out to be independent of ϵ and during this time the solution converges (in the sense of the Theorem III) to the solution of the incompressible Euler equations when $q > 1$ or to the solution of the incompressible Navier-Stokes equations when $q = 1$. For $r > 1$ the solution is regular during a time that goes to infinity as ϵ vanishes; in this situation it converges to the solution of the linearized Navier-Stokes equations when $q = 1$ or to the solution of the linearized Euler equation when $q > 1$.

The borderline consists of the case $r = q = 1$. In this case it is natural to conjecture that the DiPerna-Lions renormalized solutions of the Boltzmann equation converge (for all time and with no restriction on the size of the initial data) to a Leray solution of the incompressible Navier-Stokes equations. However, our proof of this result is incomplete without some additional compactness assumptions (cf. [BGL3]).

Leray's proof relies on the energy estimate

$$(39) \qquad \tfrac{1}{2}\|u(t)\|^2 + \int_0^t \|\nabla_x u\|^2 ds \leq \tfrac{1}{2}\|u(0)\|^2.$$

For the Boltzmann equation the classical entropy estimate plays an analogous role in the proof of DiPerna and Lions. The entropy integrand can be modified by the addition of an arbitrary conserved density; the form chosen here is well suited for comparing F_ϵ with the absolute Maxwellian M:

$$\iint \left(F_\epsilon(t) \log \left(\frac{F_\epsilon(t)}{M} \right) - F_\epsilon(t) + M \right) dv\, dx + \frac{1}{\epsilon^{1+q}} \int_0^t \int D_\epsilon\, dx\, ds$$

(40)
$$\leq \iint \left(F_\epsilon(0) \log \left(\frac{F_\epsilon(0)}{M} \right) - F_\epsilon(0) + M \right) dv\, dx\,,$$

where D_ϵ is the entropy dissipation term given by

$$D_\epsilon = \frac{1}{4} \iiint (F'_{\epsilon 1} F'_\epsilon - F_{\epsilon 1} F_\epsilon) \log \left(\frac{F'_{\epsilon 1} F'_\epsilon}{F_{\epsilon 1} F_\epsilon} \right) b\, d\omega\, dv_1\, dv\,.$$

Here $F'_{\epsilon 1}$, F'_ϵ, $F_{\epsilon 1}$ and F_ϵ are understood to mean the evaluation of the distribution at the velocity values v'_1, v', v_1 and v respectively.

The entropy integrand is a nonnegative, strictly convex function of F_ϵ with a quadratic minimum value of zero attained at $F_\epsilon = M$. This suggests that in order to obtain bounds consistent with a formal expansion of the form (38) the initial data must be taken to satisfy the bound

(41)
$$\iint \left(F_\epsilon(0) \log \left(\frac{F_\epsilon(0)}{M} \right) - F_\epsilon(0) + M \right) dv\, dx \leq K \epsilon^{2r}\,,$$

where K is a constant independent of ϵ.

The estimates (40) and (41) are enough to show that a sequence of functions g_ϵ in (38) is relatively compact in the weak topology of L^1 and that every weakly converging subsequence has a limit belonging to $L^\infty(\mathbf{R}_t^+, L^2(\mathbf{T}^3) \otimes L_M^2)$. To obtain a strong convergence of the moments for the case $q = 1$, the above estimates are used with the averaging lemma (cf. [GLPS] or [DiPL]) to obtain a compactness result for the integrand of the collision operator (36). These estimates are sufficient to show convergence to the linearized Navier-Stokes equations (29) for the case $r > q = 1$. However, for the case $q = r = 1$ an additional compactness assumption is needed to gain some time (weak) regularity for the moments $\langle vg_\epsilon \rangle$. For $\epsilon > 0$ these are not divergence free functions and therefore classical Leray type estimates provide no control on the solenoidal component of this moment. Finally, to avoid concentration phenomena as $\epsilon \to 0$, it is assumed that

$$(1 + |v|^2) \frac{g_\epsilon^2}{(3 + \epsilon g_\epsilon)} M$$

lies in a weakly compact set of $L_{\text{loc}}^1(\mathbf{R}_t^+, L^1(\mathbf{T}^3) \otimes L_M^1)$. With these two assumptions one can show ([BGL3]) the convergence (in the sense of the theorem III) of the velocity moments for renormalized solutions of the Boltzmann equation to a Leray solution of the incompressible Navier-Stokes equations.

REFERENCES

[BGL1]   C. BARDOS, F. GOLSE AND D. LEVERMORE, *Sur les limites asymptotiques de la théorie cinétique conduisant à la dynamique des fluides incompressibles*, C.R.A.S., **309-I**, (1989), 727–732.

[BGL2]   C. BARDOS, F. GOLSE AND D. LEVERMORE, *Fluid Dynamic Limits of Kinetic Equations I: Formal Derivations*, (submitted to J. Stat. Phys.).

[BGL3]   C. BARDOS, F. GOLSE AND D. LEVERMORE, *Fluid Dynamic Limits of Kinetic Equations II: Convergence Proofs for the Boltzmann Equation*, (submitted to Annals of Math.).

[BLP]   B. BAYLY, D. LEVERMORE AND T. PASSOT, *Density Variations in Weakly Compressible Flows*, (submitted to Phys. of Fluids).

[C]   R. CAFLISCH, *The Fluid Dynamical Limit of the Nonlinear Boltzmann Equation*, Comm. on Pure and Appl. Math., **33** (1980), 651–666.

[CC]   S. CHAPMAN AND T. COWLING, *The Mathematical Theory of Nonuniform Gases*, Cambridge University Press (1951).

[DMEL]   A. DE MASI, R. ESPOSITO AND J.L. LEBOWITZ, *Incompressible Navier-Stokes and Euler Limits of the Boltzmann Equation*, IHES Publication Physique 89-07.

[DiPL]   R.J. DIPERNA AND P.-L. LIONS, *On the Cauchy Problem for the Boltzmann Equation: Global Existence and Weak Stability Results*, Annals of Math., **130**, (1989), 321–366.

[GLPS]   F. GOLSE, P.-L. LIONS, B. PERTHAME AND R. SENTIS, *Regularity of the Moments of the Solution of a Transport Equation*, J. of Funct. Anal., **76** (1988), 110–125.

[KMN]   S. KAWASHIMA, A. MATSUMURA AND T. NISHIDA, *On the Fluid Dynamical Approximation to the Boltzmann Equation at the Level of the Navier-Stokes Equation*, Comm. Math. Phys., **70** (1979), 97–124.

[L]   J. LERAY, *Sur le Mouvement d'un Liquide Visqueux Emplissant l'space*, Acta Mathematica, **63**, (1934), 193–248..

[LL]   L.D. LANDAU AND E.M. LIFSHITZ, *Fluid Mechanics*, Addison-Wesley.

[N]   T. NISHIDA, *Fluid Dynamical Limit of the Nonlinear Boltzmann Equation to the Level of the Incompressible Euler Equation*, Comm. Math. Phys., **61** (1978), 119–148.

[S]   T. SIDERIS, *Formation of Singularities in Three Dimensional Compressible Fluids*, Comm. Math. Phys., **101** (1985), 475–485.

THE ESSENCE OF PARTICLE SIMULATION
OF THE BOLTZMANN EQUATION

H. BABOVSKY* AND R. ILLNER†

Abstract. We describe the mathematical structure of recently developed family of consistent and convergent particle simulation methods for the Boltzmann equation.

1. The Setting. Particle simulation methods for the Boltzmann equation have recently found wide-spread interest, in particular because new projects in spaceflight and in very high flying aircraft have revived the demand for computational rarefied gas dynamics. Because rarefied gases are believed to be well described by the Boltzmann equation, it is desirable that any such numerical scheme be consistent with this equation. In [1] and [2], we gave a convergence proof for a method which has been developed during the last 5 years and is a modification of a scheme originally suggested by Nanbu ([3], [4]). The objective of this article is to give a concise mathematical description of this method and of a more recent and more efficient variant. Our emphasis is on clarity and simplicity, and we avoid the technical details which were essential for the convergence proof in [2].

First, we clarify the concept "particle simulation". It is best explained by a simple diagram, which summarizes realistic, analytical and computational rarefied gas dynamics:

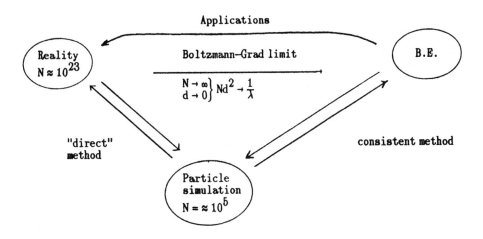

A realistic gas cloud consists of an astronomically large number of particles, far too large to track each particle individually. The Boltzmann equation is believed to

---

*Fachbereich Mathematik, Universität Kaiserslautern, 6750 Kaiserslautern, Federal Republic of Germany

†Department of Mathematics & Statistics, University of Victoria, Victoria, B.C. V8W 2Y2, Canada

be a good description because it emerges (formally) in the Boltzmann-Grad limit for the density distribution of a tagged particle. The idea of particle simulation is really to just return to the particle description, with a particle number N that is sufficiently large to guarantee "closeness" to the physical situation, but small enough to make the situation tractable from a computational point of view. Given the physical background of the problem, the particle simulation approach seems completely natural. There is, however, a second, analytical reason why particle simulation is reasonable here. We have to take a look at the Boltzmann equation to see it.

Let $f = f(t, x, v)$ be the particle density of the rarefied gas. It depends on time t, position x and velocity v, and its time evolution is given by by Boltzmann equation

$$(1) \qquad \partial_t f + v \cdot \nabla_x f = \mathcal{J}(f, f)$$

The collision operator $\mathcal{J}(f, f)$ is defined by

$$(2) \qquad \mathcal{J}(f, f)(t, x, v) = \frac{1}{\lambda} \int_{\mathbf{R}^3} \int_{S_+^2} k(|v - w|, \theta)\{f(v')f(w') - f(v)f(w)\} dn \, dw,$$

where $S_+^2 = \{n \in S^2; n \cdot (v - w) > 0\}$,

$$\theta \quad \text{is the acute angle between } n \text{ and } v - w,$$
$$v' = v - n(n \cdot (v - w))$$
$$w' = w + n(n \cdot v - w))$$

are the post-collisional velocities associated with the (ingoing) collision configuration (v, w, n), and $k(|v - w|, \theta)$ is the collision kernel (for hard spheres, $k(|v - w|, \theta) = |v - w| \cdot \cos \theta$). λ is proportional to the mean free path between collisions - for the rest of this article, we set $\lambda = 1$.

The integration in (2) is 5-dimensional, and this is the second reason why particle simulation is a sensible way to solve the Boltzmann equation numerically - except in isotropic situations where $\mathcal{J}(f, f)$ could be evaluated with low-dimensional integrals, it would be just too inefficient to evaluate (2) by quadrature formulas. Monte Carlo simulation is a well-known alternative, and we shall see that it arises quite naturally (but not necessarily) in particle simulation.

2. Reduction of the Boltzmann Equation. We now go through a series of fairly elementary steps which will reduce the Boltzmann equation to a form which is readily accessible to an approximation by point measures (particle simulation). These are

a) time disretization

b) separation of free flow and interaction ("splitting")

c) local homogenization

d) weak formulation

e) measure formulation.

To start, suppose that a particle at (x, v) in state space will go to $\phi_t(x, v)$ after time t, provided there is no collision with another particle (if the particle does not interact with the boundary of the confining container, $\phi_t(x, v) = (x + tv, v)$; otherwise, we assume that the trajectory is defined by some reasonable deterministic boundary condition, like specular reflection).

Choose a time step $\Delta t > 0$, then a first order discrete counterpart to the derivative

$$\left(\partial_t f + v \cdot \nabla_x f\right)\left(j \cdot \Delta t, \ \phi_{\Delta t}(x, v)\right)$$

is

$$(3) \qquad \frac{1}{\Delta t}\left\{f((j+1)\Delta t, \ \phi_{\Delta t}(x, v)) - f(j\Delta t, x, v)\right\}$$

Substitute (3) for the left hand side in (1), let $(y, w) = \phi_{\Delta t}(x, v)$ and evaluate. The result is

$$(4) \qquad f((j+1)\Delta t, y, w) = f(j\Delta t, \phi_{-\Delta t}(y, w)) + \Delta t \cdot \mathcal{J}(f, f)\left(j\Delta t, \phi_{-\Delta t}(y, w)\right)$$

The discretized equation (4) suggests to split the approximation into the *collision simulation*

$$(5) \qquad \tilde{f}((j+1)\Delta t, y, w) = f(j\Delta t, y, w) + \Delta t \, \mathcal{J}(f, f)(j\Delta t, y, w)$$

and the *free flow step*

$$(6) \qquad f((j+1), \Delta t, y, w) = \tilde{f}\left((j+1)\Delta t, \phi_{-\Delta t}(y, w)\right).$$

The numerical simulation of (6) will be obvious once we have understood the collision simulation. We therefore focus on that. Notice that the position variable y is not operated on in (5) - effectively, (5) is a discretized version of the spatially homogeneous Boltzmann equation.

To continue, we have to introduce a concept of "spatial cell" which already Boltzmann used in the classical derivation of his equation (his cell was just defined by $(x, x + dx)$, $(y, y + dy)$ etc.). The key idea is that such a cell is small from a macroscopic point of view, but large enough to contain many particles, and certainly large enough to keep a collision count for any particle in the cell during $\Delta t(dt)$ by just counting collisions with other particles *in the same cell*. In this collision count, the spatial variation of the gas density over the cell is neglected, i.e. spatial homogeneity over the cell is assumed (in fact, the numerical procedures we are about to describe allow to keep the exact positions of the approximating particles, but the collision partners needed for the collision simulation are assumed to be homogeneously distributed in each cell; see [2]).

Specifically, suppose that the gas in question is confined to a container $\wedge \subset \mathbb{R}^3$, and that this container is partitioned into cells by

$$\wedge = U C_i, \qquad C_i \cap C_j = \emptyset, \qquad i \neq j,$$

and we assume that the cells are such that $f(j\Delta t, \cdot)$ is on C_i well approximated by its homogenization

$$\frac{1}{\lambda^3(C_i)} \int\limits_{C_i} f \, dy$$

(we replace $f(j\Delta t, \cdot)$ by its homogenization, but keep writing $f(j\Delta t, \cdot)$). If $f(j\Delta t, \cdot)$ is locally homogeneous in this sense, so is $\tilde{f}((j+1)\Delta t, \cdot)$; however, the free flow step will destroy the homogeneity, and one has to homogenize again before the next collision simulation. We note that the cells need not be the same size and that the partition of $\wedge$ can actually be changed with time (refined, for example) to gain better adjustment to the hypothesis of local homogeneity.

We have now reduced the Boltzmann equation to the remaining key question of doing the collision simulation (5) on an arbitrary but fixed cell C_i, where $f(j \cdot \Delta t, \cdot)$ is supposed to be independent of y. To simplify notation, we write $f_j(v)$ for $f(j \cdot \Delta t, y, v)$ and $f_{j+1}(v)$ for $\tilde{f}((j+1)\Delta t, y, v)$. Then (5) reads, explicitly

(7)
$$
\begin{aligned}
f_{j+1}(v) = {} & \left(1 - \Delta t \iint k(|v-w|, \theta) dn \; f_j(w) dw\right) f_j(v) \\
& + \Delta t \iint k(|v-w|, \theta) f_j(v') f_j(w') dn \; dw.
\end{aligned}
$$

We next make the crucial assumption that there is an $A > 0$ such that

(8)
$$\int k(|v-w|, \theta) dn \le A < \infty$$

for all v, w. Unfortunately, this means that k has to be truncated even for the hard sphere case; a little thought shows that we have to modify k for large $|v-w|$, i.e. some of the collisions between particles with large relative velocity are neglected. Fortunately, for any reasonable gas cloud only few particles are affected.

Also, we renormalize $f_j(v)$ such that $\int f_j \, dw = 1$ (assuming that we have $\int_\wedge \int f_j(y, v) dv \, dy = 1$ and $\int_{C_i} \int f_j(y, v) dv \, dy = \lambda^3(C_i) \int f_j(v) dv = \gamma_{j,i}$, this means that we have to replace f_j by $f_j \cdot \frac{\lambda^3(C_i)}{\gamma_{j,i}}$; for this paper, we simply set $\frac{\lambda^3(C_i)}{\gamma_{j,i}} = 1$). Then, if $\Delta t < \frac{1}{A}$, $f_j \ge 0$ implies that $f_{j+1} \ge 0$.

Thus the truncation (8) is necessary to keep the density nonnegative, an essential feature. This is an artifact of the explicit nature of our approximation scheme; (8) can be avoided by starting from an alternative formulation of the Boltzmann equation, but this would lead to serious problems later on.

The next step is a transition to a weak formulation of (7). To this end, multiply (7) with a test function $\varphi \in C_b(\mathbf{R}_v^3)$, integrate, use the involutive property of the collision transformation and that $|v' - w'| = |v - w|$. The result is

(9)
$$\int \varphi(v) f_{j+1}(v) dv = \iint K_{v,w} \varphi f_j(v) f_j(w) dv \; dw \; ,$$

where $K_{v,w}\varphi = \left(1 - \Delta t \int k \; dn\right)\varphi(v) + \Delta t \int k \; \varphi(v')dn$ (we have also used the renormalization $\int f_j dw = 1$).

Finally, before we rewrite (9) in measure formulation, we introduce a convenient representation for $K_{v,w}\varphi$. Let v and w be given. Then, we define a continuous function $T_{v,w} : S_+^2 \to \mathbf{R}^3$ by $T_{v,w}(n) = v'$. Moreover, let $B^1 = \left\{y \in \mathbf{R}^2; \|y\| \leq \dfrac{1}{\sqrt{\pi}}\right\}$ be the circle of area 1, and assume that $\Delta t < \frac{1}{A}$.

LEMMA 2.1. *(see [1]) For all $v, w \in \mathbf{R}^3$, there is a continuous function $\phi_{v,w} :$ $B^1 \to S_+^2$ such that*

$$K_{v,w}\varphi = \iint\limits_{B^1} \varphi\left(T_{v,w} \circ \phi_{v,w}(y)\right)d^2y$$

REMARKS AND SKETCH OF THE PROOF. This lemma, which is extremely useful for the sequel, is proved in detail in [1]. The function $\phi_{v,w}$ can actually be computed in terms of the collision kernel k.

The purpose of the function $\phi_{v,w}$ is a) to decide whether the particles with velocities v nd w collide at all, and b) if they collide, with what collision parameter. We refer to y as "generalized collision parameter".

The idea of the proof is as follows. We represent B^1 by polar coordinates as $\left\{(r,\beta); 0 \leq r \leq \dfrac{1}{\sqrt{\pi}}, \; 0 \leq \beta \leq 2\pi\right\}$. There is an $r_0 < \dfrac{1}{\sqrt{\pi}}$ such that

$$\pi r_0^2 = \Delta t \int k \; dn.$$

Let $n \in S_+^2$ be represented by (θ, ψ) $(\theta \in [0, \frac{\pi}{2}], \; \psi \in [0, 2\pi))$, where θ is the polar angle with respect to the axis in direction of $v - w$, and ψ is an azimuthal angle.

For $r \geq r_0$, let $\phi_{v,w}(r, \beta) = \left(\dfrac{\pi}{2}, \beta\right)$, i.e. $\theta = \dfrac{\pi}{2}$, $\psi = \beta$. These angles correspond to a grazing collision, and therefore $T_{v,w} \circ \phi_{v,w}(r, \beta) = v$; this happens on a set of measure $1 - \Delta t \int k \; dn$.

For $r < r_0$, the collision result is nontrivial. We set again $\psi(r, \beta) = \beta$, but $\theta(r, \beta) = \theta(r)$ is defined as the inverse of a function $r(\theta)$ which satisfies

$$\frac{d}{d\theta}\left[\frac{1}{2}r^2(\theta)\right] = \Delta t \cdot k(|v - w|, \theta)\sin\theta.$$

Clearly

$$r^2\left(\frac{\pi}{2}\right) = 2\Delta t \int\limits_0^{\pi/2} k(|v-w|, \theta)\sin\theta \, d\theta = r_0^2, \text{ and } \Delta t \int \varphi(v')k \; dn = \int\limits_0^{2\pi}\int\limits_0^{r_0} \varphi\left(T_{v,w}(\theta(r), \beta)\right)r \, dr \, d\beta.$$

This completes the proof.

Now define probability measures $d\mu_j = f_j\, dv$, and let $\Psi(v, w, y) = T_{v,w} \circ \phi_{v,w}(y)$. By the lemma, (9) reduces to

$$(10) \qquad \int \varphi(v) d\mu_{j+1} = \int\limits_{v} \int\limits_{w} \int\limits_{y} \varphi \circ \Psi(y, v, w) d^2 y\, d\mu_j(v) d\mu_j(w)$$

or, if we set $dM_j := d^2 y \times d\mu_j \times d\mu_j$ (dM_j is a probability measure on $B^1 \times \mathbf{R}_v^3 \times \mathbf{R}_w^3$),

$$(11) \qquad \int \varphi d\mu_{j+1} = \int \varphi\, d\left(M_j \circ \Psi^{-1}\right)$$

In other words, we have identified the measure μ_{j+1} with $M_j \circ \Psi^{-1}$! (11) is the reduced Boltzmann equation which is most useful for the particle simulation idea.

3. The Essence. We have now arrived at the central point of particle simulation. It is best summarized as follows: Suppose we have a sequence of discrete probability measures

$$\mu_j^N = \frac{1}{N} \sum_{i=1}^{N} \delta_{V_i^N(j)}$$

such that $\mu_j^N \to \mu_j$ as $N \to \infty$ weak-* in the sense of measures, how do we find a corresponding sequence of discrete probability measures μ_{j+1}^N such that $\mu_{j+1}^N \overrightarrow{w^*} \mu_{j+1}$? μ_j^N will be called an N-particle approximation of μ_j.

Suppose next that $\Psi(v, w, y)$ is continuous as a function of v, w and y (this is certainly true if k is bounded and continuous; since we had to truncate k anyway, it is not unreasonable to enforce boundedness and continuity), and that we are given a sequence of probability measures M_j^N on $B^1 \times \mathbf{R}_v^3 \times \mathbf{R}_w^3$ such that

$$M_j^n \overrightarrow{w^*} M_j \quad \text{as } N \to \infty.$$

It is then elementary that $M_j^N \circ \Psi^{-1} \overrightarrow{w^*} M_j \circ \Psi^{-1} = \mu_{j+1}$ (see [5]). Suppose next that we have found triples (y_i^N, v_i^N, w_i^N) $(i = 1, \dots, N)$ such that

$$(12) \qquad M_j^N = \frac{1}{N} \Sigma \delta_{y_i} \times \delta_{v_i} \times \delta_{w_i} \overrightarrow{w^*} M_j,$$

then $M_j^N \circ \Psi^{-1} = \frac{1}{N} \Sigma \delta_{\Psi(y_i, v_i, w_i)}$ is an N-particle approximation of μ_{j+1}. $v_i(j+1) = \Psi(y_i, v_i, w_i)$ are the velocities of the N approximating particles after the collision simulation step, where v_i was the ingoing velocity, w_i the velocity of the collision partner, and y_i a collision parameter. The remaining question now is this: Given $v_i^N(j), i = 1, \dots, N$, how do we choose the y_i^N, w_i^N such that (12) holds?

We describe two methods. The first one, discussed in detail in [1, 2], is a Monte Carlo method, because it involves random choices: Given $v_1(j), \dots, v_N(j)$, choose sequences of independent and equidistributed random variables $(r_i)_{i \in \mathbf{N}}$ on $[0, 1]$, $(y_i)_{i \in \mathbf{N}}$ on B^1. Then, define an index $c_i \in \{1, \dots, N\}$ by $c_i = [N \cdot r_i] + 1$, let $w_i = v_{c_i}, v_i(j+1) = \Psi(y_i, v_i, w_i)$. That's all. Notice that the w_i are taken from the same set $\{v_1, \dots, v_N\}$. The same velocity can be chosen repeatedly for collision partners! In particular, we have no guarantee that the total energy and momentum are conserved in this process (however, the expected values are conserved, and see the remarks in the next section). It was shown in [1] and [2] that by the central limit theorem $M_j^N \circ \Psi^{-1}$ converges indeed w^* to μ_{j+1} almost surely.

We present a second idea that has been successfully applied. Observe that there is no reason to keep the velocities v_i^N fixed; instead, our target is to find efficient N-atomic approximations to the product measure M_j. In doing this, we may even vary N a little if this helps our approximation (the only reason to keep N approximately constant is computational efficiency).

We demonstrate the mentioned idea for the simpler case of a probability measure μ on a real interval, say $I = [0,1]$. Suppose that $\mu^N = \dfrac{1}{N} \displaystyle\sum_{i=1}^{N} \delta_{x_i}$ is a sequence of atomic measures, and that $\mu^N \xrightarrow{w^*} \mu$ as $N \to \infty$. If $N = k^2, k \in \mathbb{N}$, the following idea, due to M. Bäcker [6], yields a good approximation to $\mu \times \mu$: order the x_i such that $x_1 < x_2 < x_3 < \cdots < x_N$, let $y_0 = 0, y_1 = x_k, y_2 = x_{2k}, \dots, y_{k-1} = y_{(k-1)k}, y_k = 1$, and define intervals I_j by $I_j = [y_{j-1}, y_j)$ for $1 \le j < k, I_k = [y_{k-1}, 1]$. Then all the I_j contain exactly k of the x_i's.

The set of rectangles $R_{i,j} = I_i \times I_j$ is a cover of $[0,1]^2$ of exactly $k^2 = N$ elements. If we choose $\bar{x}_i$ as the center if I_i, then the N-atomic measure $\dfrac{1}{N} \displaystyle\sum_{i,j=1}^{k} \delta_{(\bar{x}_i, \bar{x}_j)}$ will be a reasonable discrete approximation to $\mu \times \mu$- the quality of the approximation depends of course on the quality of approximation of μ^N to μ. Figure 1 below displays the idea graphically (for $N = 9$).

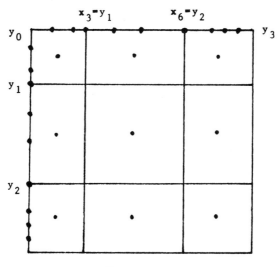

FIGURE 1

The idea readily generalizes to the higher-dimensional situation which we face with the measure M_j-of course, rather than with $N = k^2$, one has to work with numbers N which factor in a more convenient way.

It is clear that this approach should eliminate statistical fluctuations prevalent in the Monte Carlo simulation (for an unlikely possible choice of the r_i's and y_i's, one can get very poor approximations of M_j- for example, if $c_i = i$ for all i). This has been confirmed by numerical experiments. For more details on the application of this idea, and how to combine it with the first method described here, we refer the reader to [7].

4. Remarks. There are some positive and negative features of the methods outlined above. First, observe that a consistent scheme must reflect the essential properties of a rarefied gas, which enter into the derivation of the Boltzmann equation. The most fundamental of these assumptions is the propagation of molecular chaos, which is believed to hold in general, and proved to be true for special situations ([8, 9, 10]). The numerical counterpart to this is the requirement (12) that

$$M_j^N \overrightarrow{w^*} M_j$$

as $N \to \infty$, and this was exactly what we called "the essence"! Therefore, we can say that (12) means that we choose w_i, y_i such that molecular chaos is satisfied in the limit.

On the negative side, the methods described here do not guarantee the strict conservation of momentum, energy etc. (mass, of course, is conserved). However, for the Monte Carlo method, these conservation laws are satisfied in the mean, and consistency and convergence will of course guarantee that these laws hold in the limit $\Delta t \to, \Delta x \to 0(\Delta x$ is the diameter of the largest cell) and $N \to \infty$ (Clearly, N will have to grow much faster than the number of cells). We emphasize that the consistency with molecular chaos is more important than these conservation laws on the discrete level.

Also, as elaborated in detail in [1], it is possible to modify the first method such that energy and momentum are conserved exactly - the only condition is that the particle number N (per cell) be even. In this case, choose randomly a perturbation π of $\{1, \ldots, N\}$, group the N velocities in pairs

$$(v_{\pi(1)}, v_{\pi(2)}), (v_{\pi(3)}, v_{\pi(4)}) \text{ etc.,}$$

choose generalized collision parameters $(y_i)_{i=1,\ldots,N}$ randomly from B^1, and let

$$v_1' = \psi\left(y_1, v_{\pi(1)}, v_{\pi(2)}\right)$$

$$v_2' = \psi\left(y_1, v_{\pi(2)}, v_{\pi(1)}\right)$$

$$v_3' = \psi\left(y_2, v_{\pi(3)}, v_{\pi(4)}\right),$$

$$v_4' = \psi\left(y_2, v_{\pi(4)}, v_{\pi(3)}\right) \text{ etc.}$$

be the post-collisional velocities. It is clear that this version of the algorithm is designed such that momentum and energy is conserved.

5. A Technical Detail, Convergence, and Outlook. In this final section we give a few comments on the convergence proof from [2]. First, it is highly desirable to have a metric which will control the weak-*-convergence used in Section 3. The discrepancy between two measures is an efficient way to do this. We explain it for the case of probability measures on $\mathbf{R}^2$: Let Q, P be arbitrary points in $\mathbf{R}^2$, then "$\leq$" denotes the usual semi-order, i.e. $Q \leq P$ if and only if $Q_i \leq P_i, i = 1, 2$. Moreover, Let $R(P) = \{Q \in \mathbf{R}^2, Q \leq P\}$.

DEFINITION. Let μ, ν be two probability measures on $\mathbf{R}^2$. The discrepancy $D(\mu, \nu)$ between them is

$$D(u, \nu) = \sup_{P \in \mathbf{R}^2} \left| \int_{R(P)} d\mu - \int_{R(P)} d\nu \right|.$$

The useful link between D and w^*-convergence is the theorem below (for the proof see [2]).

THEOREM. *Let μ be an absolutely continuous probability measure on $\mathbf{R}^n$. Then*

$$\mu_N \overrightarrow{w^*} \mu \quad \text{if and only if } D(\mu_N, \mu) \overrightarrow{N \to \infty} 0.$$

This theorem, the central limit theorem and the Borell–Cantelli lemma were three of the most essential tools of the convergence proof in [2]. We conclude with a short synopsis of this result.

The most crucial assumption in [2] is the hypothesis that the Boltzmann equation has a unique solution f on the time interval $[0, T]$ in questions such that for some $\alpha > 0, C > 0$

$$\operatorname*{esssup}_x \int f(t, x, v) e^{\alpha v^2} \, dv \leq C, t \in [0, T].$$

There is also the truncation assumption (8) on the collision kernel k, and a regularity assumption (with respect to x) of the solution. The result is then this:

Given a sequence of particle approximations $\mu_0^N = \dfrac{1}{N} \displaystyle\sum_{i=1}^{N} \delta_{(x_i, v_i)}$ to the initial value f_0, and given sequences of time steps $(\Delta t)_n$ and cell sites $(\Delta x)_n$ such that $(\Delta t)_n \searrow 0$, $(\Delta x)_n \searrow 0$, there is a sequence $N(n) \to \infty$ such that

$$D\left(\mu_{k \cdot (\Delta t)_n}^{N(n)}, f\left(k(\Delta t)_n, x, v \right) dx \ dv \right) \longrightarrow 0$$

as $n \to \infty$, almost surely with respect to the r_i, $i = 1, 2, 3, \ldots$, for $k \cdot (\Delta t)_n \in [0, T)$. The $\mu_{k \cdot (\Delta t)_n}^{N(n)}$, of course, are the discrete measures computed from $\mu_0^{N(n)}$ via the procedures from sections 2–4.

There are several aspects of the described methods which call for further investigation. We formulate a few questions which concern some of these aspects:

Is it possible to modify the methods such that the truncation error with respect to Δt will be of higher order, without having to increase the particle number at every step?

What criteria should be applied to design an optimal partition into cells? What is an optimal number of particles per cell from a computational point of view? What can be said about convergence rates?

These and other questions leave us a lot to think about for the future. Meanwhile, the methods described here are being successfully used to model rarefied gases.

Acknowledgement. This research was supported in part by grant A7874 from the Natural Science and Engineering Research Council of Canada, and by the Institute for Mathematics and its Applications with funds provided by the National Science Foundation.

REFERENCES

[1] H. BABOVSKY, *A convergence proof for Nanbu's Boltzmann Simulation Scheme*, Eur. J. Mech., B/Fluids, 8, 1 (1989), 41–55.

[2] H. BABOVSKY, R. ILLNER, *A Convergence Proof for Nanbu's Simulation Method for the Full Boltzmann Equation*, SIAM J. Numer. Anal., 26, 1 (1989), 45–65.

[3] K. NANBU, *Theoretical Basis of the Direct Simulation Monte Carlo Method*, in: Rarefied Gas Dynamics, 1, V. Boffi, C. Cercignani (eds.), Teubner, Stuttgart (1986).

[4] K. NANBU, J. Phys. Soc. Jpn. 43 (1980), 2042.

[5] P. BILLINGSLEY, *Convergence of Probability Measures*, J. Wiley & Sons, New York (1968).

[6] M. BÄCKER, personal communication.

[7] C. LÉCOT, *A Direct Simulation Monte Carlo Scheme and Uniformly Distributed Sequences for Solving the Boltzmann Equation*, to appear in Computing.

[8] O. LANFORD III, *Time Evolution of Large Classical Systems*, in Lecture Notes in Physics, 38, Springer (1975), 1–111.

[9] R. ILLNER, M. PULVIRENTI, *Global Validity of the Boltzmann Equation for a Two-Dimensional Rare Gas in Vacuum*, Commun. Math. Phys. 105 (1986), 189–203.

[10] R. ILLNER, M. PULVIRENTI, *Global Validity of the Boltzmann Equation for Two-and Three-Dimensional Rare Gas in Vacuum: Erratum and Improved Result*, Commun. Math. Phys. 121 (1989), 143–146.

THE APPROXIMATION OF WEAK SOLUTIONS TO
THE 2-D EULER EQUATIONS BY VORTEX ELEMENTS

J. THOMAS BEALE†

Abstract. It is shown that the Euler equations of two–dimensional incompressible flow, with initial vorticity in L^p, $p > 1$, possess weak solutions which may be obtained as a limit of vortex "blobs"; i.e., the vorticity is approximated by a finite sum of cores of prescribed shape which are advected according to the corresponding velocity field. If the vorticity is instead a finite measure of bounded support, such approximations lead to a measure–valued solution of the Euler equations in the sense of DiPerna and Majda [7]. The analysis is closely related to that of [7].

Key words. incompressible flow, Euler equations, weak solutions, vortex methods

AMS(MOS) subject classifications. 76C05

Introduction. There has been renewed interest lately in the development of singularities in weak solutions of the Euler equations of two–dimensional, incompressible flow. Two examples are boundaries of patches of constant vorticity and vortex sheets. (See [16] for a general discussion of both.) In the latter case the vorticity is a measure concentrated on a curve or sheet which we take to be initially smooth. At later time, a singularity in the sheet may develop, and the nature of solution past the singularity formation is unclear. Such questions for vortex sheets were dealt with at this conference in talks by Caflisch, Krasny, Majda, and Shelley.

We focus here on discrete approximations of weak solutions of the 2-D Euler equations of the sort used in computational vortex methods. The vorticity is approximated by a finite sum of "blobs", or cores of prescribed shape, which are advected according to the corresponding velocity field. For the 2–D Euler equations with initial vorticity in L^p, $p > 1$, with bounded support, we show here that weak solutions in the usual sense are obtained in the limit as the size and spacing of the blob elements go to zero. (Weak solutions are known to be unique only if the vorticity is in L^∞.) If the initial vorticity is instead a finite measure of bounded support, as would be the case for a vortex sheet of finite length, a limit is obtained which is a measure–valued solution of the Euler equations in the sense of DiPerna and Majda [7]. The analysis is closely related to that of [7].

For both results, the number of vortex elements needed is very large compared to the radius of the vortex core, in contrast to the case of smooth flows. This seems at least qualitatively consistent with observed behavior in calculations of Krasny [13,14] and others, in which a regularization analogous to the blob elements is used to modify the vortex sheet evolution so that calculations can be continued past the time when the sheet develops singularities. Shelley and Baker [19] have used a different regularization, in which the vortex sheet is replaced by a layer of finite thickness. Both calculations are suggestive of weak solutions of the Euler equations past the time of first singularity (cf. [17]). Mathematical treatments of

†Department of Mathematics, Duke University, Durham, NC 27706. Research supported by D.A.R.P.A. Grant N00014-86-K-0759 and N.S.F. Grant DMS-8800347.

the possible nature of measured–valued solutions in 2–D, such as might occur after the singularity in the vortex sheet, have been given in [8,10]. A convergence result for vortex element approximations to vortex sheets, somewhat complementary to the results presented here, has been given by Caflisch and Lowengrub [3,15]. They show that discrete approximations to a sheet converge in a much more specific sense for a time interval before the singularity formation, provided the sheet is analytic and close to horizontal.

Recently Brenier and Cottet have obtained another convergence result with vorticity in L^p, $p > 1$. In their case the spacing of the vortex elements can be comparable to the radius of the core, unlike the first result presented here.

It is a pleasure to thank A. Majda for suggesting this investigation and for arranging a visit to the Applied and Computational Mathematics Program at Princeton University, during which this work was carried out.

1. Discussion of Results. In [6] DiPerna and Majda introduced a notion of measure-valued solution for the Euler equations of three–dimensional incompressible fluid flow, based on the conservation of energy, which was intended to incorporate possible oscillation and concentration in nonsmooth solutions. They showed that measure–valued solutions exist and can be obtained as a limit under regularization, but they may not be unique. For the two–dimensional case they used a more special definition of measure–valued solution of the Euler equations [7] which takes into account the conservation of vorticity as well as energy. They showed in [7] that various regularizations of the 2-D Euler equations converge to measure–valued solutions provided the initial vorticity is a measure on $\mathbf{R}^2$ of bounded support and finite total mass. This includes the important case of vortex sheets. They also showed that certain regularizations produce classical weak solutions (in the distributional sense) in 2-D provided the initial vorticity is in L^p for some $p > 1$. One regularization studied was an approximation by a finite number of vortex "blobs". They showed that a class of vortex blob approximations converges to a measure–valued solution of the 2-D Euler equations, provided that the total circulation is zero, but they did not determine whether classical weak solutions could be obtained in this way when the vorticity is more regular than L^1.

In this work we give another treatment of the vortex blob approximations to the 2-D Euler equations. It is similar to that of [7] but more straightforward and direct. With a slightly different choice of parameters, we show that vortex blob approximations converge to measure–valued solutions, again for initial vorticity which is a measure of bounded support and finite mass. The total circulation is arbitrary. (In two dimensions the total energy is infinite if the total circulation is nonzero.) In the case of initial vorticity in L^p for some $p > 1$, we show that a classical weak solution is obtained. A unified treatment is given for the two results; in fact, as is evident in the analysis of [7], the essential points to verify in either case are bounds for the approximate vorticity and energy and a kind of weak consistency with the Euler equations.

For smooth solutions of the Euler equations in two or three dimensions, the blob approximations of vortex methods converge with rates determined by the two length parameters, the radius of the blob elements (which can be thought of as a smoothing radius) and the spacing of the elements; the radius is usually taken larger than the spacing. Such a result was proved in two dimensions in [11]. For a summary of the theory, see, e.g., [1,2]. It has recently been shown in [9] that for smooth flows convergence is possible even with point vortices in place of the blob elements. In the nonsmooth case considered here, however, our results require that the spacing of the blobs is quite small relative to the core radius, and correspondingly the number of blob elements is large. A result of the sort presented here was given for vorticity in L^∞ in [18], as well as a treatment of stochastic differential equations of particle paths as a discretization of viscous flow. Again for vorticity in L^∞ , it has been shown [4,5] that blob approximations converge for short time with the radius comparable to the spacing. In [5] Cottet describes an elegant and appealing approach to the consistency of these methods for weak solutions in 2-D; his approach could be applied to the situation studied here.

We now describe the two–dimensional vortex blob approximation to be used. The formulation and notation correspond to Section 2 of [7]. For 2-D incompressible, inviscid flow, the vorticity ω is conserved along particle paths; this is expressed in the equation

$$\omega_t + v \cdot \nabla\omega = 0,$$

where $\omega = v_{2,1} - v_{1,2}$ is the (scalar) curl of the velocity v. We will approximate the vorticity field at a given time by a sum of vortex "blobs", i.e., translates of a core function with specified shape. The flow is then simulated by advecting these elements according to the velocity field determined by the approximate vorticity. For this purpose, we choose a core function φ depending only on the radius $r = |x|$ whose total weight is one,

$$2\pi \int_0^\infty r\varphi(r)dr = 1.$$

With scaling parameter ε, we set $\varphi_\varepsilon(x) = \varepsilon^{-2}\varphi(x/\varepsilon)$. We will assume that φ is C^1 and has bounded support, although it will be evident in our arguments that these assumptions could be relaxed. At each time t the approximate vorticity field will have the form

(1.1)
$$\omega^\varepsilon(x,t) = \sum_{j=1}^{N(\varepsilon)} \varphi_\varepsilon(x - X_j^\varepsilon(t))\Gamma_j^\varepsilon,$$

where Γ_j^ε is a strength assigned to the jth element from the initial data, and $X_j^\varepsilon(t)$ is the current location of the center of the jth blob, obtained by integrating the corresponding velocity field in time.

In two dimensions the velocity is obtained from the vorticity by convolving with the velocity kernel

(1.2)
$$K(x) = \frac{1}{2\pi|x|}(-x_2, x_1).$$

Thus corresponding to (1.1) we have the velocity field

$$(1.3) \qquad v^\varepsilon(x,t) = \sum_{j=1}^{N(\varepsilon)} K_\varepsilon(x - X_j^\varepsilon(t))\Gamma_j^\varepsilon,$$

where $K_\varepsilon = K * \varphi_\varepsilon$. It can easily be seen that

$$(1.4) \qquad K_\varepsilon(x) = K(x)f(|x|/\varepsilon),$$

where

$$(1.5) \qquad f(r) = \int_0^r s\varphi(s)ds.$$

The vortex method consists of ordinary differential equations for the blob locations obtained from (1.3),

$$(1.6) \qquad \frac{dX_k^\varepsilon}{dt} = v^\varepsilon(X_k^\varepsilon,t) = \sum_{j=1}^{N(\varepsilon)} K_\varepsilon(X_k^\varepsilon(t) - X_j^\varepsilon(t))\Gamma_j^\varepsilon,$$

together with initial conditions to be determined,

$$(1.7) \qquad X_k^\varepsilon(0) = \alpha_k^\varepsilon.$$

The regularity of K_ε for fixed ε implies that solutions of this system exist for all time.

Next we discuss the initialization of the vortex method. Since the initial vorticity ω_0 is not smooth, we define the Γ_j^ε from averages of ω_0, as in [7]. Let j be some smooth function on $\mathbf{R}^2$ of compact support with total weight one, and for $\delta > 0$, let $j_\delta(x) = \delta^{-2}j(x/\delta)$. We replace v_0 and ω_0 with the mollifications

$$(1.8) \qquad v_0^\delta = j_\delta * v_0, \quad \omega_0^\delta = j_\delta * \omega_0;$$

δ will be chosen below as a function $\delta(\varepsilon)$. For $\delta < 1$, ω_0^δ has support in some fixed bounded set. We cover this set with a grid of size h, to be chosen also as a function of ε, and introduce initial locations $\alpha_k^\varepsilon = (k_1, k_2)h$ for integer k. The kth blob is assigned the strength

$$(1.9) \qquad \Gamma_k^\varepsilon = \int_{R_k} \omega_0^\delta dx,$$

where R_k is the square with sides h centered at α_k^ε. The corresponding approximation to the initial vorticity is therefore

$$(1.10) \qquad \omega_0^\varepsilon(x) = \sum_j \varphi_\varepsilon(x - \alpha_j^\varepsilon)\Gamma_j^\varepsilon.$$

We shall say that a vector function $v(x,t)$ is a classical weak solution of the Euler equations on $0 \le t \le T$ with initial velocity v_0 if (i) for any vector test function $\Phi(x,t)$ in $C_0^\infty(\mathbf{R}^2 \times (0,T))$ with $div\ \Phi = 0$,

$$\iint (v \cdot \Phi_t + v \cdot (v \cdot \nabla)\Phi)dx\,dt = 0;$$

(ii) $div\ v = 0$ in the distributional sense; (iii) $v(\cdot, t)$ is a Lipschitz continuous function of t with values in some negative Sobolev space; and (iv) $v(\cdot, t) = v_0$. The same definition was used in [7]. We will actually verify the weak form of the equation for a much larger class of test functions than C_0^∞; see Lemma 2.1. We can now state the convergence result for classical weak solutions.

THEOREM 1. *Suppose an initial velocity field v_0 is given which is locally in L^2 of space. Assume that the vorticity $\omega_0 = \nabla \times v_0$ is in L^p for some $p > 1$ and has bounded support. For each $\varepsilon > 0$, let $\Gamma_j^\varepsilon, X_j^\varepsilon(t), v^\varepsilon(x, t)$ be determined by the vortex blob approximation described above for $0 \leq t \leq T$. Assume the parameters are chosen so that $\delta(\varepsilon) = \varepsilon^\sigma$ for some σ with $0 < \sigma < 1/4$, and $h(\varepsilon) \leq C\varepsilon^4 exp(-C_0\varepsilon^{-2})$ for a certain constant C_0 and any C. Then a subsequence of $\{v^\varepsilon\}$ converges, as $\varepsilon \to 0$, to a classical weak solution of the Euler equations with initial velocity v_0 and vorticity ω_0. The convergence takes place strongly in L^2 of any bounded region in space–time with $0 \leq t \leq T$.*

It will be evident below that if $p > 2$ the initial smoothing is not necessary, i.e., we may use ω_0 rather than ω_0^δ in (1.9). In fact, in the argument below we assume $p < 2$.

If ω_0 is a measure, the vortex blob approximation already described leads to a measure–valued solution. We do not give the definition of the measure–valued solution here, but refer instead to [7]. We now state our result in this case (cf. the result of [7], Section 2).

THEOREM 2. *Suppose ω_0 is a Radon measure $\mathbf{R}^2$ of finite mass and bounded support, and suppose that the corresponding velocity field is locally L^2 on $\mathbf{R}^2$. Define vortex blob approximations as before for $0 \leq t \leq T$, with $\delta(\varepsilon) = \varepsilon^\sigma$ for some σ with $0 < \sigma \leq 1/7$, and choose $h(\varepsilon) \leq C\varepsilon^6 exp(-C_0\varepsilon^{-2})$. Then a subsequence of $\{v^\varepsilon\}$ converges as $\varepsilon \to 0$ to a measured-valued solution of the Euler equations with specified initial condition. The convergence is strong in $L^p(\Omega)$ for $1 < p < 2$, and weak in $L^2(\Omega)$, for any bounded region Ω of space-time with $0 \leq t \leq T$.*

For a more detailed description of the limit measure-valued solution and the nature of the convergence, see Theorem 1.1 of [7] and the discussion preceding it. It will be seen below that we obtain Theorem 2 by a simple modification of the proof of Theorem 1.

2. Proof of Theorem 1. To begin the proof of Theorem 1, we discuss bounds on various quantities related to the vorticity. We will need an estimate for ω_0^δ in L^3. Since $\omega_0^\delta = j_\delta * \omega_0$ we have

$$|\omega_0^\delta|_{L^3} \leq |j_\delta|_{L^q} |\omega_0|_{L^p}, \quad \frac{1}{q} = \frac{4}{3} - \frac{1}{p}.$$

It is evident that

(2.1) $$|j_\delta|_{L^q} = 0(\delta^{-2}\delta^{2/q})$$

so that, assuming $p < 3$,

(2.2) $$|\omega_0^\delta|_{L^3} \leq C\delta^{-\beta}|\omega_0|_{L^p}$$

with

(2.3) $$\beta = 2\left(1 - \frac{1}{q}\right) = 2\left(\frac{1}{p} - \frac{1}{3}\right) \leq \frac{4}{3}.$$

We will use the passive transport $\bar{\omega}^\varepsilon(x,t)$ of ω_0^δ by the flow determined by the blob approximation, i.e., the solution of

$$\bar{\omega}_t^\varepsilon + v^\varepsilon \cdot \nabla \bar{\omega}^\varepsilon = 0, \quad \bar{\omega}^\varepsilon(\cdot, 0) = \omega_0^\delta.$$

Since $\nabla \cdot v^\varepsilon = 0$ the flow is area-preserving, and thus

(2.4)
$$|\bar{\omega}^\varepsilon(\cdot, t)|_{L^p} = |\omega_0^\delta|_{L^p} \le C|\omega_0|_{L^p}$$

(2.5)
$$|\bar{\omega}^\varepsilon(\cdot, t)|_{L^3} = |\omega_0^\delta|_{L^3} \le C\delta^{-\beta}|\omega_0|_{L^p}.$$

Next we consider the sum (1.1) as a discretization of a convolution. Let X_ε^t be the flow determined by the blob vorticity; for an initial point $\alpha \in \mathbf{R}^2$, $x(t) = X_\varepsilon^t(\alpha)$ is the solution of

$$\frac{dx}{dt} = v^\varepsilon(x,t), \quad x(0) = \alpha,$$

with V^ε given by (1.3). We will need a crude bound for the Jacobian $\partial X_\varepsilon^t/\partial\alpha$. Using the above, we have

$$\frac{dJ}{dt} = \sum_j \frac{\partial K_\varepsilon}{\partial x}(x - X_j^\varepsilon)\Gamma_j \cdot J, \quad J(0) = I.$$

It is easily seen from (1.4) that $|\nabla K_\varepsilon| \le C\varepsilon^{-2}$. Moreover

(2.6)
$$\sum_j |\Gamma_j| \le \sum_j \int_{R_j} |\omega_0^\delta| dx = |\omega_0^\delta|_{L^1} \le C|\omega_0|_{L^1}.$$

Thus for $t \le T$,

(2.7)
$$|J(\alpha,t)| \le exp(C_1\varepsilon^{-2}|\omega_0|_{L^1}T).$$

Now suppose $g(\alpha)$ is a C^1 function; we compare

$$\int g(\alpha)\omega_0^\delta(\alpha)d\alpha \quad with \quad \sum_j g(\alpha_j)\Gamma_j.$$

On R_j we have $|g(\alpha) - g(\alpha_j)| \le h|\nabla g|_{L^\infty}$ so that

$$|\int_{R_j} g(\alpha)\omega_0^\delta(\alpha)d\alpha - g(\alpha_j)\Gamma_j|$$
$$= |\int_{R_j} (g(\alpha - g(\alpha_j))\omega_0^\delta(\alpha)d\alpha| \le h|\nabla g|_{L^\infty}|\omega_0^\delta|_{L^1(R_j)}.$$

Summing over j gives

(2.8)
$$|\int g(\alpha)\omega_0^\delta(\alpha)d\alpha - \sum_j g(\alpha_j)\Gamma_j| \le h|\nabla g|_{L^\infty}|\omega_0|_{L^1}.$$

We use this to compare (1.1) with the corresponding integral. Since

$$\omega^\varepsilon(x,t) = \sum_j \varphi_\varepsilon(x - X_\varepsilon^t(\alpha_j))\Gamma_j$$

we may apply (2.8) with $g(\alpha) = \varphi_\varepsilon(x - X_\varepsilon^t(\alpha))$. We obtain

$$\omega^\varepsilon(x,t) = \int \varphi_\varepsilon(x - X_\varepsilon^t(\alpha))\omega_0^\delta(\alpha)d\alpha + E_1$$

$$= \int \varphi_\varepsilon(x - y)\bar\omega^\varepsilon(y,t)dy + E_1$$

$$= (\varphi_\varepsilon * \bar\omega^\varepsilon)(x,t) + E_1$$

with

(2.9)
$$|E_1(x,t)| \leq h|\omega_0|_{L^1}|\nabla\varphi_\varepsilon|_{L^\infty}|J|_{L^\infty}$$
$$\leq Ch\varepsilon^{-3}|\omega_0|_{L^1}exp(C_1\varepsilon^{-2}|\omega_0|_{L^1}T).$$

The error E_1 will be small if we choose h small enough relative to ε, as in Theorem 1.

Next we show that $\omega^\varepsilon(\cdot,t)$ is uniformly bounded in L^p. We saw above that ω^ε is uniformly close to $\varphi_\varepsilon * \bar\omega^\varepsilon$. We know from (2.4) that $\bar\omega^\varepsilon$, and therefore $\varphi_\varepsilon * \bar\omega^\varepsilon$, is uniformly bounded in L^p. First we have the simple estimate

(2.10)
$$|\omega^\varepsilon(\cdot,t)|_{L^1} \leq \sum_j C|\Gamma_j| \leq C|\omega_0|_{L^1},$$

using (1.9). Now since ω^ε is uniformly bounded in L^1, the measure of the set $\{x : |\omega^\varepsilon(x,t)| \geq 1\}$ is also uniformly bounded. On this set, $\omega^\varepsilon(\cdot,t)$ is close in L^∞, and therefore in L^p, to $\varphi_\varepsilon * \bar\omega^\varepsilon(\cdot,t)$, which is bounded in L^p. Thus the L^p norm of $\omega^\varepsilon(\cdot,t)$ on this set is bounded. On the remaining set, where $|\omega^\varepsilon| \leq 1$, we have $|\omega^\varepsilon|^p \leq |\omega^\varepsilon|$, so that the L^p norm is bounded in terms of the L^1 norm. In summary,

(2.11)
$$|\omega^\varepsilon(\cdot,t)|_{L^p} \leq C, \ \varepsilon > 0, \ 0 \leq t \leq T,$$

the constant depending on $|\omega_0|_{L^p}$ and $|\omega_0|_{L^1}$. In just the same way we can argue using (2.5) that

(2.12)
$$|\omega^\varepsilon(\cdot,t)|_{L^3} \leq C\delta^{-\beta}.$$

It follows from (2.11), (2.12) and the Calderon–Zygmund inequality that

(2.13)
$$|\nabla v^\varepsilon(\cdot,t)|_{L^p} \leq C,$$
(2.14)
$$|\nabla v^\varepsilon(\cdot,t)|_{L^3} \leq C\delta^{-\beta}.$$

In the first inequality we have used the fact that $p > 1$. From Sobolev's inequality we then have

(2.15)
$$|v^\varepsilon(\cdot,t)|_{L^{p^*}} \leq C, \quad \frac{1}{p^*} = \frac{1}{p} - \frac{1}{2},$$

provided $p < 2$.

In order to check the consistency of the vortex blob approximation with the Euler equations as a $\varepsilon \to 0$, we examine the error E in satisfying the vorticity evolution equation,

$$E = \frac{\partial \omega^\varepsilon}{\partial t} + v^\varepsilon \cdot \nabla \omega^\varepsilon.$$

Differentiating (1.1) and substituting from (1.6), we have, as in [7], equation (2.37),

(2.16)
$$E(x,t) = \sum_j [v^\varepsilon(x,t) - v^\varepsilon(X_j^\varepsilon, t)] \cdot \nabla \varphi_\varepsilon(x - X_j^\varepsilon) \Gamma_j^\varepsilon$$

$$= \nabla \cdot \left\{ \sum_j [v^\varepsilon(x,t) - v^\varepsilon(X_j^\varepsilon, t)] \varphi_\varepsilon(x - X_j^\varepsilon) \Gamma_j^\varepsilon \right\} \equiv \nabla \cdot F(x,t)$$

We will estimate F in L^1 of space by

(2.17)
$$|F(\cdot, t)|_{L^1} \leq \sum_j |F_j(\cdot, t)|_{L^1} |\Gamma_j|,$$

where

$$F_j(x,t) = \left[v^\varepsilon(x,t) - v^\varepsilon(X_j^\varepsilon, t) \right] \varphi_\varepsilon(x - X_j^\varepsilon)$$

$$= \int_0^1 \nabla v^\varepsilon(sx + (1-x)X_j^\varepsilon) \cdot (x - X_j^\varepsilon) \varphi_\varepsilon(x - X_j^\varepsilon) ds$$

$$\equiv \int_0^1 f_j(x, s, t) ds.$$

We set $\mu_\varepsilon(z) = z \varphi_\varepsilon(z)$, so that

$$f_j(x, s, t) = \nabla v^\varepsilon(sx + (1-s)X_j^\varepsilon) \cdot \mu_\varepsilon(x - X_j^\varepsilon).$$

We will estimate the x–integral of $|f_j|$ using Holder's inequality. Since $\mu_\varepsilon(z) = \varepsilon^{-1} \mu(z/\varepsilon)$,

$$|\mu_\varepsilon|_{L^r} \leq C \varepsilon^{-1+2/r},$$

which is small if $r < 2$. We choose $r = 3/2$ and bound the other factor in L^3. We saw in (2.14) that $|\nabla v^\varepsilon|_{L^3} \leq C \delta^{-\beta}$ with some $\beta \leq 4/3$. Thus after rescaling, the L^3 norm of $\nabla v^\varepsilon(sx + (1-s)X_j^\varepsilon)$, as a function of x, is bounded by $C s^{-2/3} \delta^{-\beta}$. Therefore,

$$\int |f_j(x, s, t)| ds \leq C s^{-2/3} \delta^{-\beta} \varepsilon^{1/3},$$

and integrating in s,

(2.18)
$$\int |F_j(x,t)| dx \leq C \delta^{-\beta} \varepsilon^{1/3}.$$

Combining the last inequality with (2.17), we have

$$|F(\cdot,t)|_{L^1} \le C \delta^{-\beta} \varepsilon^{1/3} \sum_j |\Gamma_j| \le C \delta^{-\beta} \varepsilon^{1/3} |\omega_0|_{L^1}.$$

If $\delta = \varepsilon^{\sigma}$, we have a power of ε of $1/3 - \sigma\beta > (1 - 4\sigma)/3 \equiv a$; this is positive provided $\sigma < 1/4$, and we have

(2.19) $$|F(\cdot,t)|_{L^1} \le C\varepsilon^a, \quad some \ a > 0.$$

We now use (2.19) to check the weak consistency of the vortex blob approximation as $\varepsilon \to 0$; that is, we show that for suitable test functions $\Phi(x,t)$ with $div \ \Phi = 0$,

(2.20) $$\int_0^T \int (v^\varepsilon \cdot \Phi_t + v^\varepsilon \cdot (v^\varepsilon \cdot \nabla)\Phi)dxdt \to 0$$

as $\varepsilon \to 0$. More specifically, we have the following, which implies (2.20).

LEMMA 2.1. *For every vector test function* Φ *with divergence zero in* $W^{s,q}(\mathbf{R}^2)$, *where* $1 < q < \infty$, $s \ge 1$ *and* $s > 2/q$, *we have for* $0 \le t \le T$,

(2.21) $$\left| \int (v_t^\varepsilon \cdot \Phi - v^\varepsilon \cdot (v^\varepsilon \cdot \nabla)\Phi dx \right| \le C\varepsilon^a |\Phi|_{W^{s,q}}.$$

Proof. We first check that the integral is defined. From (1.3), (1.6) we have

$$v_t^\varepsilon = -\sum_j v^\varepsilon(X_j^\varepsilon,t) \cdot \nabla K_\varepsilon(x - X_j^\varepsilon)\Gamma_j.$$

Under our present assumptions, $\nabla K_\varepsilon \in L^r$ for $r > 1$ and v^ε is bounded, so that $v_t^\varepsilon \in L^r$, $r > 1$. Also $v^\varepsilon \in L^b$ for $b > 2$ so that $v^\varepsilon \cdot v^\varepsilon \in L^r$ for $r > 1$. Thus each of the two terms makes sense. We have chosen $s > 2/q$ so that

(2.22) $$|\Phi|_{L^\infty} \le C|\Phi|_{W^{s,q}}$$

by Sobolev's Lemma.

Next we interpret the equation

(2.23) $$\omega_t^\varepsilon + v^\varepsilon \cdot \nabla\omega^\varepsilon = E$$

in terms of v^ε. As in the remarks above, $u^\varepsilon \equiv v_t^\varepsilon + v^\varepsilon \cdot \nabla v^\varepsilon$ is in L^r for each $r > 1$. The projection P onto divergence–free vector fields is bounded on L^r, and we may write $u^\varepsilon = Pu^\varepsilon + \nabla p^\varepsilon$ with $\nabla p^\varepsilon \in L^r$. Moreover, $\nabla \times (Pu^\varepsilon) = \nabla \times u^\varepsilon = E$, using (2.23), so that $Pu^\varepsilon = K * E$. Thus

(2.24) $$v_t^\varepsilon + v^\varepsilon \cdot \nabla v^\varepsilon + \nabla p^\varepsilon = K * E.$$

Now suppose $\Phi \in C_0^\infty$, not necessarily with divergence zero. Multiplying (2.24) by Φ and integrating by parts, we have

$$\int (v_t^\varepsilon \Phi - v^\varepsilon \cdot (v^\varepsilon \cdot \nabla)\Phi + \nabla p^\varepsilon \cdot \Phi) dx$$
$$= \int (K * E) \cdot \Phi dx = - \int E \cdot (K * \Phi) dx$$
$$= - \int (\nabla \cdot F)(K * \Phi) dx = \int F \cdot \nabla (K * \Phi) dx,$$

where $K * \Phi = K_1 * \Phi_1 + K_2 * \Phi_2$. The operator $\nabla(K * \cdot)$ is bounded on $W^{s,q}$ since it is a second derivative of the Green's function. The last integral is therefore bounded by

$$|F|_{L^1}|\nabla(K * \Phi)|_{L^\infty} \leq |F|_{L^1}|\nabla(K * \Phi)|_{W^{s,q}}$$
$$\leq |F|_{L^1}|\Phi|_{W^{s,q}} \leq C\varepsilon^a|\Phi|_{W^{s,q}}.$$

Finally if $\Phi \in W^{s,q}$ and $\nabla \cdot \Phi = 0$, we may approximate Φ in $W^{s,q}$ by a sequence of test functions Φ_N for which the above argument shows

$$\left| \int (v_t^\varepsilon \cdot \Phi_N - v^\varepsilon \cdot (v^\varepsilon \cdot \nabla)\Phi_N + \nabla p^\varepsilon \cdot \Phi_N) dx \right| \leq C\varepsilon^a|\Phi_N|_{W^{s,q}}.$$

Regarding $v_t^\varepsilon, v^\varepsilon \cdot v^\varepsilon$ and ∇p^ε as elements of $L^{q'}$, with q' dual to q, we may let $\Phi_N \to \Phi$ in this inequality; the pressure term goes to zero since $\nabla \cdot \Phi = 0$. Thus (2.21) is established, and the proof of the Lemma is complete.

We shall also need an estimate for v_t^ε. From (2.21) we have, for Φ as in the Lemma,

$$(2.25) \qquad \left| \int v_t^\varepsilon \cdot \Phi dx \right| \leq \int |v^\varepsilon \cdot (v^\varepsilon \cdot \nabla)\Phi| dx + C\varepsilon^a|\Phi|_{W^{s,q}}.$$

We saw in (2.15) that v^ε is bounded in L^{p*}, $p* > 2$. Thus $v^\varepsilon \cdot v^\varepsilon \in L^r$ with $r = p*/2 > 1$. It is convenient to choose q dual to r, and we then find

$$(2.26) \qquad \left| \int v_t^\varepsilon \cdot \Phi dx \right| \leq C|\Phi|_{W^{s,q}}$$

provided $div\, \Phi = 0$, $\Phi \in W^{s,q}$, $s \geq 1$, and $s > 2/q$. Since $div\, v^\varepsilon = 0$ and the projection onto divergence–free vector fields is a bounded operator on $W^{s,q}$, (2.26) holds for arbitrary $\Phi \in W^{s,q}$. We may therefore conclude that

$$(2.27) \qquad |v_t^\varepsilon|_{W^{-s,r}} \leq C, \quad r = p*/2.$$

Finally we verify that a classical weak solution is obtained in the limit $\varepsilon \to 0$. The argument is similar to the proofs of Theorem 1.1 and 1.2 in [7]. By (2.15) we know that

$$(2.28) \qquad v^\varepsilon \text{ is uniformly bounded in } L^{p^*}(\mathbf{R}^2 \times (0,T))$$

with $p^* = 2p/(2-p) > 2$. We may select a subsequence $\varepsilon_n \to 0$ so that v^ε converges weakly in L^{p^*} to a limit v. We also have bounds on the x– and t–derivatives of v^ε: from (2.13)

(2.29) $$\nabla v^\varepsilon \text{ is bounded in } L^\infty(0,T;L^p(\mathbf{R}^2))$$

and from (2.27),

(2.30) $$v_t^\varepsilon \text{ is bounded in } L^\infty(0,T;W^{-s,r}(\mathbf{R}^2)).$$

for a certain choice of (s,r). We may as well assume that we have chosen ε_n so that also

(2.31) $$v_t^\varepsilon \to v_t \text{ weakly in } L^2(0,T;W^{-s,r}(\mathbf{R}^2)).$$

The estimates (2.28), (2.29), (2.30) enable us to use compactness on finite subregions. Let R be any finite radius, and ρ any test function in $C_0^\infty(B_R)$, where $B = \{x : |x| < R\}$. We may regard $\{\rho v^\varepsilon\}$ as a bounded set in $L^2(0,T;W^{1,p}(B_R))$. The Sobolev imbedding of $W^{1,p}(B_R)$ in $L^m(B_R)$ is compact for $1 < m < p^*$; since $p^* > 2$, we may choose $m = 2$. Moreover, the subspace of $W^{-s,r}(\mathbf{R}^2)$ with support in B_R contains $L^2(B_R)$ if s is large enough so that $W^{s,r^*}(B_R) \subseteq L^2(B_R)$. Thus we can conclude from the Lions– Aubin Lemma (e.g., see [20], Theorem III.2.1) that $\{\rho v^\varepsilon\}$ has compact closure in $L^2(B_R \times (0,T))$. Along our subsequence v^{ε_n} which converges weakly in $L^{p^*}(\mathbf{R}^2 \times (0,T))$, any convergent subsequence in $L^2(B_R \times (0,T))$ must have the same limit. Therefore

(2.32) $$v^{\varepsilon_n} \to v \text{ in } L^2(B_R \times (0,T)), \text{ any } R > 0.$$

It now follows immediately from the weak consistency of (2.20), (2.21) that

$$\int_0^T (v \cdot \Phi_t + v \cdot (v \cdot \nabla)\Phi)dxdt = 0$$

for any test function Φ with $div\ \Phi = 0$; i.e., v is a classical weak solution of the Euler equations. Moreover the bound (2.30) for v_t^ε and (2.31) imply that v is Lipschitz continuous in time in $W^{-s,r}$ and $v(0) = v_0$. This completes the proof of the theorem.

In the case $p > 2$ we may carry out the estimate of F_j in (2.18) differently so that there is no dependence on δ. In applying Holder's inequality to $f_j = \nabla v^\varepsilon \cdot \mu_\varepsilon$, we can estimate ∇v^ε on L^p and μ_ε in L^r, with r dual to p. Since ω^ε is bounded in L^p, ∇v^ε is also, according to (2.13). Thus (2.18) is replaced by

$$\int |F_j(x,t)|dx \le C|\mu_\varepsilon|_{L^r} = C\varepsilon^{-1+2/r} = C\varepsilon^{1-2/p},$$

which is small. Therefore in this case our argument applies without the initial smoothing by j_δ.

3. Proof of Theorem 2. According to Theorem 1.1 of [7], the conclusions of Theorem 2 will be established if we show that

(1) v^ε has locally bounded kinetic energy,

$$\int_{|x|\leq R} |v^\varepsilon(x,t)|^2 dx \leq C(R), \quad 0 \leq t \leq T,$$

(2) ω^ε is uniformly bounded in $L^1(\mathbf{R}^2)$,

(3) v^ε is weakly consistent with the Euler equations, in the same sense as before,

(4) v^ε is uniformly Lipschitz with respect to t in some negative Sobolev space, and

(5) v^ε vanishes uniformly at ∞, in the sense of Remark 1.1 of [7].

We will decompose $v^\varepsilon(x,t)$ as $v^\varepsilon = \tilde{v}^\varepsilon(x,t) + \bar{v}(x)$, where $\bar{v}$ is a steady, smooth solution of the Euler equations and $\tilde{v}^\varepsilon$ is in $L^2(\mathbf{R}^2)$ at each time with zero total circulation. To do this we set

(3.1) $$\bar{\omega}(x) = \Gamma_0 \varphi_1(x), \quad \Gamma_0 = -\sum_j \Gamma_j = -\int \omega_0,$$

and $\bar{v} = K * \bar{\omega}$, $\tilde{v}^\varepsilon = v^\varepsilon - \bar{v}$, $\tilde{\omega}^\varepsilon = \omega^\varepsilon - \bar{\omega}$. For large x, $\tilde{v}^\varepsilon$ is a linear combination of translates of K; since the coefficients sum to zero, $\tilde{v}^\varepsilon = O(|x|^{-2})$ as $x \to \infty$, and therefore $\tilde{v}^\varepsilon \in L^2$. We will show below that

(3.2) $$|\tilde{v}^\varepsilon(\cdot,t)|_{L^2} \leq C, \quad 0 \leq t \leq T;$$

then (1) and (5) above follow directly. (2) was shown earlier in (2.10), with $|\omega_0|_{L^1}$ replaced by the total mass $\|\omega_0\|$ in the present case. The argument leading to (2.19) and the proof of Lemma 2.1 apply without change since they only used bounds on ω_0^δ in L^1, and thus (3) holds. Finally, (4) will be verified after (3.2) is established. As a first step toward (3.2), we modify the previous treatment of weak consistency to estimate the error F in L^2 norms:

LEMMA 3.1. *With the choice of parameters of Theorem 2, $F(x,t)$, defined by (2.16), satisfies*

$$|F(\cdot,t)|_{L^2} \leq C.$$

Proof. With notation as before, we write

(3.3) $$F(x,t) = \sum_j F_j(x,t)\Gamma_j = \sum_j F_j(x,t)\gamma_j h^2$$

with $\gamma_j = j^{-2}\Gamma_j$. We note first that

(3.4) $$\sum_j \gamma_j^2 h^2 = \sum_j h^{-2}\left(\int_{R_j} \omega_0^\delta dx\right)^2 \leq \sum_j h^{-2} \cdot h^2 \int_{R_j} |\omega_0^\delta|^2 dx = |\omega_0^\delta|_{L^2}^2.$$

We may estimate $|\omega_0^\delta|_{L^2}$ just as in (2.1), (2.2) and find

(3.5)
$$|\omega_0^\delta|_{L^2} \leq \delta^{-1}\|\omega_0\|.$$

Now according to a standard criterion for boundedness of integral operators, we will have

(3.6)
$$|F(\cdot,t)|_{L^2}^2 \leq M^2 \sum_j \gamma_j^2 h^2$$

provided we show that

(3.7)
$$\int |F_j(x,t)| \leq M, \quad each \quad j,$$

and

(3.8)
$$\sum_j |F_j(x,t)| h^2 \leq M, \quad each \quad x,$$

The first of these was established in (2.18) with $M = C\delta^{-4/3}\varepsilon^{1/3}$. To estimate the second we will replace the sum with an integral. With x fixed, let

$$g(\alpha) \; [v^\varepsilon(x,t) - v^\varepsilon(y,t)]\varphi_\varepsilon(x - y), \quad y = X_\varepsilon^t(\alpha).$$

Then
$$\sum_j |F_j(x,t)| h^2 = \sum_j |g(jh)| h^2,$$

which is a discretization of

(3.9)
$$\int |g(\alpha)| d\alpha = \int |v^\varepsilon(x,t) - v^\varepsilon(y,t)| \; |\varphi_\varepsilon(x-y)| dy.$$

As in our earler discussion of quadrature, the difference between the sum and the integral is bounded by $h \cdot |\nabla g|_{L^\infty}$. Now $|\nabla K_\varepsilon| \leq C\varepsilon^{-2}$ so that

$$|\nabla v^\varepsilon(x,t)| \leq \sum_j C\varepsilon^{-2}|\Gamma_j| \leq C\varepsilon^{-2}\|\omega_0\|.$$

Estimating as before we see that the discretization error is bounded by

$$Ch(\varepsilon^{-1} + \varepsilon^{-2}\|\omega_0\|exp(C_1\varepsilon^{-2}\|\omega_0\|T)) \cdot (\varepsilon^{-3}exp(C_1\varepsilon^{-2}\|\omega_0\|T))$$
$$\leq Ch\varepsilon^{-5}\|\omega_0\|exp(2C_1\varepsilon^{-2}\|\omega_0\|T).$$

We choose h in terms of ε as in Theorem 2 so that this quantity is $O(\varepsilon)$. We have now reduced (3.6) to showing that (3.9) is bounded by M for each x. But this is essentially the same as estimating (3.7), with (y,x) taking the place of (x, X_j^ε) previously. Thus the argument leading to (3.7) also shows that (3.9) is bounded by the M defined above. In view of the discretization estimate, (3.8) is now verified. Combining (3.4), (3.5), (3.6) we have

(3.10) $$|F(\cdot,t)|_{L^2} \le C\delta^{-7/3}\,\varepsilon^{1/3} \le C\varepsilon^\gamma, \quad \gamma \ge 0,$$

provided $\delta = \varepsilon^\sigma$ with $\sigma \le 1/7$. This completes the proof of the lemma.

We will now obtain the bound (3.2) by an energy estimate for $\tilde{v}^\varepsilon$. Using the decomposition for v^ε, we have the equation for $\tilde{v}^\varepsilon$, as in (2.24)

$$\tilde{v}^\varepsilon_t + (v^\varepsilon \cdot \nabla)\tilde{v}^\varepsilon + (\tilde{v}^\varepsilon \cdot \nabla)\bar{v} + \nabla p = K * (\nabla \cdot F).$$

We estimate in the usual way, obtaining

$$\frac{1}{2}\frac{d}{dt}|\tilde{v}^\varepsilon|^2_{L^2} \le c|\nabla\bar{v}|_{L^\infty}|\tilde{v}^\varepsilon|^2_{L^2} + |K^*(\nabla \cdot F)|_{L^2}|\tilde{v}^\varepsilon|_{L^2}.$$

Now $F \to K^*(\nabla \cdot F)$ is a bounded operator on L^2, since it is a composition of two derivatives with the Green's function. Thus the inequality gives exponential growth of the norm, and

$$|\tilde{v}^\varepsilon(\cdot,t)|_{L^2} \le C(T)|\tilde{v}^\varepsilon_0|_{L^2}, \quad 0 \le t \le T.$$

The proof of (3.2) is now reduced to bounding the initial $\tilde{v}^\varepsilon$ in L^2. We have

$$\tilde{v}^\varepsilon_0(x) = \sum_j K_\varepsilon(x - \alpha_j)\Gamma_j + K_1(x)\Gamma_0.$$

Since ω_0 has compact support, the α_j lie in some fixed radius, say $R \ge 1$. The sum is a discretization of the integral

$$\int K_\varepsilon(x - \alpha)\omega_0^\delta(\alpha)d\alpha = \varphi_\varepsilon * (j_\delta * v_0)$$

which, by hypothesis, is bounded in L^2. The discretization error can be bounded pointwise, using (2.8), by $Ch\varepsilon^{-2}$, which is small for our choice of h. It follows that v^ε_0, and therefore $\tilde{v}^\varepsilon_0$, is uniformly bounded in L^2 of $\{|x| < 2R\}$. For $|x| > 2R$, each K_ε is just K, and we can write

$$\tilde{v}^\varepsilon_0(x) = \sum_j [K(x - \alpha_j) - K(x)]\Gamma_j.$$

It is easy to see that this is bounded by

$$\sum_j C|x|^{-2}|\Gamma_j| \le C\|\omega_0\|\ |x|^{-2}$$

for $|x| > 2R$, and thus bounded in L^2. This completes the verification of (3.2).

It remains to verify the Lipschitz continuity of v^ε in time. We return to (2.25), which was a consequence of Lemma 2.1. We will choose $q = 3$ and $s = 2$, so that $\Phi \in W^{s,q}$ implies $\nabla\Phi \in L^\infty$ and therefore $\nabla\varphi \in L^r$ with $3 \le r \le \infty$. We use this fact to estimate the integral on the right in (2.25). We had $\tilde{v}^\varepsilon$ bounded in L^2, and $\bar{v} \in L^3$, for example. Thus the product $v^\varepsilon \cdot v^\varepsilon$ is a sum of terms in L^p for $1 \le p \le 3/2$. Since $\nabla\Phi$ is bounded in the dual spaces for such L^p, we may estimate the integral using Holder's inequality. We conclude as before that

$$|v^\varepsilon_t(\cdot,t)|_{W^{-2,3/2}} \le C,$$

so that v^ε is Lipschitz continuous in $W^{-2,3/2}$. We have now verified all the conditions (1)–(5) for the convergence to a measure-valued solution.

REFERENCES

[1] J. T. BEALE AND A. MAJDA, *High order accurate vortex methods with explicit velocity kernels*, J. Comput. Phys., 58 (1985), pp. 188–208.

[2] J. T. BEALE AND A. MAJDA, *Vortex methods for fluid flow in two or three dimensions*, Contemp. Math., 23 (1984), pp. 221–229.

[3] R. CAFLISCH AND J. LOWENGRUB, *Convergence of the vortex method for vortex sheets*, preprint.

[4] J. P. CHOQUIN, G. H. COTTET, AND S. MAS-GALLIC, *On the validity of vortex methods for nonsmooth flows*, in *Vortex Methods* (C. Anderson and C. Greengard, editors), Lecture Notes in Mathematics, Springer–Verlag, pp. 56–67.

[5] G.H. COTTET, Thèse d'Etat, Université Pierre et Marie Curie.

[6] R. DiPERNA AND A. MAJDA, *Oscillations and concentrations in weak solutions of the incompressible fluid equations*, Commun. Math. Phys., 108 (1987), pp. 667–689.

[7] R. DiPERNA AND A. MAJDA, *Concentrations and regularizations for 2-D incompressible flow*, Comm. Pure Appl. Math., 60 (1987), pp. 301–45.

[8] R. DiPERNA AND A. MAJDA, *Reduced Hausdorff dimension and concentration- cancellation for 2-D incompressible flow*, J. Amer. Math. Soc., 1 (1988), pp. 59–95.

[9] J. GOODMAN, T. HOU, AND J. LOWENGRUB, *Convergence of the point vortex method for the 2-D Euler equations*, preprint.

[10] C. GREENGARD AND E. THOMANN, *On DiPerna–Majda concentration sets for two-dimensional incompressible flow*, Comm. Pure Appl. Math., 41 (1988), pp. 295–303.

[11] O. HALD, *The convergence of vortex methods, II*, SIAM J. Numer. Anal., 16 (1979), pp. 726–755.

[12] R. KRASNY, *Desingularization of periodic vortex sheet roll-up*, J. Comput. Phys., 65 (1986), pp. 292–313.

[13] R. KRASNY, *Computation of vortex sheet roll-up in the Trefftz plane*, J. Fluid Mech., 184 (1987), p. 123.

[14] R. KRASNY, *Computation of vortex sheet roll-up*, in *Vortex Methods* (C. Anderson and C. Greengard, editors), Lecture Notes in Mathematics, Springer- Verlag, pp. 9–22.

[15] J. LOWENGRUB, *Convergence of the vortex method for vortex sheets*, Thesis, New York University, 1988.

[16] A. MAJDA, *Vorticity and the mathematical theory of incompressible fluid flow*, Comm. Pure Appl. Math., 39 (1986), pp. S187–S220.

[17] A. MAJDA, *Mathematical fluid dynamics: the interaction of nonlinear analysis and modern applied mathematics*, to appear in the Proc. of the Centennial Celebration of the Amer. Math. Society.

[18] C. MARCHIORO AND M. PULVIRENTI, *Hydrodynamics in two dimensions and vortex theory*, Commun. Math. Phys., 84 (1982), pp. 483–503.

[19] M. SHELLEY AND G. BAKER, *On the relation between thin vortex layers and vortex sheets: Part 2, numerical study*, preprint.

[20] R. TEMAM, *The Navier-Stokes Equations*, North-Holland, Amsterdam, 1977.

LIMIT BEHAVIOR OF APPROXIMATE SOLUTIONS
TO CONSERVATION LAWS

CHEN GUI-QIANG*

Abstract. We are concerned with the limit behavior of approximate solutions to hyperbolic systems of conservation laws. Several mathematical compactness theories and their role are described. Some recent and ongoing developments are reviewed and analyzed.

AMS(MOS) subject classifications. 35–02, 41–02, 35B25, 35D05, 35L65, 46A50, 46G10, 65M10.

1. Introduction. We are concerned with the limit behavior of approximate solutions to hyperbolic systems of conservation laws. The Cauchy problem for a system of conservation laws in one space dimension is of the following form:

$$(1.1) \qquad u_t + f(u)_x = g(x, t, u),$$

$$(1.2) \qquad u \Big|_{t=0} = u_0(x),$$

where $u = u(x, t) \in \mathbf{R}^n$ and both f and g are smooth nonlinear functions $\mathbf{R}^n$ to $\mathbf{R}^n$. The system is called strictly hyperbolic in a domain $\mathcal{D}$ if the Jacobian $\nabla f(u)$ has n real and distinct eigenvalues

$$(1.3) \qquad \lambda_1(u) < \lambda_2(u) < \ldots < \lambda_n(u)$$

at each state $u \in \mathcal{D}$. If the Jacobian $\nabla f(u)$ has n real and indistinct eigenvalues $\lambda_i(u)(i = 1, 2, \ldots, n)$ in $\mathcal{D}$, one calls the system nonstrictly hyperbolic in $\mathcal{D}$. An eigenfield corresponding to λ_j is genuinely nonlinear in the sense of Lax [LA2] if λ_j's derivative in the corresponding eigendirection never vanishes, i.e.,

$$(1.4) \qquad r_j \cdot \nabla \lambda_j \neq 0,$$

where

$$\nabla f r_j = \lambda_j r_j.$$

The system is called genuinely nonlinear if all of its eigenfields are genuinely nonlinear. Otherwise, one calls the system linearly degenerate.

The quasilinear systems of conservation laws result from the balance laws of continuum physics and other fields (e.g., conservation of mass, momentum, and energy) and, therefore, describe many physical phenomena. In particular, important

---

*Partially supported by U.S. NSF Grant # DMS–850403, by CYNSF, and by the Applied Mathematical Sciences subprogram of the Office of Energy Research, U.S. Department of Energy, under Contract W-31-109-Eng-38.

Courant Institute of Mathematical Sciences, 251 Mercer Street, New York, NY 10012 U.S.A.

Current address: Department of Mathematics, The University of Chicago, Chicago, IL 60637.

examples occur in fluid dynamics (see Section 5), solid mechanics (see Section 4), petroleum reservoir engineering (see Section 4), combustion theory and game theory [CR].

Since f is a nonlinear function, solutions of the Cauchy problem (1.1)–(1.2) (even starting from smooth initial data) generally develop singularities in a finite time, and then the solutions become discontinuous functions. This situation reflects the physical phenomenon of breaking of waves and development of shock waves. For this reason, attention focuses on solutions in the space of discontinuous functions, where one cannot directly use the classical analytic techniques that predominate in the theory of partial differential equations of other types.

To overcome this difficulty, one constructs approximate solutions $u^\epsilon(x,t)$ to the following perturbations:

a. Perturbation of equations: One of the perturbation prototypes is the viscosity method; that is, $u^\epsilon(x,t)$ are generated by the corresponding parabolic system of the form

(1.5)
$$\begin{cases} u_t + f(u)_x = g(x,t,u) + \epsilon(D(u)u_x)_x, \\ u\big|_{t=0} = u_0(x), \end{cases}$$

where D is a properly selected and nonnegative matrix. Usually one chooses D to be the unit matrix.

b. Perturbation of Cauchy data: $u^\epsilon(x,t)$ are generated by the following Cauchy problem:

(1.6)
$$\begin{cases} u_t + f(u)_x = g(x,t,u), \\ u\big|_{t=0} = u_0^\epsilon(x). \end{cases}$$

c. Perturbation of both equations and Cauchy data: Besides the viscosity method with perturbated Cauchy data $u\big|_{t=0} = u_0^\epsilon(x)$ (see (1.5)), another perturbation prototype is the difference method; that is, $u^\epsilon(x,t)$ ($\epsilon = \Delta x$, space step length) are generated by the difference equations

$$\begin{cases} D_t u + D_x f(u) = g^\epsilon(x,t,u), \\ u\big|_{t=0} = u_0(x;\epsilon), \end{cases}$$

and then one studies limit behaviors of the approximate solutions $u^\epsilon(x,t)$ as $\epsilon \to 0$: convergence and oscillation. Examples of this approach are the Lax-Friedrichs scheme [LA1], the Glimm scheme [GL], the Godunov scheme [GO], higher-order schemes (e.g., [LW], [SZ], [TA]) and the fractional step schemes (e.g., [DCL2]).

The motivation for using approximate solutions comes from continuum physics, numerical computations, and mathematical considerations. The system of gas dynamics generally involves viscosity terms, although the viscosity coefficient is very small and can be ignored; the initial value function is determined only by using statistical data and some averaging methods described by a weak topology. Numerical computations of systems of conservation laws are limited to calculations of difference equations and discrete Cauchy mesh data. In game theory with non–zero sum, derivative functions of the stochastic game values and the deterministic game values satisfy systems of conservation laws with and without viscosity terms, respectively. Therefore, studying the relationship between the stochastic game and the deterministic game when "noise" disappears is equivalent to studying the limit behavior of the approximate solutions as $\epsilon \to 0$ (see [CR]). Moreover, one expects to use "good" Cauchy data (e.g., total variation functions) to approximate "bad" Cauchy data (e.g., L^∞ functions) to obtain a solution to the Cauchy problem with "bad" Cauchy data. Thus, such a study enables us to understand how the behavior of the system at the microscopic level affects the behavior of the system at the macroscopic level and, therefore, understand the well–posedness of the Cauchy problem (1.1)–(1.2) in a weak topology.

The remainder of this paper has the following organization. Section 2 focuses on compactness theories. Sections 3 and 4 discuss the limit behavior of approximate solutions to the Cauchy problem for the scalar conservation law and for hyperbolic systems of conservation laws, respectively. For concreteness, we focus our attention on homogeneous systems (i.e., $h(u) \equiv 0$) in Sections 3 and 4. Section 5 focuses on approximate solutions generated by the Lax–Friedrichs scheme, the Godunov scheme, and the viscosity method for the homogeneous system of isentropic gas dynamics and on the fractional–step Lax–Friedrichs scheme and Godunov scheme for the inhomogeneous system of isentropic gas dynamics. Section 6 concludes our review with some remarks about distinguishing features of multidimensional conservation laws. The techniques and strategies developed in this direction should be applicable to other interesting problems of nonlinear analysis and their regularizations.

This paper is dedicated to the memory of Ronald J. DiPerna. His life and his work is an inspiration to the author.

2. Compactness Theories. One of the main difficulties in studying nonlinear problems is that, after introducing a suitable sequence of approximations, one needs enough a priori estimates to ensure the convergence of a subsequence to a solution. This argument is based on compactness theories. Here we describe several important compactness theories that have played a significant role in the field of conservation laws.

2.1. Classical Theories

The two important compactness theorems provide natural norms for the field of conservation laws in classical theories of compactness: BV and L^1 compactness theorems.

2.1.1. BV Compactness Theorem

THEOREM 2.1. *There exists a subsequence converging pointwise a.e. in any function sequence that has uniform control on the L^∞ and total variation norms.*

In the context of a strictly hyperbolic system of conservation laws, the L^∞ norm and the total variation norm provide a natural pair of metrics to study the stability of approximate solutions in the sense of L^∞. The L^∞ norm serves as an appropriate measure of the solution amplitude, while the total variation norm serves as an appropriate measure of the solution gradient. The role of these norms is indicated by Glimm's theorem [GL] concerning the stability and convergence of the Glimm approximate solutions, provided that the total variation norm of the initial data $u_0(x)$ is sufficiently small for systems of conservation laws, and by results of Oleinik [OL], Conway and Smoller [CS], and others concerning the stability and convergence of the Lax–Friedrichs and the Godunov approximate solutions with large initial data $u_0(x)$ for the scalar conservation law. The families of approximate solutions $\{u^\epsilon\}$ are stable in the sense that

$$\begin{cases} |u^\epsilon(\cdot, t)|_\infty & \leq \text{const.} \, |u_0|_\infty, \\ \mathrm{TV} \, u^\epsilon(\cdot, t) & \leq \text{const.} \, \mathrm{TV} \, u_0, \end{cases}$$

where constants are independent of ϵ and depend only on the flux function f. Furthermore, there exists a subsequence that converges pointwise a.e. to a globally defined distribution solution u. Until the end of the 1970s, almost all results concerning the stability and convergence of approximate solutions for conservation laws were obtained with the aid of the BV compactness theorem (e.g., [BA, DZ, LLO, NI, NS, SR1, TE1, ZG]).

2.1.2. L^1 Compactness Theorem

A more general compactness framework for conservation laws is the L^1 compactness theorem.

THEOREM 2.2. *A function sequence* $\{u^\epsilon(x)\} \subset L^1(\Omega)$, $\Omega \subset\subset \mathbf{R}^n$, *is strongly compact in* L^1 *if and only if*

(i) $\|u^\epsilon\|_{L^1} \leq M$, M *is independent of* ϵ.

(ii) $\{u^\epsilon(x)\}$ *is equicontinuous in the large, i.e.,* $\forall \, h > 0$, $\exists \delta(h) > 0$ *such that for all* $u^\epsilon \in \{u^\epsilon(x)\}$,

$$\int_\Omega |u^\epsilon(x+y) - u^\epsilon(x)| dx \leq h$$

if only $|y| < \delta(h)$.

In the context of conservation laws, the role of the L^1 norm is indicated by Kruskov's theorem [KR] concerning the stability and convergence of the viscosity approximate solutions and the uniqueness of generalized solutions for the scalar conservation law and by Temple's theorem [TE3] concerning the weak stability of generalized solutions with respect to the initial value for systems of conservation laws.

2.2. The Theory of Compensated Compactness

Weak topology has played an important role in studying linear problems where weak continuity can be used; however, lack of weak continuity in nonlinear problems has long restricted the use of weak topology. The theory of compensated compactness established by Tartar [T1–T3] and Murat [M1–M4] is intended to render weak topology more useful in solving nonlinear problems. In other words, the theory deals with the behavior of nonlinear functions with respect to weak topology, for instance, the weak continuity and the weak lower semicontinuity of nonlinear functions. Here we restrict our attention to that partion of the theory relating to conservation laws.

As is well known, it is difficult to clarify the conditions to ensure weak continuity and weak lower semicontinuity for general nonlinear functions (e.g., [DA, M1–M4, T1–T4]). However, for a 2×2 determinant, a satisfactory result can be obtained [M1–M2, T2].

THEOREM 2.3. *Let* $\Omega \subset \mathbf{R} \times \mathbf{R}^+ = \mathbf{R}^2_+$ *be a bounded open set and* $u^\epsilon : \Omega \to \mathbf{R}^4$ *be measurable functions satisfying*

$$
\left\{
\begin{array}{l}
w - \lim u^\epsilon = u, \quad \text{in } L^2_4(\Omega), \\
\left\{
\begin{array}{l}
\dfrac{\partial u^\epsilon_1}{\partial t} + \dfrac{\partial u^\epsilon_2}{\partial x}, \\
\dfrac{\partial u^\epsilon_3}{\partial t} + \dfrac{\partial u^\epsilon_4}{\partial x},
\end{array}
\right. \quad \text{compact in } H^{-1}_{\text{loc}}(\Omega).
\end{array}
\right.
$$

Then there exists a subsequence (still labeled) u^ϵ *such that*

$$
w - \lim \begin{vmatrix} u^\epsilon_1 & u^\epsilon_2 \\ u^\epsilon_3 & u^\epsilon_4 \end{vmatrix} = \begin{vmatrix} u_1 & u_2 \\ u_3 & u_4 \end{vmatrix}, \quad \text{in the sense of distribution,}
$$

where $w - \lim$ *denotes the weak limit in the sense of weak topology.*

This is one prototype in the theory of compensated compactness. A simpler proof of the theorem can be found in [CH1]. Two questions must be addressed: (1) what in general is the exact relationship between $w - \lim f(u^\epsilon)$, and the nonlinear function f and the sequence u^ϵ ? and (2) can $w - \lim f(u^\epsilon)$ be represented by f and the sequence u^ϵ ? The following theorem answers these questions.

THEOREM 2.4. *Let* $K \subset \mathbf{R}^m$ *be a bounded open set and* $u^\epsilon : \mathbf{R}^n \to \mathbf{R}^m$ *be arbitrary measurable functions,* $u^\epsilon(y) \in K$, *a.e., for* $\epsilon > 0$. *Then there exists a family of probability measures* $\nu_y \in \text{Prob}(\mathbf{R}^n)$, $y \in \mathbf{R}^n$, *such that*

$$
\text{supp}\, \nu_y \subset \overline{K}, \quad y \in \mathbf{R}^n,
$$

and, for any continuous function f, *there is a subsequence (still labeled)* u^ϵ *satisfying*

$$
w^* - \lim f(u^\epsilon) = \langle \nu_y(\lambda), f(\lambda) \rangle = \int f(\lambda) d\nu_y(\lambda),
$$

where $w^* - \lim$ *denotes the weak limit in the sense of weak–star topology in the* L^∞ *space.*

COROLLARY 2.1. *The uniformly bounded and measurable functions u^ϵ converge to u a.e. if and only if the corresponding family of Young measures ν_y reduces at most all points y to a family of Dirac measures concentrated at $u(y)$, i.e., $\nu_y = \delta_{u(y)}$.*

In the context of conservation laws, one could expect that the family of Young measures ν_y, $y = (x, t)$, uniquely determined by the approximate solutions u^ϵ, satisfies two restrictions:

(i) Static behavior: Tartar's functional equation

$$(2.1) \qquad \left\langle \nu_y, \begin{vmatrix} \eta_1 & q_1 \\ \eta_2 & q_2 \end{vmatrix} \right\rangle = \begin{vmatrix} \langle \nu_y, \eta_1 \rangle & \langle \nu_y, q_1 \rangle \\ \langle \nu_y, \eta_2 \rangle & \langle \nu_y, q_2 \rangle \end{vmatrix}, \qquad \text{a.e.}$$

for entropy pairs $(\eta_i, q_i) \in E_1 (i = 1, 2)$.

(ii) Dynamic behavior: The quasientropy condition

$$(2.2) \qquad \partial_t \langle \nu_y, \eta(\lambda) \rangle + \partial_x \langle \nu_y, q(\lambda) \rangle \leq \langle \nu_y, \nabla \eta(\lambda) g(y, \lambda) \rangle,$$

for entropy pair $(\eta, q) \in E_2$.

Remark. A pair of mappings $(\eta, q) : \mathbf{R}^n \to \mathbf{R}^2$ is called an entropy pair of system (1.1) if it satisfies $\nabla q = \nabla \eta \nabla f$. For the strictly hyperbolic case, one could expect that the entropy pair space E_1 contains all continuous entropy pairs, while the entropy pair space E_2 contains all convex entropy pairs.

The problem of clarifying the convergence and oscillation of the approximate solutions $u^\epsilon(x, t)$ is equivalent to solving the following functional problem.

Problem (FP). Is the family of Young measures ν_y, $y = (x, t) \in \mathbf{R}_+^2$, satisfying the functional conditions (2.1)–(2.2) a family of Dirac measures or not?

For genuinely nonlinear systems, one can conclude that the Young measure satisfying Tartar's functional equation is a Dirac measure [T2, DI1–DI2, SR2, CH1–CH2, DCL2]. However, for linearly degenerate systems, Tartar's functional equation cannot ensure that ν_y is a Dirac measure. Thus one needs to find new restrictions on ν_y to reduce $\nu_y = \delta$ (e.g., [CH3]). The inequality (2.2) is motivated by the entropy condition; that is, an L^∞ solution $u = u(y)$ to conservation laws (1.1) is called admissible if

$$\partial_t \eta(u(y)) + \partial_x q(u(y)) \leq 0$$

for all convex entropy pairs (η, q).

In general, the weaker the nonlinearity of systems becomes, the larger the space E_2 becomes; that is, the stronger the restriction of (2.2) on ν_y becomes, the weaker the restriction of (2.1) on ν_y becomes. The results in subsequent sections are obtained by solving the functional problem (FP). For details, we refer the reader to [CH1–CH3, CL, DCL1–DCL2, DI1–2, TA2] and references cited therein.

As is well known, for the scalar conservation law, Oleinik [OL], Conway and Smoller [CS], and others proved that the Lax–Friedrichs approximate solutions satisfy the BV compactness framework and, therefore, obtained the convergence of the

Lax–Friedrichs scheme. The next step is to determine whether the Lax–Friedrichs approximate solutions to systems such as that of isentropic gas dynamics still satisfy the BV compactness framework. The answer should be negative, based on the analysis of Liu and Smoller [LS]. One does not expect that the global solutions to the Cauchy problem for the system of isentropic gas dynamics containing a vacuum are always locally finite total variation functions. This factor motivates people to seek new compactness frameworks that are satisfied by approximate solutions and that still ensure the existence of subsequence converging pointwise a.e.. Such compactness frameworks can indeed be established with the aid of the theory of compensated compactness (see Section 5).

3. Scalar Conservation Law. In this section we discuss the limit behavior of approximate solutions to the Cauchy problem (1.1)–(1.2) for the scalar conservation law $(n = 1)$.

Consider the approximate solutions $u^\epsilon(x,t)$ with the initial data $u_0^\epsilon(x)$ that satisfy Framework (A):

A$_1$: $\|u^\epsilon(x,t)\|_{L^\infty} \leq M$ (M is independent of ϵ).

A$_2$: For entropy pair $(\eta_0, q_0) = \left(f(u) - f(k), \int_k^u (f'(\xi))^2 d\xi\right)$, k the constant,

$$\eta_0(u^\epsilon)_t + q_0(u^\epsilon)_x, \quad \text{compact in } H_{\text{loc}}^{-1}.$$

Then we have the following result.

THEOREM 3.1. [CL]. *(Also see [CH1, T2].) There are two different cases.*

(i) $f''(u) \neq 0$, *a.e.* $u \in [-M, M]$. *The initial oscillations will be cancelled instantaneously by the nonlinearity. That is, the sequence $u^\epsilon(x,t)$ satisfying* (A_1)–(A_2) *with arbitrary initial data $u_0^\epsilon(x)$ converges pointwise a.e. to a uniquely admissible L^∞ solution to the corresponding Cauchy problem (1.1)* $(n = 1)$ *with initial data $u_0(x) = w^* - \lim u_0^\epsilon(x)$.*

(ii) f *linearly degenerate. If the initial sequence $u_0^\epsilon(x)$ is a highly oscillatory sequence, the initial oscillation will propagate as t evolves. That is, the sequence $u^\epsilon(x,t)$ satisfying (A_1)–(A_2) is also an oscillatory sequence that does not converge in the sense of strong topology.*

This theorem shows that the limit behavior of the sequence $u^\epsilon(x,t)$ with the highly oscillatory data sequence $u_0^\epsilon(x)$ uniquely depend on the nonlinear extent of the flux function. If f has some degenerate intervals, one cannot expect the convergence of $u^\epsilon(x,t)$ with a highly oscillatory data sequence $u_0^\epsilon(x)$.

Notice that the approximate solutions generated from the viscosity method, the Lax–Friedrichs scheme, the Godunov scheme, and some higher–order schemes, as well as the exact solutions, satisfy Framework (A) (see [CH1, DI1, SZ, TA]). We conclude immediately the following corollaries.

COROLLARY 3.1. *Suppose that the approximate solutions $u^\epsilon(x,t)$ are generated from the viscosity method, the Lax–Friedrichs scheme, the Godunov scheme,* and some higher–order schemes (e.g., the spectral method and the shock–capturing streamline–diffusion finite–element method) with initial data $u_0^\epsilon(x)$ satisfying

$$\|u_0^\epsilon(x) - \bar{u}\|_{L^\infty \cap L^2} \leq M \quad (M \text{ is independent of } \epsilon).$$

Then the results of Theorem 3.1 hold for the sequence $u^\epsilon(x, t)$.

COROLLARY 3.2. Suppose that $u^\epsilon(x, t)$ is a sequence of exact solutions, especially the Glimm solutions [GL], uniformly bounded in L^∞ to the Cauchy problem (1.1)–(1.2) with initial data $u_0^\epsilon(x)$ that are of locally bounded variation for fixed $\epsilon > 0$ and uniformly L^∞ bounded for all $\epsilon > 0$. Then the results of Theorem 3.1 hold for the sequence $u^\epsilon(x, t)$.

In fact, cancellation of the initial oscillations is an essential feature of the nonlinearity, while propagation of the initial oscillations is a feature of the linearity. For example, consider the solutions $u^\epsilon(x, t)$ of

$$\begin{cases} u_t + (au)_x & = \epsilon u_{xx}, \quad a \text{ constant}, \\ u \big|_{t=0} & = u_0^\epsilon(x). \end{cases}$$

Then

$$u^\epsilon(x, t) = -\frac{1}{\sqrt{\pi}} \int_{-\infty}^{\infty} u_0^\epsilon(x + at - 2\epsilon\sqrt{t}y)e^{-y^2} \, dy,$$

and $u^\epsilon(x, t)$ have the same limit behaviors as $u_0^\epsilon(x)$.

4. 2×2 Strictly Hyperbolic Systems. Now we are concerned with limit behavior of the approximate solutions $u^\epsilon(x, t)$ to the Cauchy problem (1.1) ($n = 2$). First we consider two extreme cases.

4.1. Linear Case

In this case, $f(u) = Au$, where A is the 2×2 constant matrix. We immediately observe the following fact.

THEOREM 4.1. The approximate solutions $u^\epsilon(x, t)$ generated from the viscosity method, the Lax–Friedrichs scheme, and the Godunov scheme, as well as the exact solutions, have the same limit behaviors as the initial data $u_0^\epsilon(x)$.

Theorem 4.1 shows that the convergence or oscillation of the approximate solutions $u^\epsilon(x, t)$ is uniquely determined by the limit behavior of the initial data $u_0^\epsilon(x)$. This fact is also true for systems comprising more than two equations.

4.2. Genuinely Nonlinear Case

Consider the approximate solutions $u^\epsilon(x, t)$ with the initial data $u_0^\epsilon(x)$ that satisfy Framework (B):

B_1: $u^\epsilon(x, t) \in K$, a.e., where $K \subset \mathbf{R}^2$ is a bounded set.

B_2: For any C^2 entropy pair (η, q),

$$\eta(u^\epsilon)_t + q(u^\epsilon)_x, \quad \text{compact in } H_{\text{loc}}^{-1}(\Omega).$$

Then we have the following theorem.

THEOREM 4.2. [DI1]. *For a* 2×2 *strictly hyperbolic and genuinely nonlinear system, Framework (B) is compact.*

Thus, in any approximate solutions $u^\epsilon(x,t)$ satisfying Framework (B), there exists a subsequence converging pointwise a.e. to an admissible solution u no matter how the initial data $u_0^\epsilon(x)$ oscillate or not; that is, the initial oscillations can be canceled instantaneously.

COROLLARY 4.1. *Consider a* 2×2 *strictly hyperbolic and genuinely nonlinear system. Suppose that the approximate solutions* $u^\epsilon(x,t)$ *are generated from the viscosity method, the Lax–Friedrichs scheme, and the Godunov scheme, uniformly bounded in* L^∞, *with initial data* $u_0^\epsilon(x)$ *satisfying*

$$\|u_o^\epsilon(x) - \bar{u}\|_{L^\infty \cap L^2} \le M \quad (M \text{ is independent of } \epsilon).$$

Then there exists a subsequence converging pointwise a.e. to an admissible solution u.

From the view of numerical computation, this result means that, even if the initial mesh data are oscillatory, the values determined by the finite difference schemes are no longer oscillatory as $t = m\Delta t$, $m >> 1$. This ensures the validity of the computational program.

4.3. Systems with One Linearly Degenerate Field

In this section we study the limit behavior of approximate solutions to systems between the linear and genuinely nonlinear cases. This study is intrinsically important from the view of both mathematical theory and numerical analysis because of the complexity of coupling linearity and nonlinearity. Specifically we examine systems with one linear degenerate field. Such systems arise in many areas (e.g., [KK, PO]).

4.3.1. Framework

Consider the approximate solutions $u^\epsilon(x,t)$ with initial data $u_0^\epsilon(x)$ that satisfy Framework (C):

C_1: $u^\epsilon(x,t) \in K$, a.e., where $K \subset \mathbf{R}^2$ is a bounded set.

C_2: For any C^2 entropy pair (η, q),

$$\eta(u^\epsilon)_t + q(u^\epsilon)_x, \quad \text{compact in } H_{\text{loc}}^{-1}(\Omega).$$

C_3: There exists a continuous function set Λ such that

(i) For any $a(z) \in \Lambda$,

$$E_a(u^\epsilon)_t + F_a(u^\epsilon)_x \mapsto A(a; x, t) \le 0,$$

in the weak topology of $H_{\text{loc}}^{-1}(\Omega)$. In particular, there exists $a_0(z) \in \Lambda$ such that

(4.1)
$$\begin{cases} w^* - \lim E_{a_0}(u^\epsilon) & \ge 0, \\ A(a_0; x, t) & = 0. \end{cases}$$

(ii) The set Λ is sufficiently large that, for any bounded variation function $g(z)$,

$$\int_\alpha^\beta a(z)dg(z) \leq 0$$

implies that

$$g(z) \equiv 0, \qquad z \in (\alpha, \beta),$$

where (E_a, F_a) is an entropy pair corresponding to the linearly degenerate field with the eigenvalue λ_2 and has the formula

$$\begin{cases} E_a = \Phi_0 a(z), \\ F_a = \lambda_2 \Phi_0 a(z), \end{cases}$$

with

$$\Phi_0 = \exp\left(\int_{\bar{w}}^w \frac{\lambda_2 w(\xi)}{\lambda_1(\xi, z) - \lambda_2(\xi)} d\xi\right),$$

and (w, z) are the Riemann invariants corresponding to the first and second eigenfields, respectively.

Remark. For the dissipative or exact approximate solutions, the function $a(z)$ such that E_a is a convex entropy belongs to Λ, while the condition (4.1) is in general satisfied by the system under consideration.

THEOREM 4.3. [CH3]. Consider a strictly hyperbolic system of conservation laws with one linearly degenerate field. Suppose that the approximate solutions $u^\epsilon(x,t)$ with uniformly bounded initial data $u_0^\epsilon(x)$ satisfy Framework (C). Then

(i) If another field is genuinely nonlinear and the initial data sequence satisfies

$$z_0^\epsilon(x) = z(u_0^\epsilon(x)) \to z_0(x), \qquad \text{a.e.},$$

there exists a subsequence (still labeled) $u^\epsilon(x,t)$ converging pointwise a.e. to an admissible L^∞ solution to the corresponding hyperbolic system (1.1) with the Cauchy data $u_0(x) = w^* - \lim u_0^\epsilon(x) \in L^\infty$.

(ii) If the composite sequence of the approximate initial data $z_0^\epsilon(x) = z(u_0^\epsilon(x))$ is a highly oscillatory sequence, the oscillation will propagate along the linearly degenerate field; that is, the corresponding Riemann invariant $z^\epsilon(x,t) = z(u^\epsilon(x,t))$ is an oscillatory sequence that does not converge in the sense of strong topology.

This theorem is established by a study of the static and dynamic behavior of the family of probability measures $\{\nu_{x,t}\}_{(x,t)\in\mathbf{R}_+^2}$, which corresponds to the approximate solutions. The basic motivation is that the static relation (2.1) represents an imbalance of regularity: the operator on the left is more regular than the one on the right as a result of cancellation, which forces the support of the measures $\nu_{x,t}$ to lie in the linearly degenerate field (see Serre's theorem [SR2]). Moreover, the dynamic relation (2.2) represents a behavior of the measures $\nu_{x,t}$ as t evolves. The key idea is to make a coordinate transformation so as to clarify that a family of measures derived from $\nu_{x,t}$ is the same as the corresponding family of initial measures in the new coordinate system. The idea is fulfilled in [CH3].

Remark 1. Theorem 4.3 shows that convergence of the approximate solutions completely depends on the behavior of the Cauchy data. Efficient calculation of the numerical solutions as the mesh length goes to zero completely depends on the behavior of the initial mesh data. From the view of theoretical analysis, if the initial data are highly oscillatory, then the singular limits of the approximate solutions to system (1.1) with linearly degeneracy must be represented by the measure-valued solutions introduced by DiPerna [DI3] or by corresponding homogenization equations of well–posed Cauchy problems with respect to the weak-star topology of L^∞. We omit the discussion in this paper.

Remark 2. Theorem 4.1 (ii) shows that the initial oscillations can propagate along the linearly degenerate field, which is not affected by the nonlinear field. This phenomenon differs significantly from the phenomena in [DI1, LL, CH1].

4.3.2. Systems with Two Linearly Degenerate Fields

Using Theorem 4.3, we obtain the following theorem.

THEOREM 4.4. [CH3]. *Consider a strictly hyperbolic system of conservation laws with two linearly degenerate fields. If the initial data sequence $u_0^\epsilon(x) \in L^\infty \cap BV$ satisfies*

$$\begin{cases} \|u_0^\epsilon\|_{L^\infty} & \le M \quad (M \text{ is independent of } \epsilon), \\ \sup \lambda_1(u_0^\epsilon(x)) & < \inf \lambda_2(u_0^\epsilon(x)). \end{cases}$$

Then the corresponding BV solutions $u^\epsilon(x,t)$, especially the Glimm solutions, have the following limit behavior:

(i) *Let the corresponding initial Riemann invariants*

$$w_0^\epsilon(x) = w(u_0^\epsilon(x)), \quad z_0^\epsilon(x) = z(u_0^\epsilon(x))$$

be compact in the strong topology of L^∞. Then there exists a subsequence converging pointwise a.e. to an admissible L^∞ solution to the completely degenerate system (1.1) with the Cauchy data $u_0(x) = w^* - \lim u_0^\epsilon(x)$.

(ii) *Let one or two of the corresponding initial Riemann invariants $w_0^\epsilon(x)$ and $z_0^\epsilon(x)$ be a highly oscillatory sequence. Then the oscillation will propagate along the corresponding linearly degenerate field or fields.*

In fact, in this case, Λ is the continuous function space. Therefore Framework (C) always holds. Condition (C$_1$) is a corollary of the L^∞ uniformly bounded estimate of the Riemann solutions.

COROLLARY 4.2. *There exists an admissible L^∞ solution to the completely degenerate system with the Cauchy data $u_0(x) \in L^\infty$, $\sup \lambda_1(u_0^\epsilon(x)) < \inf \lambda_2(u_0^\epsilon(x))$,* *that satisfies*

$$\left(\frac{a(w)}{\lambda_2(w) - \lambda_1(z)}\right)_t + \left(\frac{\lambda_1(z)a(w)}{\lambda_2(w) - \lambda_1(z)}\right)_x = 0, \qquad \forall\, a \in C,$$

$$\left(\frac{b(z)}{\lambda_2(w) - \lambda_1(z)}\right)_t + \left(\frac{\lambda_2(w)b(z)}{\lambda_2(w) - \lambda_1(z)}\right)_x = 0, \qquad \forall\, b \in C.$$

Remark. A partial result of Theorem 4.4 is also obtained by Serre [SR3], in which a corresponding homogenization system is found.

4.3.3. Systems with One Constant Field and One Line Field

Here we discuss systems with one contact field and one line field

(4.2) $$u_t + (u\phi(u))_x = 0$$

where $\phi(u) : \mathbf{R}^2 \to \mathbf{R}^1$ is a nonlinear mapping. This class of systems, which was identified by Temple [TE2], is a subclass of systems for which the shock wave and rarefaction wave curves coincide. Such systems arise in elasticity theory [KK] and enhanced oil recovery [PO], for example.

THEOREM 4.5. [CH3]. *Consider a system of conservation laws with one contact field and one line field. Suppose that the exact solutions $u^\epsilon(x,t)$ to system (4.2) with the initial data $u_0^\epsilon(x) \in L^\infty$ lie in a bounded set K on which the system is strictly hyperbolic. Then the results of Theorem 4.3 hold.*

In this case, the function space Λ is also the continuous function space.

Example 1. The elasticity model system $\phi = \phi(u)$ satisfies

(4.3) $$\begin{cases} D_1: & \phi(u) > 0, \\ D_2: & \phi(u) = \text{const.} > 0 \text{ is a simple close curve that contains} \\ & \text{the origin } (0,0). \end{cases}$$

For example, $\phi(u) = \phi(r) = 1 + \delta\frac{(r-1)^2}{r}$, $r = \sqrt{u_1^2 + u_2^2}$, $\delta > 0$ constant, satisfies D_1–D_2.

One can construct admissible solutions $u^\epsilon(x,t)$ (e.g., [CH3]) satisfying

(4.4) $$0 < \alpha \le \phi(u^\epsilon(x,t)) \le \beta < \infty$$

to the system (4.2)–(4.3) with the initial data sequence $u_0^\epsilon(x)$ satisfying (4.4) and

$$\text{TV}_{-\infty}^{\infty}(\phi(u_0^\epsilon(x)), \quad \theta(u_0^\epsilon(x))) \le M_\epsilon < \infty,$$

where $\phi(u)$ and $\theta(u) = \arctan\frac{u_1}{u_2}$ are the Riemann invariants of the system (4.2). We conclude that the initial oscillations will propagate along the linearly degenerate field corresponding to the Riemann invariant θ. As a byproduct, we have the following theorem.

THEOREM 4.6. [CH4]. *Suppose that $u_0(x)$ satisfies*

$$0 < \alpha \leq \phi(u_0(x)) \leq \beta < \infty$$

on which the system (4.2)–(4.3) is strictly hyperbolic. Then there exists a global L^∞ solution $u(x,t)$ to the system (4.2)–(4.3) with the Cauchy data $u_0(x)$ satisfying

$$0 < \alpha \leq \phi(u(x,t)) \leq \beta < \infty$$

and

$$(ra(\theta))_t + (ra(\theta)\phi(r,\theta))_x = 0.$$

A detailed discussion of general ϕ can also be found in [CH4].

Example 2. This example involves a system arising in the polymer flooding of an oil reservoir [PO],

(4.5)
$$\phi(u_1, u_2) = \frac{f\left(u_1, \frac{u_2}{u_1}\right)}{u_1},$$

where $0 \leq u_1 \leq 1$, $0 \leq u_2 \leq u_1$ are the concentration of water and the overall concentration of a polymer at any x and t, respectively, and $f(u_1, c)$ is a smooth function such that $f(u_1, c)$ increases from zero to one with one inflection point where c is constant and such that $f(u_1, c)$ decreases with increasing c for fixed u_1 (see [TE1]). The essential feature of system is that the strict hyperbolicity fails along a certain curve in state space.

Using Temple's theorem [TE1], we have a global solution sequence $u^\epsilon(x,t)$ satisfying (4.2) and (4.5) with the Cauchy data $u_0^\epsilon(x)$ of bounded variation.

THEOREM 4.7. [CH3]. *The initial oscillations will still propagate along the linearly degenerate field for the nonstrictly hyperbolic system (4.2) and (4.5).*

Thus if the initial data sequence $u_0^\epsilon(x)$ is a highly oscillatory sequence, the exact solution sequence $u^\epsilon(x,t)$ is a highly oscillatory sequence, too. One cannot expect convergence of the Glimm approximate solutions with highly oscillatory initial data.

Remark. The arguments in Frameworks (A)–(C) could be extended to L^p approximate solutions. We refer the reader to [DAF1, SH, LP].

5. The System of Isentropic Gas Dynamics. Here we describe the limit behavior of the approximate solutions $u^\epsilon(x,t)$, generated from the fractional step Lax–Friedrichs scheme and Godunov scheme to the inhomogeneous system of isentropic gas dynamics in the Euler coordinate

(5.1)
$$\begin{cases} \rho_t + (\rho u)_x & = U(\rho, u; x, t), \\ (\rho u)_t + (\rho u^2 + p(\rho))_x & = V(\rho, u; x, t), \end{cases}$$

(5.2)
$$(\rho, u)\Big|_{t=0} = (\rho_0(x), u_0(x)),$$

where u, ρ, and p are the velocity, the density, and the pressure, respectively. For a polytropic gas, $p(\rho) = k^2 \rho^\gamma$, where k is the constant and $\gamma > 1$ is the adiabatic exponent (for usual gas, $1 < \gamma \leq 5/3$). The system (5.1) with $(U, V) \neq (0, 0)$ is a gas dynamics model of nonconservative form with a source. For instance, $(U, V) = (0, \alpha(x, t)\rho)$, where $\alpha(x, t)$ represents a body force (usually gravity) acting on all the fluid in any volume. An essential feature of the system is a nonstrict hyperbolicity; that is, a pair of wave speeds coalesce on the vacuum $\rho = 0$.

We also describe the limit behavior of the approximate solutions $u^\epsilon(x, t)$ (especially generated from the Lax-Friedrichs scheme, the Godunov scheme, and the viscosity method) to the homogeneous system of isentropic gas dynamics. The homogeneous system of (5.1) is

(5.3)
$$\begin{cases} \rho_t + (\rho u)_x & = 0, \\ (\rho u)_t + (\rho u^2 + p(\rho))_x & = 0. \end{cases}$$

For the Cauchy problem of the homogeneous system (5.3), many existence theorems of global solutions have been obtained [RI, ZG, BA, NI, NS, DZ, LLO, DI2]. The first large-data existence theorem of global solutions was established by Nishida [NI] for $\gamma = 1$ by using the Glimm method [GL]. DiPerna [DI2] established a large data existence theorem for $\gamma = 1 + 2/(2m+1)$, $m \geq 2$ integers, by using the viscosity method and the theory of compensated compactness. Both results assume that the initial density $\rho_0(x)$ is away from the vacuum. In this section we describe recent achievements for the problem.

5.1. Compactness Framework

THEOREM 5.1. *Suppose that the approximate solutions*

$$v^\epsilon(x, t) = (\rho^\epsilon(x, t), m^\epsilon(x, t)) = (\rho^\epsilon(x, t), \rho^\epsilon(x, t)u^\epsilon(x, t))$$

to the Cauchy problem (5.1)–(5.2) ($1 < \gamma \leq 5/3$) satisfy the following framework.

(i) *There is a constant $C > 0$ such that*

$$0 \leq \rho^\epsilon(x, t) \leq C, \quad |u^\epsilon(x, t)| \leq C, \quad \text{a.e.}$$

(ii) *On any bounded domain $\Omega \subset \mathbf{R}_+^2$ and for any weak entropy pair (η, q) (i.e., $\eta(0, u) = 0$), the measures*

$$\eta(v^\epsilon)_t + q(v^\epsilon)_x \qquad \text{compact in } H_{\text{loc}}^{-1}(\Omega).$$

Then there exists a subsequence (still labeled) v^ϵ such that

$$(\rho^\epsilon(x, t), m^\epsilon(x, t)) \to (\rho(x, t), m(x, t)), \quad \text{a.e.}$$

This compactness framework is established by an analysis of weak entropy and a study of regularities of the family of probability measure $\{\nu_{x,t}\}_{(x,t)\in\mathbf{R}_2^+}$, which corresponds to the approximate solutions. The basic motivation is that the commutativity relation (2.1) represents an imbalance of regularity: the operator on the left is more regular than the one on the right as a result of cancellation, which forces the measure $\nu_{x,t}$ to reduce to a point mass. We recall that the derivative of a Radon measure in the Lebesque sense vanishes except at one point, implying that the measure is a point mass. The challenge is to choose the entropy pairs whose leading term is coercive with respect to the 2×2 determinant and to show that the coercive behavior guarantees that the derivative of $\nu_{x,t}$ vanishes except at one point. The essential difficulty is that only a subspace of entropy pairs, weak entropy pairs, can be used in the relation (2.1). The strategy is fulfilled in [CH1–CH2, DCL1–DCL2].

5.2. Convergence of the Lax–Friedrichs Scheme and the Godunov Scheme

Using Theorem 5.1 and making several estimates, we obtain the following theorem.

THEOREM 5.2. [DCL1, CH1–2]. *Suppose that the initial data* $v_0(x) = (\rho_0(x), \rho_0(x)u_0$ *satisfy*

(5.4) $$\left\{ \begin{array}{l} 0 \le \rho_0(x) \le M, \quad |u_0(x)| \le M, \\ \int_{-\infty}^{\infty}(\eta_*(v_0(x)) - \eta_*(\bar{v}) - \nabla\eta_*(\bar{v})(v_0(x) - \bar{v}))dx \le M, \end{array} \right.$$

for some constant state $\bar{v}$ *and the mechanical energy* $\eta_* = \frac{1}{2}\rho u^2 + \frac{\rho^\gamma}{\gamma(\gamma-1)}$. *Then there exists a convergent subsequence in the Lax–Friedrichs approximate solutions and the Godunov approximate solutions* $v^\epsilon(x,t)$ ($\epsilon = \Delta x$, *the space step length*), *respectively, that have the same local structure as the random choice approximations of Glimm [GL], such that*

$$(\rho^{\epsilon k}(x,t), m^{\epsilon k}(x,t)) \to (\rho(x,t), m(x,t)), \quad \text{a.e.}$$

Define $u(x,t) = m(x,t)/\rho(x,t)$, *a.e. Then the pair of functions* $(\rho(x,t), u(x,t))$ *is a generalized solution of the Cauchy problem (5.2)–(5.3) satisfying*

$$0 \le \rho(x,t) \le C, \quad |u(x,t)| \le \frac{|m(x,t)|}{\rho(x,t)} \le C, \quad \text{a.e.}$$

5.3. Convergence of the Viscosity Method

Consider the viscosity approximations $v^\epsilon(x,t)$ determined by

(5.5) $$\left\{ \begin{array}{ll} \rho_t + m_x & = \epsilon\rho_{xx}, \\ m_t + \left(\frac{m^2}{\rho} + p(\rho)\right)_x & = \epsilon m_{xx}, \end{array} \right.$$

(5.6) $$(\rho, m)\Big|_{t=0} = (\rho_0^\epsilon(x), m_0^\epsilon(x)),$$

where $v_0^\epsilon(x) = (\rho_0^\epsilon(x), m_0^\epsilon(x))$ is an approximate sequence of the initial data $v_0(x) = (\rho_0(x), \rho_0(x)u_0(x))$.

LEMMA 5.1. *Suppose that the initial data* $(\rho_0(x), u_0(x))$ *satisfy*

$$\begin{cases} (\rho_0(x) - \bar{\rho}, u_0(x) - \bar{u}) \in L^2 \cap L^\infty, \\ \rho_0(x) \geq 0. \end{cases}$$

Then there is an approximate sequence $v_0^\epsilon(x)$ *satisfying*

$$\begin{cases} v_0^\epsilon(x) \to v_0(x), \quad \text{in } L^2, \\ v_0^\epsilon(x) - \bar{v} \in C_0^1(-\infty, \infty), \\ 0 \leq \rho_0^\epsilon(x) \leq M_0, \quad |u_0^\epsilon(x)| \leq M_0, \end{cases}$$

such that there exist global solutions $(\rho^\epsilon, u^\epsilon)$ *to the Cauchy problem (5.2)–(5.3) satisfying*

$$(\rho^\epsilon(\cdot, t) - \bar{\rho}, u^\epsilon(\cdot, t) - \bar{u}) \in C^1 \cap H^1,$$
$$0 \leq \rho^\epsilon \leq M, \quad |u^\epsilon| \leq M,$$

where both M_0 *and* M *are constants independent of* ϵ.

THEOREM 5.3. [CH1]. *Suppose that the initial data* $v_0(x) = (\rho_0(x), \rho_0(x)u_0(x))$ *satisfy (5.4). Then there exists a convergent subsequence in the viscosity approximations* $v_0^\epsilon(x, t)$ *such that*

$$(\rho^{\epsilon_k}(x, t), m^{\epsilon_k}(x, t)) \to (\rho(x, t), m(x, t)).$$

Define $u(x, t) = \frac{m(x, t)}{\rho(x, t)}$, *a.e. Then the pair of functions* $(\rho(x, t), m(x, t))$ *is a generalized solution to the Cauchy problem (5.2)–(5.3) satisfying*

$$0 \leq \rho(x, t) \leq C, \quad |u(x, t)| \leq \frac{|m(x, t)|}{\rho(x, t)} \leq C, \quad \text{a.e.}$$

5.4. Convergence of the Fractional Step Lax–Friedrichs Scheme and Godunov Schemes

In Section 5.1, the L^∞ uniformly bounded estimate of the approximate solutions plays an important role in ensuring the compactness of the approximate solutions. As a general rule, one uses the principle of invariant region or the (weak) maximal principle to get the L^∞ estimate. For the Cauchy problem (5.1)–(5.2) and $(U, V) = (0, \alpha\rho u)$, $\alpha \leq 0$, there exist bounded invariant regions. For the general inhomogeneous term (e.g., $(U, V) = (0, \alpha\rho u), \alpha > 0$), however, there are usually no bounded invariant regions. Nevertheless, we use two difference schemes—the fractional–step Lax–Friedrichs scheme and Godunov scheme (see [DCL2]), which are generalizations of those of Lax–Friedrichs [LA1] and Godunov [GO]—to construct approximate solutions $v^\epsilon(x, t)$ ($\epsilon = \Delta x$, the space step length). If the inhomogeneous terms satisfy conditions C1°–C3° (see [DCL2]), which contains cases of $(0, \alpha(x, t)\rho), (0, \alpha(x, t)\rho u), (\alpha(x, t)\rho, \alpha(x, t)\rho u)$, and $(0, \alpha(x, t)\rho u \ln(|u|+1))$, $\alpha(x, t) \in C(\mathbf{R} \times \mathbf{R}^+)$, we can overcome the difficulty by analyzing the solution of a nonlinear ordinary differential equation for the fractional–step Lax–Friedrichs scheme and Godunov scheme.

THEOREM 5.4. [DCL2]. *Suppose that the inhomogeneous term satisfies conditions C1°–C3° (see [DCL2]) and the initial data $(\rho_0(x), u_0(x))$ satisfy (5.4). Then there exists a convergent subsequence in the approximations $v^\epsilon(x,t)$ such that*

$$(\rho^{\epsilon_k}(x,t), m^{\epsilon_k}(x,t)) \to (\rho(x,t), m(x,t)) \quad \text{a.e.}$$

Define $u(x,t) = \frac{m(x,t)}{\rho(x,t)}$, a.e. Then the pair of functions $(\rho(x,t), m(x,t))$ is a generalized solution to the Cauchy problem (5.1)–(5.2) satisfying

$$0 \le \rho(x,t) \le C, \qquad |u(x,t)| \le C.$$

Remark. For a study of the convergence of approximate solutions for transonic flow problems, we refer the reader to the work of Morawetz and Kohn [MO, MK]. For a study of the limit behavior of approximate solutions to hyperbolic systems of conservation laws with an isolated umbilic point, we also refer the reader to [KA].

6. Remarks on Multidimensional Conservation Laws. From discussions in the literature, it should be evident that the multidimensional hyperbolic conservation laws are extremely complex and that rigorous theory of such conservation laws is in its infancy. In particular, the study of the limit behavior of approximate solutions to multidimensional conservation laws is one of the major challenge problems in the field of nonlinear analysis.

An essential feature in the multidimensional case is degeneracy, i.e., failure of strict hyperbolicity or genuine nonlinearity (e.g., [LA3, CC]). Another distinguishing feature in this case is the rollup of the compressible vortex sheet through the non-linear interaction of the kink mode and the development of the vorticity in the compressible vortex sheet (e.g., [AM]), which is quite different from the one–dimensional case. Therefore, one should first study degenerate systems of one–dimensional conservation laws and systems of two–dimensional steady conservation laws to understand and solve multidimensional conservation laws. One should also understand the limit behavior of approximate solutions for the incompressible Navier–Stokes equation, because of the appearance of the vorticity in the compressible vortex sheet. We refer the reader to DiPerna and Majda's work [DM 1-3](also see [GT]) for two–dimensional incompressible flow, the work of Morawetz and Kohn [MO, MK] for transonic flow, and recent work on degenerate systems of one-dimensional conservation laws (e.g., [CH1-4, KA, SR2-3]).

REFERENCES

[AM] ARTOLA, M., MAJDA, A.J., *Nonlinear development of instabilities in supersonic vortex sheets, I: The basic kink modes*, Physica 28D (1987), pp. 253–281; II: Resonant interaction among kink modes. SIAM Appl. Math.

[BA] BAHKVAROV, N., *On the existence of regular solutions in the large for quasilinear hyperbolic systems*, Ah. Vychisl. Mat. Fiz. 10 (1970), pp. 969–980.

[CC] CHANG, T., CHEN, G.-Q., *Some fundamental concepts about system of two spatial dimensional conservation laws*, Acta Mathematica Scientia 6 (1986), pp. 463–474.

[CH1]   CHEN, G.-Q., *The theory of compensated compactness and the system of isentropic gas dynamics*, Preprint, February 1989.

[CH2]   CHEN, G.-Q., *Convergence of the Lax-Friedrichs scheme for isentropic gas dynamics (III)*, Acta Math. Sci. 5 (1986), pp. 75–120; 8, (1988) (in Chinese).

[CH3]   CHEN, G.-Q., *Propagation and cancellation of oscillations in hyperbolic systems of conservation laws*, Preprint, August 1989.

[CH4]   CHEN, G.-Q., *The systems hyperbolic conservation laws with a symmetry*, Preprint, May 1989.

[CL]   CHEN, G.-Q., LU, Y., *A study to applications of the theory of compensated compactness*, Chinese Science Bulletin 34 (1989), pp. 15–19.

[CR]   CHEN, G.-Q., RUSTICHINI, A., *Global solution to a system of conservation laws and a non-zero sum game*, to appear.

[CCS]   CHUEH, K., CONLEY, C., SMOLLER, J., *Positively invariant regions for systems of nonlinear diffusion equations*, Indiana U. Math. J. 26 (1977), pp. 373–392.

[CS]   CONWAY, E., SMOLLER, J., *Global solutions of the Cauchy problem for quasilinear first-order equations in several space variables*, Comm. Pure Appl. Math. 19 (1966), pp. 95–105.

[DAC]   DACOROGNA, B., *Weak continuity and weak lower semicontinuity of nonlinear functionals*, in *Lecture Notes in Math.*, Vol. 922, Springer-Verlag, 1982, pp. 1-120.

[DAF1] DAFERMOS, C.M., *Estimates for conservation laws with little viscosity*, SIAM J. Math. Anal. 18 (1987), pp. 409-421.

[DAF2] DAFERMOS, C.M., *Hyperbolic systems of conservation laws*, LCDS No. 83-5, March 1983.

[DCL1] DING, X., CHEN, G.-Q., LUO, P., *Convergence of the Lax-Friedrichs scheme for isentropic gas dynamics (I), (II)*, Acta Math. Sci. 5, 483-500, 501-540 (1985); 7, (1987) (in Chinese).

[DCL2] DING, X., CHEN, G.-Q., LUO, P., *Convergence of the fractional step Lax-Friedrichs scheme and Godunov scheme for isentropic gas dynamics*, Comm. Math. Phys. 121 (1989), pp. 63–84.

[DZ]   DING SHIA-SHI, ZHANG-TUNG, WANG CHING-HUA, HSIAO-LING, LI TSAI-CHUNG, *A study of the global solutions for quasilinear hyperbolic systems of conservation laws*, Scientia Sinica 16 (1973), pp. 317–335.

[DI1]   DIPERNA, R., *Convergence of approximate solutions to conservation laws*, Arch. Rat. Mech. Anal. 88 (1985), pp. 223–270.

[DI2]   DIPERNA, R.J., *Convergence of the viscosity method for isentropic gas dynamics*, Comm. Math. Phys. 91 (1983), pp. 1–30.

[DI3]   DIPERNA, R., *Measure-valued solutions to conservation laws*, Arch. Rat. Mech. Anal. 8 (1985), pp. 223–270.

[DM1]   DIPERNA, R., MAJDA, A., *Oscillations and concentrations in weak solutions of the incompressible fluid equations*, Comm. Math. Phys. 108 (1987), pp. 667–689.

[DM2]   DIPERNA, R., MAJDA, A., *Concentrations in regularizations for 2-D incompressible flow*, Comm. Pure Appl. Math. 40 (1987), pp. 301–345.

[DM3]   DIPERNA, R., MAJDA, A., *Reduced Hausdorff dimension and concentration-cancellation for 2-D incompressible flow*, J.A.M.S..

[GL]   GLIMM, J., *Solutions in the large for nonlinear hyperbolic systems of equations*, Comm. Pure Appl. Math. 18 (1965), pp. 95–105.

[GO]   GODUNOV, S.K., *A difference method for numerical calculation of discontinuous solutions of the equation of hydrodynamics*, Mat. Sb. 47, 89 (1959), pp. 271-306.

[GT]   GREENGARD, C., THOMANN, E., *On DiPerna-Majda concentration sets for two-dimensional incompressible flow*, Preprint, June 9, 1987.

[KK]   KEYFITZ, B., KRANZER, H., *A system of nonstrictly hyperbolic conservation laws arising in elasticity theory*, Arch. Rat. Mech. Anal. 72, (1980), pp. 219–241.

[KA]   KAN, P.-T., *On the Cauchy problem of a 2 × 2 system of nonstrictly hyperbolic conservation laws*, Ph.D. thesis, Courant Institute of Mathematical Sciences, NYU, 1989.

[KR]   KRUSKOV, S., *First-order quasilinear equations with several space variables*, Mat. Sb. 123, 228-255 (1970).

[LA1]   LAX, P.D., *Weak solutions of nonlinear hyperbolic equations and their numerical computation*, Comm. Pure Appl. Math. 7 (1954), pp. 159–193.

[LA2]   LAX, P.D., *Hyperbolic systems of conservation laws II*, Comm. Pure Appl. Math. 10 (1957), pp. 537–566.

[LA3]   LAX, P.D., *Hyperbolic systems of conservation laws in several space variable. Courant Mathematics and Computing Laboratory*, DOE/ER/03077-246, UC-32, May 1985.

[LL]   LAX, P., LEVERMORE, C.D., *The small dispersion limit for the KdV equation. I, II, III*, Comm. Pure Appl. Math. 36 (1983), pp. 253–290, 571-594, 809-829.

[LW]   LAX, P., WENDROFF, B., *Systems of conservation laws*, Comm. Pure Appl. Math. 13 (1960), pp. 217–237.

[LLO]   LIN LONGWEI, *A study of the global solutions for system of gas dynamics*, Acta Scien. Nat. Jilin Univ. 1 (1978), pp. 96–106.

[LP]   LIN, P.-X., *Young measures and application of compensated compactness to systems of nonlinear elasticity*, Preprint, June 30, 1989.

[LS]   LIU, T.-P., SMOLLER, J., *The vacuum state in isentropic gas dynamics*, Adv. Appl. Math. 1 (1980), pp. 345–359.

[LU]   LIU, T.-P., *Admissible solutions of hyperbolic conservation laws*, Amer. Math. Soc. Memoir no. 240, Providence, 1981.

[MO]   MORAWETZ, C.S., *On a weak solution for a transonic flow problem*, Comm. Pure Appl. Math. 38 (1985), pp. 797–818.

[MK]   MORAWETZ, C.S., KOHN, R., *Transonic flow past a bump*, (to appear).

[M1]   MURAT, F., *Compacité par compensation*, Ann. Scula Norm. Sup. Pisa Sci. Math. 5 (1978), pp. 489-507.

[M2]   MURAT, F., *Compacite par compensation II*, Proceedings of the International Meeting on Recent Methods in Nonlinear Analysis, Rome, ed. E. De Giorgi, E. Magenes, and U. Mosco, Pitagora Editrice, Bologna (1979), pp. 245–256.

[M3]   MURAT, F., *Compacite par compensation: Condition necessaire et suffisante de continuité faible sous une hypothese de rang constant*, Ann. Scula Norm. Sup. Pisa 8 (1981), pp. 69–102.

[M4]   MURAT, F., *L'injection du cone positif de H^{-1} dans $W^{-1,q}$ est compacte pour tout $q < 2$*, J. Math. Pure Appl. 60 (1981), pp. 309–322.

[MPT]   MCLAUGHIN, D.W., PAPANICOLAOU, G., TARTAR, L., *Weak limits of semilinear hyperbolic systems with oscillating data*, Lecture Notes in Phys. Vol. 230, Springer-Verlag, 1985, 277-289.

[NI]   NISHIDA, T., *Global solution for an initial-boundary-value problem of a quasilinear hyperbolic system*, Proc. Jap. Acad. 44 (1968), pp. 642–646.

[NS]   NISHIDA, T., SMOLLER, J., *Solutions in the large for some nonlinear hyperbolic conservation laws*, Comm. Pure Appl. Math. 26 (1973), pp. 183–200.

[OL]   OLEINIK, O., *Discontinuous solutions of nonlinear differential equations*, Usp. Mat. Nauk. (N.S.) 12 (1957), pp. 3–73.

[PO]   POPE, G.A., *The application of fractional flow theory to enhanced oil recovery*, Soc. of Petroleum Eng. Journal 20 191 (1980).

[RS]   RASCLE, M., SERRE, D., *Compacité par compensation et systèmes hyperboliques de lois de conservation*, Applications, Compt. Rend. Acad. Sci. Paris, Séries I, 299, 673–676 (1984).

[RI]   RIEMANN, B., *Uber die Fortpflanzung ebener Luftwellen von endlicher Schwingungsweite*, Abhandl. Koenig. Gesell. Wiss., Goettingen, Vol. 8, 43 (1860).

[SH]   SHEARER, J., *Global existence and compactness for a system of conservation laws in L^p*, Ph.D. thesis, University of California at Berkeley, 1989.

[SR1]   SERRE, D., *Solutions à variation bornée pour certains systèmes hyperboliques de lois de conservation*, J. Diff. Equs. 68, 137-168 (1987).

[SR2]   SERRE, D., *Compacité par compensation et systèmes hyperboliques de lois de conservation*, Compt. Rend. Acad. Sci. Paris, Série I, 299, 555-558 (1984).

[SR3]   SERRE, D., *Propagation des oscillations dans les systèmes hyperboliques non linéaires*, No. 54. Université de Saint-Etienne, 1986.

[SZ]   SZEPESSY, A., *Convergence of a shock-capturing streamline diffusion finite element method for a scalar conservation law in two space dimensions*,, Math. Comp. (1989), to appear.

[TA]   TADMOR, E., *Semi-discrete approximations to nonlinear systems of conservation laws: Consistency and L^∞ -stability imply convergence*, Preprint, 1988.

[T1]   TARTAR, L., *Une nouvelle méthode de résolution d'équations aux dérivées partielles non-linéaires*, Lecture Notes in Mathematics, Vol. 665, Springer-Verlag, 1977, 228-241.

[T2]   TARTAR, L., *Compensated compactness and applications to partial differential equations*, in Research notes in mathematics, nonlinear analysis and mechanics: Heriot-Watt Symposium, Vol. 4, ed. R.J. Knops, Pitman Press, 1979.

[T3]   THE COMPENSATED COMPACTNESS METHOD APPLIED TO SYSTEM OF CONSERVATION LAWS, in Systems of nonlinear partial differential equations, ed. J.M. Ball, NATO ASI Series, D. Reidel Publishing Co., 1983.

[T4]   TARTAR, L., *Discontinuities and oscillations*, in Direction in Partial Diff. Eqns., Mathematics Research Center Symposium, ed. M.G. Crandall, P.H. Rabinowitz, and R.E.L. Turner, Academic Press, 1987, pp. 211–233.

[TE1] TEMPLE, B., *Global solution of the Cauchy problem for a class of 2 x 2 nonstrictly hyperbolic conservation laws*, Adv. in Appl. Math. 3 (1982), pp. 335-375.

[TE2] TEMPLE, B., *Degenerate systems of conservation laws*, Contemporary Math. 60 (1987), pp. 125–133.

[TE3] TEMPLE, B., *Weak stability in the global L^1-norm for system of hyperbolic conservation laws*, Preprint (1989).

[ZG] ZHANG TONG, GUO YU-FA, *A class of initial-value problems for systems of aerodynamic equations*, Acta Math. Sinica 15 (1965), pp. 386–396.

MODELING TWO–PHASE FLOW OF REACTIVE
GRANULAR MATERIALS

Pedro F. Embid* and Melvin R. Baer[†]

Abstract. In this study, we examine a two–phase model proposed by Baer and Nunziato to describe the modes of combustion from deflagration to detonation in reactive granular materials. The model treats all phases in nonequilibrium and fully compressible. A compaction evolutionary equation, describing changes in volume fraction, provides model closure. In contrast to a pressure equilibrium model that has elliptic regions, the system of equations is hyperbolic except at points where the relative flow is locally sonic.

Key words. Combustion, multiphase flow, hyperbolic equations

AMS(MOS) subject classifications. 80A25, 76T05, 58G16

1. Introduction. Deflagration–to–detonation transition (DDT) in gas–permeable reactive granular materials is a highly complex phenomena. Extensive experimental and theoretical studies of the various thermal, mechanical, and chemical mechanisms governing the initiation and acceleration of the combustion processes (see, *i.e.* [1],[3]) have been conducted.

It is well known that the combustion of granular materials is inherently unsteady and evolves as various stages characterized by well-defined speeds of propagation. Initially, the combustion is slow and dominated by heat conduction. As heat is convected by the hot product gases, permeating into the granular reactant ahead of the flame, substantial acceleration of the flame is induced. During this accelerated combustion, disparity of the solid and gas pressures causes compaction on the granular bed ahead of the flame and a reduction of the permeability produces rapid gas pressurization. This, in turn, accelerates combustion until a shock wave develops. Thereafter, other mechanisms, such as the heat release from localized hot spots, enhances the wave growth to detonation. Thus, four major flame propagation regimes occur during DDT: conductive burning, convective burning, compressive burning and detonation.

Based on these physical observations, it is apparent that an appropriate model describing the reactive flow field must incorporate numerous thermal and mechanical processes associated with combustion. In particular, the model must describe flows at low and high speeds, with differences in thermodynamic properties between the phases (*e.g.* pressure) and take into account the compressibility of both phases consistent with compaction of the granular reactant.

In this study, we examine a model formulated by Baer and Nunziato [1] for reactive granular flow and contrast it with a traditional equal pressure model [7]. Since the equal pressure model is based on the assumption of pressure equilibrium between both phases, a characteristics analysis shows that system of equations is hyperbolic only at high relative speeds and elliptic regions occur at low relative speeds. This implies the initial value problem for the pressure equilibrium model is ill–posed and is inappropriate during the early stages of combustion prior to detonation. During combustion in granulated materials, field variables, such as phase pressures and velocities,

*Mathematics Department, University of New Mexico, Albuquerque, New Mexico 87131

[†]Fluid and Thermal Sciences Department, Sandia National Laboratories, Albuquerque, New Mexico 87185. This work supported by the U. S. Department of Energy under Contract Number DE-AC04-76DP00789.

take widely and disparate ranges of values. In addition, such systems are known to produce nonphysical instabilities in numerical solutions.

The model posed by Baer and Nunziato [1] is based on the continuum theory of mixtures and describes two–phase nonequilibrium flows. The formulation employs balance laws allowing the phases to exchange mass, momentum and energy. Constraints on these interactions are imposed so that overall conservation of mass, momentum and energy, as well as the entropy inequality, are satisfied. The characteristics for this system are real and elliptic regions never appear. Thus, the system is hyperbolic except at points where the relative flow is sonic. Also, numerical calculations based on this model (see [1],[2]) predict combustion waves that agree well with experimental observations.

In the next section, we present the formulation and the characteristics analysis of the equal pressure model and the Baer–Nunziato model. Comparison of numerical calculations using the Baer–Nunziato model to experimental wave trajectories are shown to illustrate the predictability of the model.

It is noted that the elementary wave analysis of the Baer–Nunziato model including the characteristic analysis, classification of wave fields, Riemann invariants, simple waves and special discontinuous solutions are given in [4]. Continued studies with Prof. A. Majda, are underway toward determining a more fundamental mathematical explanation of the DDT mechanism described by the Baer–Nunziato model.

2. **Continuum Models for Reactive Two Phase Flows.** There are many approaches in developing the conservation equations describing reactive two–phase flows. A commonly used derivation is based on the continuum theory of mixtures, as given in [1]. In this approach, one assumes the coexistence of both phases at every point in the flow field. Each phase is then assigned independent thermodynamic and hydrodynamic states: density ρ_a, pressure p_a, internal energy e_a, volume fraction ϕ_a, and velocity v_a, where the subscript a takes the values s (solid) or g (gas). For a two–phase flow, the description requires a total of ten state variables. Using the conservation laws of mass, momentum and energy for each phase, six field equations are defined. Additionally, two equations of state, $p_a = p_a(\rho_a, e_a)$, and a saturation constraint $\phi_s + \phi_g = 1$ increase the total number of equations to nine. Since the number of unknowns exceeds the number of equations, a problem of closure exists, and must be resolved by producing another constraint or an additional evolution equation. It is this issue which has led to much controversy in the multiphase flow literature. In the remainder of the paper, we discuss the consequences of both ways of resolving the closure difficulty in the context of an equal pressure model and the Baer–Nunziato nonequilibrium model.

Model 1. Equal Pressure Model

The equations describing the equal pressure model are given by (see *i.e.* [7]):

Conservation of mass

$$\frac{\partial}{\partial t}(\phi_a \rho_a) + \frac{\partial}{\partial x}(\phi_a \rho_a v_a) = 0 \quad , \quad a = s, g \quad , \tag{1}$$

Conservation of momentum

$$\frac{\partial}{\partial t}(\phi_a \rho_a v_a) + \frac{\partial}{\partial x}(\phi_a \rho_a v_a^2) + \phi_a \frac{\partial p_a}{\partial x} = 0 \quad , \tag{2}$$

Conservation of energy

$$\frac{\partial}{\partial t}(\phi_a \rho_a \eta_a) + \frac{\partial}{\partial x}(\phi_a \rho_a \eta_a v_a) = 0 \quad , \tag{3}$$

Saturation constraint

$$\phi_g = 1 - \phi_s \quad , \tag{4}$$

Pressure equilibrium constraint

$$p_s = p_g = p \quad , \tag{5}$$

Equations of state

$$\rho_a = \rho_a(p, \eta_a) \quad , \tag{6}$$

where η_a is the entropy of phase a.

In this model, the pressure equilibrium constraint (5) is enforced. For simplicity, the source and diffusion terms in (1-6) have been excluded since these terms are not required in a characteristics analysis.

The characteristics for the system (1-6) have been determined in [6] and [7]. In these studies the focus of the analysis centered on determining regions where the system equation has complex eigenvalues. Here, we provide a simpler analysis based on the previous work to show the existence of elliptic regions.

Using the variables $\mathbf{U} = (p, \phi_s, v_s, v_g, \eta_s, \eta_g)^T$, one recasts the system (1-3) in the form

$$\frac{\partial \mathbf{U}}{\partial t} + \mathbf{A}(\mathbf{U})\frac{\partial \mathbf{U}}{\partial x} = 0 \quad . \tag{7}$$

As given in [6],[7], the characteristic equation $det(\lambda \mathbf{I} - \mathbf{A}(\mathbf{U})) = 0$ is:

$$(\lambda - v_s)(\lambda - v_g)\left[\left(\frac{\lambda - v_s}{c}\right)^2\left(\frac{\lambda - v_g}{c}\right)^2 - \phi_g\frac{\rho}{\rho_g}\left(\frac{\lambda - v_s}{c}\right)^2 - \phi_s\frac{\rho}{\rho_s}\left(\frac{\lambda - v_g}{c}\right)^2\right] = 0 \quad , \tag{8}$$

where ρ and c are defined as

$$\rho = \phi_s \rho_s + \phi_g \rho_g \quad , \tag{9}$$

$$\text{and} \quad \frac{1}{\rho c^2} = \frac{\phi_s}{\rho_s c_s^2} + \frac{\phi_g}{\rho_g c_g^2} \quad . \tag{10}$$

The following notation is introduced

$$\delta = \frac{(v_s - v_g)}{2c} \quad , \tag{11}$$

$$\theta = \frac{1}{c}\left[\lambda - \frac{(v_s + v_g)}{2}\right] \quad , \tag{12}$$

$$G_\delta(\theta) = \frac{\phi_g \rho}{\rho_g}(\theta - \delta)^2 + \frac{\phi_s \rho}{\rho_s}(\theta + \delta)^2 \quad , \tag{13}$$

$$F_\delta(\theta) = (\theta + \delta)^2(\theta - \delta)^2 \quad , \tag{14}$$

and (8) is written as:

$$(\lambda - v_s)(\lambda - v_g)[F_\delta(\theta) - G_\delta(\theta)] = 0. \tag{15}$$

Therefore, the eigenvalues of the 6x6 matrix $A(U)$ are given by $\lambda = v_s, v_g$ and by the solutions of $F_\delta(\theta) = G_\delta(\theta)$. We first examine $F_\delta(\theta)$ and $G_\delta(\theta)$ for different values of δ for conditions when both phases are present.

The quadratic function $\psi = G_\delta(\theta)$ is always positive and its minimum is given by

$$\theta. = \frac{(\phi_g \rho_s - \phi_s \rho_g)}{(\phi_g \rho_s + \phi_s \rho_g)} \delta, \tag{16}$$

and the corresponding value

$$\psi_* = G_\delta(\theta.) = C\rho^2 , \tag{17}$$

where $C > 0$ is given by

$$C = 4\rho\phi_s\phi_g / (\phi_g \rho_s + \phi_s \rho_g) \le 4 . \tag{18}$$

Note that $|\theta.| < \delta$ and C is independent of δ.

The quadratic function $\psi = F_\delta(\theta)$ is nonnegative and has roots $\theta = -\delta, -\delta, \delta, \delta$. It is readily observed that

$$\min_{|\theta| \le \delta} F_\delta(\theta.) = \tilde{C}\delta^4 , \tag{19}$$

where $\tilde{C} = 16(\phi_s\phi_g\rho_s\rho_g)^2 / (\phi_s\rho_g + \phi_g\rho_s)^4 > 0$ and is independent of δ. The maximum value over this interval is

$$\max_{|\theta| \le \delta} F_\delta(\theta) = F(0) = \delta^4. \tag{20}$$

The solutions of $F_\delta(\theta) = G_\delta(\theta)$ for small and large values of δ are then determined. The nondimensional quantity δ given in (11) can be interpreted as a relative Mach number, c being the average sound speed given in (10).

Case 1. $\delta \ll 1$. Since $G_\delta(\theta) > 0$, $F_\delta(\theta) \ge 0$ for all θ, $F_\delta(\pm\delta) = 0$ and $deg\, F_\delta = 4 > 2 = deg\, G_\delta$ then $F_\delta(\theta) = G_\delta(\theta)$ has two real roots in $|\theta| > \delta$. On the other hand, from (17) and (20) $\max_{|\theta| \le \delta} F_\delta(\theta) = \delta^4 \ll C\delta^2 = \min_{\theta \in \mathbb{R}} G_\delta(\theta)$. Therefore, $F_\delta(\theta) = G_\delta(\theta)$ has only two real roots and the remaining two roots are complex conjugates. This situation is depicted in Figure 1(a).

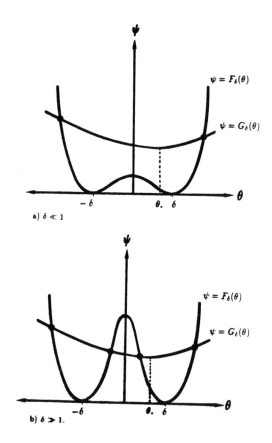

Figure 1. Graphical representations of the intersections of $F_\delta(\theta)$ and $G_\epsilon(\theta)$
a) $\delta \ll 1$ and b) $\delta \gg 1$.

Case 2. $\delta \gg 1$. Following the previous arguments $F_\delta(\theta) = G_\delta(\theta)$ has two real roots for $|\theta| > \delta$. On the other hand, $G_\delta(\theta_-) = C\rho^2 \ll \tilde{C}\delta^4 = F_\delta(\theta_-)$ and since $F_\delta(\pm\delta) = 0 < G_\delta(\pm\delta)$, it is concluded that $F_\delta(\theta) = G_\delta(\theta)$ has two real roots in $|\theta| < \delta$. Hence, in this case, the four roots of $F_\delta(\theta) = G_\delta(\theta)$ are real. This situation is depicted in Figure 1(b).

In summary, this system has four real characteristics when $\delta \gg 1$, however, for the case $\delta \ll 1$ two characteristics are complex. It is noted that when one phase vanishes, the system of equation corresponds to a single phase system and all characteristics are real as given in Equation (18).

A modification of the equal pressure model has been proposed in [5], with the pressure equilibrium assumption (5) replaced by

$$p_g - p_s = p_\sigma \tag{21}$$

where the additional p_σ term represents surface tension effects. This modification renders the characteristics of the system real. However, this is not completely satisfactory because some of the eigenvalues are infinite, implying that some modes have infinite propagation speed.

Model 2. Baer–Nunziato nonequilibrium model.

The reactive multiphase flow equations formulated in [1] can be written as:

Conservation of Mass

$$\frac{\partial}{\partial t}(\phi_a \rho_a) + \frac{\partial}{\partial x}(\phi_a \rho_a v_a) = c_a^+, \tag{22}$$

Conservation of Momentum

$$\frac{\partial}{\partial t}(\phi_a \rho_a v_a) + \frac{\partial}{\partial x}(\phi_a \rho_a v_a^2 + \phi_a p_a) = m_a^+, \tag{23}$$

Conservation of Energy

$$\frac{\partial}{\partial t}(\phi_a \rho_a E_a) + \frac{\partial}{\partial x}((\phi_a \rho_a E_a + \phi_a p_a)v_a) = e_a^+, \tag{24}$$

Compaction Evolutionary Equation

$$\frac{\partial \rho_s}{\partial t} + \frac{\partial}{\partial x}(\rho_s v_s) = -\frac{\rho_s F}{\phi_s}. \tag{25}$$

Saturation Constraint

$$\phi_s + \phi_g = 1. \tag{26}$$

Equations of State

$$p_a = p_a(\rho_a, e_a). \tag{27}$$

The total phase energy $E_a = e_a + v_a^2/2$ is introduced.

In this case, the closure problem is resolved upon recognizing ϕ_s as an additional degree of freedom and an equation for its evolution is derived by combining (22) and (25).

The interaction terms, c_a^+, m_a^+ and e_a^+, represent the exchange of mass, momentum and energy between the phases, respectively. To assure conservation of mass, momentum and energy of the total mixture, the following constraints are required:

$$c_s^+ = -c_g^+, \quad m_s^+ = -m_g^+, \quad e_s^+ = -e_g^+. \tag{28}$$

The forcing function for the compaction equation, F, dictates that compaction is rate dependent and is driven by pressure differences, thus:

$$F = \frac{\phi_s \phi_g}{\mu_c}(p_s - p_g - \beta_s), \tag{29}$$

where β_s is the configuration stress (reflecting distortion of the solid granular material) and μ_c is the compaction viscosity.

Explicit constitutive equations for the interaction terms are given in [1]. In the momentum and energy phase interactions, it is convenient to separate the volume fraction derivative terms describing geometric coupling between the phases according to:

$$m_s^+ = p_s \frac{\partial \phi_s}{\partial x} + m_s^* \ , \tag{30}$$

$$e_s^+ = v_s p_s \frac{\partial \phi_s}{\partial x} + e_s^* \ . \tag{31}$$

The m_s^* and e_s^* terms are now strictly algebraic (the diffusion terms are omitted). Note that the system (22-25) is not in divergence free form.

The definition of F and the interaction terms are shown to be consistent with the entropy inequality for the mixture

$$\sum_a \left(\frac{\partial}{\partial t}(\phi_a \rho_a \eta_a) + \frac{\partial}{\partial x}(\phi_a \rho_a v_a \eta_a) \right) \geq 0 \tag{32}$$

A characteristic analysis for the system (22-25) has also been conducted. For the sake of brevity, the mathematical details are omitted and the interested reader can find a complete development in [4].

The field equations (22-25) represent a first order system of equations written in the form:

$$\frac{\partial \mathbf{U}}{\partial t} + \mathbf{A(U)} \frac{\partial \mathbf{U}}{\partial x} = \mathbf{S(U)} \ , \tag{33}$$

where $\mathbf{U} = (\phi_s \rho_s, \phi_s \rho_s v_s, \phi_s \rho_s E_s, \rho_s, \phi_g \rho_g v_g, \phi_g \rho_g E_g)^T$.

The 7x7 matrix $\mathbf{A(U)}$ has eigenvalues $v_s \pm c_s, v_s, v_s, v_g \pm c_g, v_g$. All the eigenvalues are real but v_s is a double eigenvalue and in general there is no possible ordering for the eigenvalues since those corresponding to one phase may coincide with others associated to the other phase, hence the system is not strictly hyperbolic. Nevertheless, the system is hyperbolic provided that $\mathbf{A(U)}$ has a complete set of eigenvalues. The analysis of the eigenvectors reveals that this is the case except when the eigenvalue v_s coalesce with either $v_g + c_g$ or $v_g - c_g$. At this point, v_s becomes a triple eigenvalue and one of the eigenvectors associated with v_s aligns with the eigenvector associated with $v_g + c_g$ or $(v_g - c_g)$. Therefore, at points where $v_s - v_g = \pm c_g$, the Jordan form of $\mathbf{A(U)}$ has a nontrivial 2x2 block, and the system becomes degenerate. We term $v_s - v_g = \pm c_g$ the choked flow conditions. As a physical interpretation, flow at the pore level is analogous to flow in a moving variable cross-sectional area duct (due to local variation of porosity) and the choked flow conditions correspond to sonic points of the relative gas flow.

It is noted that two pressure models equilibrium and a closure evolutionary equation for the volume fraction has been previously studied [6]. However, the formulation of these systems, the physical significance of the characteristic values and the degenerate points has not been clarified.

To demonstrate the predictability of the Baer-Nunziato model in describing DDT in reactive granular flows, we present comparisons of numerical calculations to experimental measurements.

The model was used to simulate flame spread in a 1.0 cm–long column of explosive CP with uniform grains of 15 μm surface–mean diameter and at different initial partial densities. Combustion was initiated by a high pressure/temperature gas pulse at one end of the tube. The numerical grid consisted of 201 computational nodes uniformly distributed along the column and the system of equations was solved numerically using the method–of–lines. The trajectory of the luminous front was taken from image–enhanced photographic streak records and compared to the computed location of the burn front.

Shown in Figure 2 are numerical–experimental comparisons of the burn front path for varied initial densities 1.2, 1.4, and 1.6 g/cm^3. Consistent with experimental observations, calculations predict the convective and compressive burning stages prior to the onset of detonation. Moreover, as the density increases from 1.2 to 1.6 g/cm^3, the detonation speed increases from 5.2 to 6.5 mm/μs, in agreement with the experimental measurements.

3. **Summary.** In this paper we presented the characteristics analysis of the reactive multiphase flow equations formulated by Baer and Nunziato, and contrast its mathematical properties with those of a pressure equilibrium model. The system is hyperbolic except at choked flow points, where the coupling of the hydrodynamic fields of both phases render the system degenerate. The system is well suited for numerical calculations and adequately describes the complex combustion processes involved in DDT of granular explosives.

Acknowledgement. We gratefully acknowledge fruitful discussions with P. Lax (Courant Inst.), A. Majda (Princeton University) and J. Nunziato (SNLA).

66

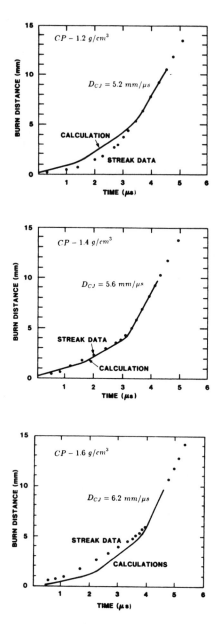

Figure 2. A comparison of calculations and experimental data of the combustion front trajectories for granular CP at various densities a) 1.2 g/cm^3, b) 1.4 g/cm^3, and c) 1.6 g/cm^3.

REFERENCES

1. M. R. Baer and J. W. Nunziato, "A Two–Phase Mixture Theory for Deflagration–to–Detonation Transition (DDT) in Reactive Granular Materials ," *Intl. J. Multiphase Flow*, 12 (1986) pp. 861–889.

2. M. R. Baer, R. J. Gross, J. W. Nunziato, and E. A. Igel, "An Experimental and Theoretical Study of Deflagration–to–Detonation Transition (DDT) in the Granular Explosive, CP," *Combustion and Flame* 65 (1986) pp. 15–30.

3. R. R. Bernecker, "The Deflagration–to–Detonation Transition of High Energy Propellants – A Review," *AIAA Journal* 24 (1986) pp. 82-91.

4. P. F. Embid and M. R. Baer, "Mathematical Analysis of a Two–Phase Model for Reactive Granular Material," SAND88-3302, Sandia National Laboratories, 1989, in press.

5. J. D. Ramshaw and J. A. Trapp, "Characteristics, Stability and Short–Wavelength Phenomena in Two–Phase Flow Equation Systems," *Nucl. Sci. Eng.* 66 (1978) 93–102.

6. V. H. Ransom and D. L. Hicks, "Hyperbolic Two–Pressure Models for Two–Phase Flow," *J. Comput. Phys.* 53 (1984) 124-151.

7. H. B. Stewart and B. Wendroff, "Two–Phase Flow: Models and Methods," *J. Comput. Phys.* 56 (1984) 363–409.

SHOCKS ASSOCIATED WITH ROTATIONAL MODES

HEINRICH FREISTÜHLER*

Abstract. Many interesting systems of conservation laws in several space variables are isotropic in the sense of equivariance under rotations acting on the spatial components of the independent variable and on appropriate components of the state variable (e.g. the velocity vector in gas dynamics, or the velocity vector and the magnetic field vector in magnetohydrodynamics). As consequences of the isotropy, the conservation law which governs the propagation of plane waves in a given direction is not only independent of this direction, but also inherits a rotational symmetry, where now rotations act only on the transverse parts of the said components of the state variable. Generically — e.g. in magnetohydrodynamics, but not in gas dynamics — these transverse parts exhibit an interesting wave pattern including isolated linearly degenerate rotational modes. The talk is especially on the following three facts: These modes have not only contact discontinuities, but also shocks associated with them; for an arbitrarily given strictly stable and rotationally equivariant viscosity many of these shocks have viscous profiles; for an arbitrary choice of the four viscosity parameters in the usual dissipative version of magnetohydrodynamics there exist intermediate magnetohydrodynamic shocks which have a viscous profile.

Some details. A detailed analysis of the wave pattern for systems with rotationally symmetric flux has been given in [2]. This includes the observation of the rotational modes and of shocks associated with them, and an existence and uniqueness theorem on locally posed Riemann problems. The theory given in [2] has also been applied there to important cases from magnetohydrodynamics and isotropic elasticity.

Shocks associated with rotational modes and their viscous profiles have been observed in [4] in a very general setting, assuming standard action of the orthogonal group on the state space.

For a simple model illustrating these phenomena see [3] and [5].

Those results of [4] which are of interest here have been generalized in [5] to the situation of compact linear degeneracy, i.e. a linearly degenerate mode whose eigenspace bundles have compact integral manifolds (for the special case of rotational modes corresponding to standard action of the orthogonal group they are spheres). This covers e.g. more general actions of the orthogonal group.

Note that even this is not directly applicable to certain physical examples, the obstacle being that the viscosity matrices that are commonly used there are not strictly stable since they are singular. But e.g. in the case of magnetohydrodynamics this technical difficulty can be overcome by appropriate adaptations, see [5]. Here, the symmetry approach yields the existence of the abovementioned profiles for (certain) subfast-superslow shocks, often denoted as $u_1 \to u_2$. (Starting from a rotationally symmetric case occuring for the system of o.d.e. governing the traveling waves, also profiles for certain superfast-subslow intermediate shocks $u_0 \to u_3$ and for certain superfast-superslow intermediate shocks $u_0 \to u_2$ are established in [5].)

With respect to viscous profiles for intermediate magnetohydrodynamic shocks, I would like to point out the recent papers [1], [8], and especially [6], [7] where the

---

*RWTH Aachen, Templergraben 55, 5100 Aachen, West Germany

first proof has been given that such shocks of type $u_1 \to u_2$ have profiles for certain values of the flux and viscosity parameters.

REFERENCES

[1] M. BRIO, *Propagation of weakly nonlinear magnetoacoustic waves*, Wave Motion, 9 (1987), pp. 455–458.

[2] H. FREISTÜHLER, *Rotational degeneracy of hyperbolic systems of conservation laws*, Arch. Rat. Mech. Anal (to appear).

[3] ——————————, *A standard model of generic rotational degeneracy*, in *Nonlinear Hyperbolic Equations — Theory, Computation Methods, and Applications*, Vieweg, Braunschweig, 1989.

[4] ——————————, *Instability of vanishing viscosity approximation to hyperbolic systems of conservation laws with rotational invariance*, J. Diff. Eqs (to appear).

[5] ——————————, *On compact linear degeneracy*, IMA preprint (1989).

[6] H. HATTORI & K. MISCHAIKOW, *An application of connection matrix to magnetohydrodynamic shock profiles*, in *Multidimensional hyperbolic problems and computations*, IMA Volumes in Mathematics and its Applications, 29, Springer-Verlag, New York, 1989.

[7] K. MISCHAIKOW & H. HATTORI, *On the existence of intermediate magnetohydrodynamic shock waves*, Dynamics and Diff. Eqs (to appear).

[8] C.C. WU, *On MHD intermediate shocks*, Geophys. Res. Lett., 14 (1987), pp. 668–671.

SELF-SIMILAR SHOCK REFLECTION
IN
TWO SPACE DIMENSIONS

HARLAND M. GLAZ*

1. Introduction. Self-similar oblique shock wave reflection for the equations of unsteady inviscid gas dynamics is among the outstanding unsolved problems in nonlinear PDE. This is so due not only to its' standing as the simplest possible nontrivial shock wave interaction problem for these equations and the importance of the problem in engineering applications, but to the tantalizingly large amount of experimental and computational data which is available after half a century of intensive research since WW II.

The purpose of this paper is to precisely state the problem, along with other problems related to various aspects of stability. Additionally, we provide a historical perspective and discuss some of the results which have become available during the past decade or so; this recent work is of relatively high quality due to advances in shock tube technology, data analysis (especially infinite-fringe interferometry), and computational fluid dynamics (CFD).

The equations of unsteady, inviscid gasdynamics in Cartesian coordinates are

$$\rho_t + (\rho u)_x + (\rho v)_y = 0 \tag{1a}$$

$$(\rho u)_t + (\rho u^2 + p)_x + (\rho u v)_y = 0 \tag{1b}$$

$$(\rho v)_t + (\rho u v)_x + (\rho v^2 + p)_y = 0 \tag{1c}$$

$$(\rho E)_t + (\rho u E + u p)_x + (\rho v E + v p)_y = 0 \tag{1d}$$

where ρ is the density, $\mathbf{u} = (u,v)$ is the velocity field, $E = \frac{1}{2}(u^2 + v^2) + e$ is the total specific energy, e is the specific internal energy, and p is the pressure. The system is closed by specifying an equation-of-state (EOS), $p = p(\rho, e)$; an important example is the polytropic EOS, $p = (\gamma - 1)\rho e$, where $\gamma > 1$ is constant. Equations (1) are a system of conservation laws

$$U_t + F(U)_x + G(U)_y = 0$$

where $U = (\rho, \rho u, \rho v, \rho E)^t$, etc. The initial conditions for oblique shock wave reflection are depicted in Figure 1 for a standard computational setup. An incident shock wave, I, of shock wave Mach number M_s traverses an ambient (i.e., $\mathbf{u}_0 = (0,0)$) medium and approaches a wedge surface inclined at θ_w degrees; since the upstream flow is ambient, $M_s = \sigma/c_0$ where $\sigma = $ shock speed and c_0 is the upstream sound speed. The coordinate system is set by taking the corner at $(x, y) = (0, 0)$ and the

*Department of Mathematics, University of Maryland, College Park, Maryland 20742. The support of the National Science Foundation, Grant DMS-8703971, and the Defense Nuclear Agency, Contract DNA001–87–0303 is gratefully acknowledged.

time $t = 0$ at the instant the shock reaches the corner. There are no length scales in the initial data, the equations (1), or the EOS, so it is reasonable to look for self–similar or pseudosteady solutions for $t > 0$ in the similarity variables

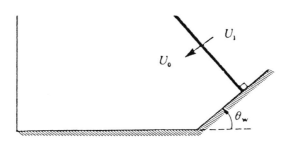

Figure 1. Initial conditions in a computational frame for self-similar oblique shock wave reflection. The laboratory frame is obtained by a clockwise rotation of θ_w degrees.

(2) $$(\xi, \eta) = (x/t, y/t).$$

In conservation form, the transformed equations are

(3a) $$(\rho\tilde{u})_\xi + (\rho\tilde{v})_\eta = -2\rho$$
(3b) $$(\rho\tilde{u}^2 + p)_\xi + (\rho\tilde{u}\tilde{v})_\eta = -3\rho\tilde{u}$$
(3c) $$(\rho\tilde{u}\tilde{v})_\xi + (\rho\tilde{v}^2 + p)_\eta = -3\rho\tilde{v}$$
(3d) $$(\rho\tilde{u}\tilde{H})_\xi + (\rho\tilde{v}\tilde{H})_\eta = -\rho(\tilde{u}^2 + \tilde{v}^2) - 2\rho\tilde{H}$$

where $\tilde{\mathbf{u}} = (\tilde{u}, \tilde{v})$ is the self-similar velocity field, $\tilde{u} = u - \xi$, $\tilde{v} = v - \eta$, $\tilde{H} = \frac{1}{2}(\tilde{u}^2 + \tilde{v}^2) + h$ is the self-similar total enthalpy, and $h = e + p/\rho$ is the specific enthalpy. Additionally,

(4) $$\tilde{M}^2 = (\tilde{u}^2 + \tilde{v}^2)/c^2$$

defines the self-similar Mach number $\tilde{M}$, where c = sound speed. In vector form, the equations (3) are written

$$\tilde{F}_\xi + \tilde{G}_\eta = \tilde{S}$$

where $\tilde{F} = (\rho\tilde{u}, \rho\tilde{u}^2 + p, \rho\tilde{u}\tilde{v}, \rho\tilde{u}\tilde{H})^t$, etc. One notes that the first-order part of this system, $\tilde{F}_\xi + \tilde{G}_\eta = 0$, are exactly the equations of steady gas dynamics in the (ξ, η) coordinates.

The wall boundary condition for equations (1) is perfect reflection, i.e., flow is parallel to the wall. It is easily seen that this condition transforms analogously for equations (3). The boundary condition at infinity for equations (1) is the time-dependent Dirichlet condition provided by the Rankine-Hugoniot conditions connecting states U_0 and U_1 in Figure 1. Since no disturbance can reach infinity in finite time, this boundary condition can also be easily transformed. Further, the boundary may be moved to a finite distance from the origin so long as it is outside the region of disturbed flow; the intersection of the incident shock with such a boundary is easily calculated in the (ξ, η)-frame for $t > 0$ by using the known shock speed, σ.

Unlike equations (1), which are always hyperbolic under very general conditions on the EOS, the equations (3) of steady gas dynamics with source terms are hyperbolic only when the flow is supersonic in the (ξ, η)-frame. For subsonic flow, the streamline characteristic remains real but the sound wave characteristics become complex. Thus, equations (3) are of mixed type. Note, however, that for the boundary conditions under consideration here, the flowfield becomes supersonic as $(\xi, \eta) \to \infty$ for any initial data; in particular, any finite farfield boundary chosen as above should satisfy the condition that the flow is supersonic nearby.

Summarizing, we are interested in solving the boundary value problem given by equations (3) subject to the given boundary conditions. Evidently, the solution depends parametrically on M_s, θ_w, the EOS and the ambient data (ρ_0, e_0). For the important case of a polytropic gas, the parameters are M_s, θ_w and γ; the ambient state is no longer an independent parameter. To the best knowledge of the author, there do not exist any mathematical results pertaining to the question of global solutions for this problem; the issues of existence, uniqueness, and continuous dependence on the parameters remain open for all parameter ranges.

Any conjectures that we have concerning these questions must be inferred from shock tube experiments and CFD simulations. Additionally, mathematical analysis is often applied to infinitesimal or local flowfield regions, especially shock wave interactions, once such regions are apparent from experiments or calculations. Of course, such experiments and calculations are subject to experimental error (e.g., the incident shock has a finite thickness and may not move at precisely constant speed, the wall will not be perfectly smooth, data analysis errors, etc.) and numerical error. Also, any experimental result will represent a solution, not of equations (1), but of an augmented version of equations (1) which takes into account real gas effects such as vibrational relaxation, dissociation-recombination chemistry, and viscosity. In some examples (especially at high M_s), these effects are significant; in all cases, it is necessary to carefully analyze them. Numerical simulations may not be fully resolved, may introduce 'artificial' viscosity, and may contain other systematic errors; furthermore, it is not possible to prove convergence or obtain useful error estimates at the present time. Despite all of these caveats, most workers in the field would at least maintain that the data is very useful in deducing properties of solutions of equations (3). Still, it must be emphasized that the answers to these questions are not at all obvious.

The phenomenology of oblique shock wave reflections will be outlined in the next section, based on experimental and computational work of the last decade or so. A brief historical survey and guide to the literature is also included in this section. In the following section, several issues are taken up in greater detail: the effects of viscous and nonequilibrium terms, the question of whether realized flows are indeed self-similar, and mathematical questions related to solutions of equations (3). The final section presents our conclusions.

2. Phenomenology of oblique shock wave interaction. The four types of self-similar oblique shock wave reflections observed in experiments and calculations are depicted schematically in Figure 2 and consist of (a) regular reflection (RR), (b) single Mach reflection (SMR), (c) complex Mach reflection (CMR), and (d) double Mach reflection (DMR). Some standard terminology is also defined by this figure: the triple point trajectory angles χ, χ'; the shock waves I (incident), R,R' (reflected), M,M' (Mach); the first and second triple points T,T'; the kink K; the slip (and contact) surfaces S,S'; the angles δ, ω and ω' between I and R, R and the first triple point trajectory, and R and the wall, respectively; the bow shock standoff distance s; and the states U_0–U_5 produced by these reflections. We will not distinguish between the two segments R,R' of the reflected shock, in general. Notice that the triple point trajectory must be a straight line by self-similarity, which also implies that the ratio s/L is independent of time where L is as indicated in the figure.

A fundamental objective of research in the field is to predict the reflection pattern as a function of the problem parameters. A typical transition (or bifurcation) diagram is shown in Figure 3; the curves are constructed here from analytic formulas based on heuristic transition criteria (to be discussed below) and applied to a polytropic gas with $\gamma = 1.4$. The depicted (M_s, θ_w) – parameter space is complete for a polytropic gas and the results for different values of γ (say, $1.09 \leq \gamma \leq \frac{5}{3}$) are qualitatively similar. As $M_s \rightarrow \infty$, the transition lines become asymptotically flat (so $M_s \leq 10$ is sufficient for studying a $\gamma = 1.4$ gas; different maximums would be required for very low γ gases); of course, this is not necessarily the case for a general EOS. Also, there is a dependence on the initial state for a general EOS but this dependence is very weak, especially for initial data near atmospheric conditions.

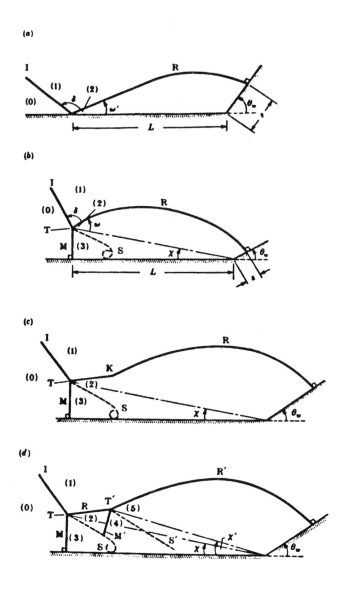

Figure 2. The four main types of self-similar reflection: (a) RR, (b) SMR, (c) CMR and (d) DMR. The various symbols are defined in the text.

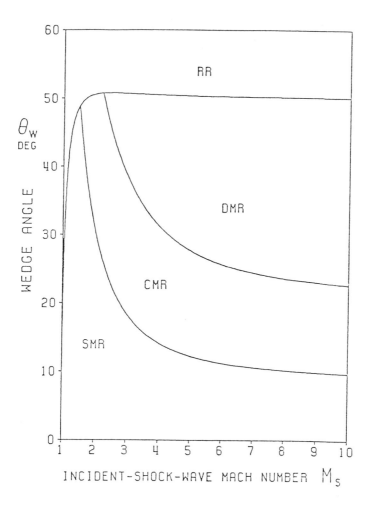

Figure 3. (M_s, θ_w) transition diagram for a $\gamma = 1.4$ gas.

Considerable effort has been expended in constructing transition diagrams based on experimental fits. The most thorough work has been undertaken by I.I. Glass and coworkers at UTIAS; results of infinite – fringe interferometry are available for CO_2 [1]; N_2 [5],[7]; Ar [5],[8]; real air [16],[37]; SF_6 [31],[48],[49]; and isobutane [48],[49]. Interesting Soviet shadowgraphs for these gases and Freon–12 as well may be found in [45]. Since these results incorporate real gas effects such as viscosity, vibrational relaxation and high temperature chemistry, there will inevitably be disagreements with any analytic theory (even if assumed correct). Indeed, the disagreements can be quite substantial at high M_s and low γ, although differences are quantitative and not qualitative.

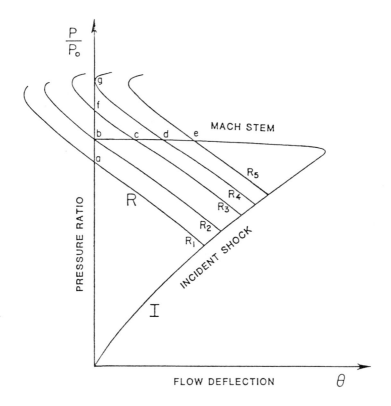

Figure 4. Schematic pressure ratio – flow deflection shock polar diagram for steady flow shock jump conditions; five reflected shock polars are drawn from the same incident shock polar.

Numerous criteria have been developed for the RR–MR transition. The main analytic criteria are called the detachment condition and the mechanical equilibrium condition. These criteria are determined from shock polar diagrams such as that of Figure 4. Here, the curve I denotes the set of postshock states attainable behind the incident shock wave; taking some point on this curve and constructing the set of states attainable from this point by a backwards–facing shock leads to the reflected shock curves R_k. The curves are constructed by solving the Rankine–Hugoniot conditions for steady gas dynamics in the shock–fixed frame. Curves such as R_1 meet the $\theta = 0$ axis (i.e., $\theta_1 + \theta_2 = 0$ which is required to satisfy the inviscid boundary conditions at the wall) in two points ('a' and a point above 'g'). The lower (upper) intersection point is called the weak (strong) shock solution; experiments always recover the weak solution although neither solution can be rejected on theoretical grounds at the present time. Mach reflection occurs for curves such as R_5 which do not intersect the $\theta = 0$ axis. The point 'e' describes the state just behind the Mach stem at the triple point, since it clearly satisfies the three - shock jump conditions: $\theta_1 - \theta_2 = \theta_3$ and $p_2 = p_3$, i.e., U_2 and U_3 are separated by a slip/contact surface. The detachment condition is given by R_4 which is the unique curve meeting the $\theta = 0$ axis in exactly one point; alternatively, $\theta_2 = \theta_{2m}$ (= the maximum deflection angle) which implies $\theta_1 = -\theta_{2m}$. Another way of looking at the detachment condition is that it states that RR will persist until it is impossible to solve the two - shock jump conditions at the reflection point. One notes that the detachment condition implies that the postshock pressure undergoes a large jump across the transition curve (remark: the point 'g' may lie inside or outside the I curve in general and 'g' = 'b' is a possible singular case for which this jump does not occur). The mechanical equilibrium condition (which goes back to von Neumann) is partially motivated by viewing this transition pressure jump as physically unnatural. It is described by the unique curve R_2 (so, in this theory if one starts on R_3 then the result is Mach reflection at 'c') which intersects both the I curve and the $\theta = 0$ axis and, thereby, results in a continuous pressure variation at transition. The papers [25],[26],[27] are especially important in understanding the possibilities for the RR–MR transition.

The differences between the two criteria are large and are illustrated for a $\gamma = 1.4$ gas in Figure 5; the region between the two curves is referred to as the overlap region. The experimental record on this subject is both substantial and quite inconclusive. Indeed, the 'persistence of RR' below the detachment line has often been observed. Questions in this area have been recognized since before 1950 and have been studied intensively since then. During the late 1970's and later, many researchers have proposed that experimental transitions must be governed by phenomena involving length scales at the reflection/triple point (and, therefore, such theories are not directly useful in a self – similar theory) such as the internal shock structure and various relaxation lengths and, especially, viscosity. Explanations involving a 'boundary layer defect' [29],[30] have been especially prominent. These ideas will not be developed here except to note that it is natural to expect that a boundary layer will shift the effective wedge angle and, therefore, the transition line (which can be used to explain the phenomenon of 'RR persistence' below the detachment line). Some very interesting recent experimental work, exploiting the dependence of viscosity on the gas density, has been reported by the UTIAS group [49],[50]; using a limiting argument based on their data, they support the mechanical equilibrium condition in the inviscid limit for some gases.

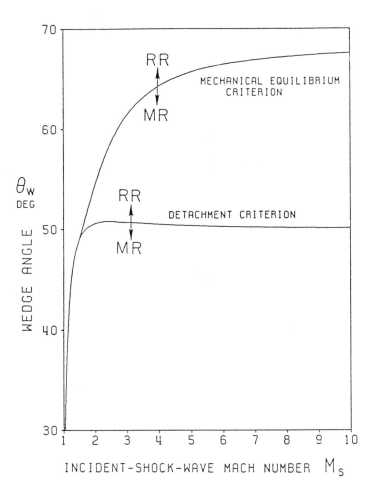

Figure 5. Mechanical equilibrium and detachment conditions as RR–MR transition criteria illustrated i the (M_s, θ_w) – plane.

Another important distinction is that between a weak and strong incident shock. The above discussion applies only to the strong case. If the inverse pressure ratio p_1/p_2 is sufficiently high then it becomes impossible to solve the three - shock jump conditions at the triple point and one is in the weak case; this occurs for $M_s \leq 1.1$ or so. Again, the older experimental record is ambiguous and triple points have been observed in some cases, but not in others for which the incident shock/Mach stem undergoes a continuous curvature. The effects of shock wave thickness are considered important in this regime [32]. This and other possible explanations are discussed in the excellent newer study [17]. Recent work [14],[43] involving sophisticated numerical computations may have resolved the difficulties here in the context of the self – similar theory; the reader is referred to these references for details.

The SMR–CMR and CMR–DMR transitions have received relatively less attention in the literature than the RR–MR transition. Today, most theories for these transitions are entirely inviscid; an example of a much earlier theory involving internal degrees of freedom of the gas may be found in [24]. Define M_{2T} = flow Mach number in region (2) relative to T and M_{2K} = flow Mach number in region (2) relative to the kink, K. Then, it is observed in both experiments and computations that $M_{2T} \geq 1$ is necessary for CMR and that $M_{2K} \geq 1$ is necessary for DMR. Additionally, the presence of a kink probably requires the existence of a band of compression waves near K (which may not be visible in experiments but are often clear in computations) and this assumption has been used to imply that $\delta \geq 90°$ (see Figure 2) for CMR. If so, the orientation of the Mach stem at the triple point – which is usually not perpendicular to the wall – becomes of interest; see [47] for information. To complete the theory for these necessary conditions it is necessary to locate the position of the kink; the theory used here is somewhat more speculative and we do not describe it here. We refer to [2],[26],[31],[36],[47] for further details concerning these criteria and the kink location. Comparison with the experimental record is mixed with the disagreements growing with M_s, in general. This may be expected since there is essentially no theory for sufficient conditions for these transitions (and real gas effects are significant at high M_s). As is known from numerical simulations (see below) the inviscid dynamics of CMR and DMR flowfields is quite complex in the Mach stem region; an interesting description of the dynamics of this region from a semianalytic point–of–view may be found in [41].

A potentially important feature of the transition diagram (Figure 3) is the near confluence of the transition lines at approximately $(M_s, \theta_w) = (2.0, 49°)$. Indeed, workers in the field have long conjectured that these lines are actually coincident. In the experimental studies of the SMR–CMR and CMR–DMR transitions [1],[2], the authors noted that for θ_w approaching the RR–MR transition line from below, the triple point T approaches coincidence with the kink K or the second triple point T'; if this were so (and it should be remarked that a careful experimental or numerical study would be quite difficult), it would imply that $M_{2T} = M_{2K}$ in the limit. If the aforementioned criteria for the SMR–CMR and CMR–DMR transitions are also correct, this would imply the desired result. The mathematical implications of such an 'organizing center' are interesting and are discussed in the paper of Majda in this volume. The discussion here is somewhat amplified in [37].

A new experimental phenomenon attaining prominence since 1981 is the so-called terminal double Mach reflection (TDMR), a subtype of DMR characterized by the condition $\chi' = 0$. Interferograms and shadowgraphs may be reviewed in [6],[31],[33],[40]. Since such a flowfield consists of two disconnected components (except for one point) this author finds their existence in a self–similar theory very unlikely; extensive numerical experiments support this view. However, it should be noted that TDMR is only observed for very low γ gases and only at high M_s. Such gases are quite complex in the sense that they contain many internal degrees of freedom. Thus, relaxation processes along with viscosity acting near the second triple point for χ' near 0 may explain TDMR in an unsteady theory. Another outside possibility is that the equilibrium EOS is nonconvex at high temperatures for such gases and it is then possible that the reflected shock is unstable (to unsteady perturbations) [39] under certain conditions; such instabilities may be observed in the literature, but can be explained by phenomena (e.g., sidewall boundary layers in the shock tube) more mundane than a true inviscid instability.

The earliest numerical simulations of oblique shock wave reflection known to the author are in the references [28],[35],[44],[46]. These calculations were already able to predict the overall waveform but the hardware capabilities of the time were wholly inadequate to resolve flowfield details. With the advent of supercomputers and the coincident development of high-order upwind schemes for conservation laws the situation changed. Some early (i.e., about 1980 – 1982) calculations with an FCT scheme and a prototype second-order Godunov scheme are discussed in [16]. A Soviet contribution in this period is reported in [15].

The most thorough numerical studies of oblique shock wave reflection have utilized the second–order Godunov scheme described in [13]. The results of a comparison with UTIAS experimental data for real air are reported in [20] with further details in [21]. Among the most interesting aspects of these calculations is the rollup of the main slip surface (emanating from the first triple point) into a strong vortex which can push forward the Mach stem, for high M_s DMR flowfields. Roughly simultaneously with these results, the UTIAS group obtained interferograms which also demonstrated the effect clearly. Since the calculations were inviscid, it followed that the phenomenon had to be due to dynamic effects alone with the boundary layer being insignificant. It may be noted that until the early 1980's, the usual theory for the slip surface was that it terminated at the self–similar stagnation point on the wall – and this is in fact the case for many weaker SMR flowfields. This vortical structure and associated effects substantially complicates the formulation of a mathematical theory for DMR flow, primarily because the DMR region of the flowfield can be broken up into multiple subsonic and supersonic pockets; also, new triple points form in the region for some cases (although these effects only become dominant at low γ and require a different test gas). Despite this success and the overall excellent qualitative agreement between corresponding experiments and calculations, the comparisons at high temperature were well off quantitatively. This is so because the high M_s experiments must be conducted using low density ambient air in order to protect the shock tube apparatus; the low densities accentuate vibrational nonequilibrium and other high temperature effects which cannot be modelled with a $\gamma = 1.4$ model and may or may not be well modelled with equilibrium real air EOS software such as those used in [13],[20],[21]. The next study undertaken was similar in design but used SF_6 as the test gas and related equilibrium EOS software in the calculations, see [22]. Here, the quantitative agreement was as strong as possible; since SF_6 attains equilibrium quickly an equilibrium EOS is a much better model here than for real air.

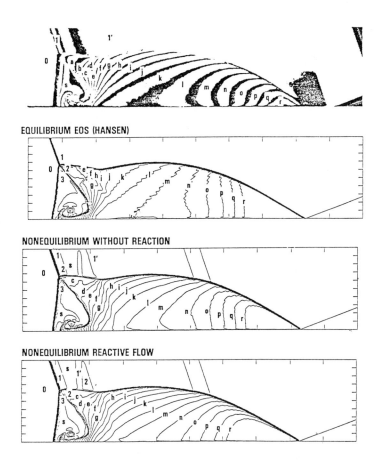

Figure 6. $M_s = 7.19, \theta_w = 20°$ flowfield. From top to bottom: (a) interferrogram from UTIAS experiment, (b) calculation with an equilibrium air EOS, (c) calculation with vibrational nonequilibrium but frozen chemistry and (d) calculation with vibrational nonequilibrium and 5 – species chemistry. Figures (b) – (d) show density contours.

In later work, a numerical scheme based on the formulation of [13] but incorporating extra equations and source terms sufficient to model real air up to temperatures encountered in the experimental literature was developed and used to study the high M_s DMR flowfields of the preceding real air study [23]; the new model used a five species chemistry model plus vibrational nonequilibrium with rate constants taken from the recent literature. Two sets of the resulting comparisons are shown here in Figures 6 and 7. In each case, it may be noted that the qualitative and quantitative agreement improves when the full nonequilibrium reactive model is used. Even fine details such as the relaxation zone beneath R and behind T are in strong agreement; prior to the new calculation, it would have been quite reasonable to ascribe this region in the interferograms to optical effects. Another obvious conclusion is that the equilibrium real air EOS used is inadequate for these flows and that they are truly unsteady; at very short times the flowfield would be nearly self–similar with a frozen EOS, and at very large times the flowfield would once again approach self–similarity with an appropriate equilibrium EOS. It is interesting that the scale of the experiments happens to lie in the intermediate nonequilibrium range. It may also be noted that these figures clearly demonstrate the vortex rollup and Mach stem toe–out discussed above.

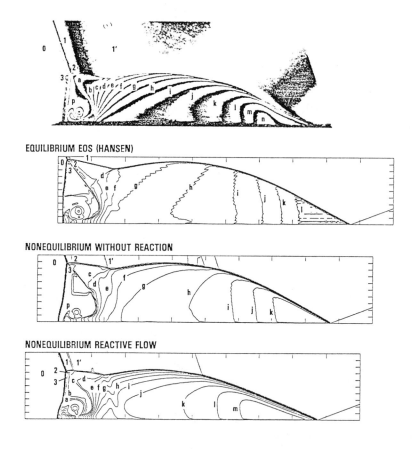

Figure 7. $M_s = 8.86, \theta_w = 20°$ flowfield. See Figure 6 caption.

An unsplit version of the second–order Godunov scheme (the scheme of [13] is directionally split) has also been developed [12] and later implemented in a code with local adaptive mesh refinement capability [11]. This allows nearly an order-of-magnitude better resolution in the DMR region; an example is shown in Figure 8. The main point of this result for our purposes is the apparent Kelvin–Helmholtz instability of the slip surface. It is essentially inconceivable that this feature could appear in a mathematical self–similar solution. It is, of course, possible that a self–similar solution exists with a smooth slip surface but that this solution is unstable with respect to unsteady perturbations.

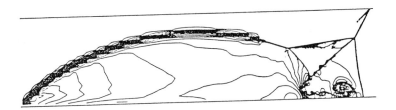

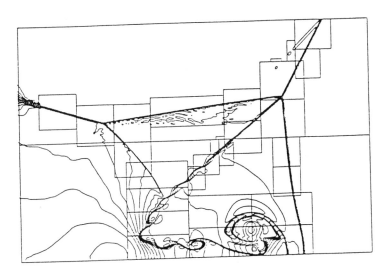

Figure 8. Local adaptive mesh refinement calculation of a DMR flowfield. A density contour plot of the entire flowfield is shown in the top figure where the coarse downstream mesh is apparent. The bottom figure of the DMR region superposes the adaptive mesh structure on the density contour plot.

The above calculation raises the question of whether or not calculated and experimental flowfields are truly self–similar for the initial data of Figure 1 and assuming no chemical reactions, etc. In this author's experience the answer is yes in an overall sense; unavoidable inviscid instabilities such as that above are obvious counterexamples for CFD simulations. The analysis of the experimental record is more subtle and the reader is referred to [6] for details.

A brief historical survey of oblique shock wave reflection is now presented; the references give much fuller accounts. The phenomenon of Mach reflection was first observed by Ernst Mach around 1878. The subject then remained dormant until the 1940's when von Neumann, Friedrichs, Bethe and many others began intensive research into all aspects of shock wave theory; a substantial portion of von Neumann's collected works deal with shock wave reflections. After the war, a group at Princeton under the leadership of Bleakney formed; CMR and DMR were soon discovered and the problems with the RR–MR transition noted. During the next two decades, several groups became prominent; we mention Henderson in Australia, Glass in Toronto and Bazhenova in Moscow. In the 1970's a group in W. Germany under Hornung made important contributions. Until the 1980's all advances in understanding came from experimental studies and analysis. Experimental research continues today with the main centers (and researchers) being Sendai, Japan (Takayama); Beersheba, Israel (Ben–Dor); Moscow (Bazhenova and Gvozdeva); Victoria, B.C., Canada (Dewey); and Henderson, Hornung, and Glass are active as well. Of course, CFD studies are of equal importance now and they take place in many centers.

Self–similar reflections are but one aspect of the subject of shock wave reflections in general. Research in the field tends to be conducted from this more general viewpoint, primarily because applications are always unsteady. Among the related oblique shock wave reflection problems we mention reflection from curved wedges and the problem of a (sperically or cylindrically symmetric) blast wave at height–of–burst above a reflecting surface. The papers [9],[10] are recent references in this area; applications of the second–order Godunov scheme to unsteady flows may be found in [19]. Finally, the concave wedge case only has been discussed here; the concave case has been extensively studied in the Soviet Union and [3] contains typical results.

From a phenomenological point–of–view the field is fairly mature and a number of review papers have appeared. A survey of the early work may be found in [42]. More recent survey papers include [4],[6],[18],[25],[29]. The books (in Russian) [34],[38] are also useful.

3. Theoretical issues. Several questions concerning solutions of equations (3) (and related solutions of equations (1) and augmented systems) for self–similar oblique shock wave reflection are discussed in the light of the preceding phenomenological review.

The issue of continuous dependence of the solutions on the parameters is taken up first. As previously noted, there are abrupt transitions in the (M_s, θ_w)–plane, across which the flowfield may change type discontinuously. Considering the RR–MR transition, one observes that the DMR region occupies very little of the disturbed flowfield just after transition (although it grows very quickly away from the transition curve). Therefore, one might reasonably expect continuous dependence in a norm such as L^2. On the other hand, the postshock states for RR and DMR (whether measured at the wall or at the triple point) differ by a large amount, unless the mechanical equilibrium condition is the assumed transition criterion. Thus, continuous dependence in a norm such as L^∞ will depend on such criteria, and is doubtful.

Augmenting equations (1) in order to account for vibrational nonequilibrium, chemical reactions, ionization and other high temperature effects destroy the self–similarity of the solutions by introducing multiple length scales into the problem. It has already been noted that both EOS variations and the above effects can change the resulting solution to a considerable extent; in particular, the transition lines are all shifted. However, there is no clear evidence that these changes are not continuous with respect to the magnitude of the source terms in a nonequilibrium/reacting flow or to the variations of the pressure function in a more general EOS (these terms are all functions of temperature which are known for a given test gas; one may imagine varying the magnitude of these terms by varying the test gas), although this issue has not been closely studied. Thus, one expects structural stability of the solutions of equations (1) with respect to such perturbations (and of equations (3) with respect to EOS perturbations). One possible contradiction lies in the RR–MR transition, where large shifts in the transition line have been proposed based on length scales introduced by such effects.

Turning now to the Navier-Stokes extension of equations (1), it is again the case that viscous terms introduce length scales and solutions are not self–similar. In particular, experimental data is only approximately self–similar at best. There are at least four flowfield effects which should be considered:

(1) Finite shock thickness. This effect becomes important as $M_s \to 1$ and is certainly so for the weak shock case. Also, triple point configurations can be affected if shock thickness needs to be accounted for. In some theories, the length scale introduced by viscosity at the reflection/triple point plays a role in the RR–MR transition criterion, even for large M_s; such theories sometimes combine viscosity with the effects of relaxation scales as above.

(2) The boundary layer at the wall. Since the Navier–Stokes boundary conditions must be satisfied, any N–S solution will differ dramatically from the corresponding Euler solution in the boundary layer, which grows downstream from zero height at the reflection point/base of Mach stem. Far enough upstream of the corner, these effects are visually negligible since the laminar boundary layer does not have enough time (on scales of typical experimental simulations) to either grow significantly, transition to turbulence or separate. However, many authors believe that the 'boundary layer defect' has an effect on the RR–MR transition.

(3) The corner $\lambda - shock$. For unattached wave configurations (i.e., R is not attached to the wedge corner), a $\lambda - shock$ – which is a characteristic feature of shock wave/boundary layer interactions – often (but not always) forms in the corner region in experiments. These solutions differ dramatically from the corresponding Euler solutions, but only locally near the corner; in particular, numerical/experimental comparisons seem to indicate that the important Mach stem region is not affected.

(4) Small scales. No matter how small the viscosity, it will destroy any complicated inviscid structure which exists at small enough length scales. The relevance of this factor is debatable. At this point, no numerical simulation has been carried out with sufficient resolution to capture viscous length scales for the corresponding experimental test gas. For relatively simple flowfields such as RR and some SMR cases it is very unlikely that any complex structure exists at scales smaller than those already computed for either equations (1) or (3). However, numerical experimentation with adaptive mesh refinement has revealed considerable small–scale structure for DMR flowfields at low γ and high M_s. One is led to wonder whether or not relevant inviscid length scales might become arbitrarily small as $\gamma \rightarrow 1.0$ at high M_s.

In summary, the evidence indicates that the zero viscosity limit is singular for equations (1), but the significance of this fact is not clear.

Stability of self–similar flowfields with respect to unsteady perturbations has already been considered for the Kelvin–Helmholtz instability of the main slip surface in DMR flowfield calculations at high enough resolution. This result alone indicates that a wide variety of self–similar solutions are not stable to unsteady perturbations. A more interesting scenario concerns the RR–MR transition. It is possible that equations (3) have multiple solutions near the transition curve comprising a hysterisis loop with stable and unstable branches; a possible mechanism for jumping between branches in an experiment or CFD simulation might be a judiciously chosen unsteady perturbation. Finally, very small scale structures in complex situations such as low γ DMR flow might well be created or destroyed by small perturbations in an experiment or calculation. To distinguish such events from true self–similar structure one would have to continue the experiment/calculation long enough so that the events were no longer small scale.

4. Conclusions. From the discussion of the preceeding section it is seen that the questions of existence and uniqueness for self–similar oblique shock wave reflection have no obvious answers. However, certain regions of parameter space can be selected where the situation is more clear; we refer to RR cases away from the transition line and to weak SMR cases. Here, there is no reason to believe other than that solutions exist and are unique. The RR situation is somewhat easier to consider from the PDE point–of–view, primarily because the slip surface of MR is absent. SMR cases for which the slip surface terminates at the wall boundary stagnation point should not be significantly more difficult.

Future CFD studies are indicated in several areas. For example, the possibility of an 'organizing center' should be carefully looked into. Another area would be to try unsteady perturbations near the RR–MR transition line. Also, the study of RR–MR transition would be greatly facilitated by a high resolution Navier-Stokes capability for modelling the boundary layer; it is quite likely that a few such calculations would substantially clear up some of the ambiguities in the experimental record. Finally, more work is needed in the low γ regime.

REFERENCES

[1] S. ANDO, *Pseudo-stationary oblique shock-wave reflections in carbon dioxide - domains and boundaries*, University of Toronto Institute for Aerospace Studies (UTIAS) Tech. Note No. 231 (1981).

[2] T.V. BAZHENOVA, V.P. FOKEEV AND L.G. GVOZDEVA, *Regions of various forms of Mach reflection and its transition to regular reflection*, Acta Astronautica, 3 (1976), pp. 131–140.

[3] T.V. BAZHENOVA, L.G. GVOZDEVA AND YU.V. ZHILIN, *Change in the shape of the diffracting shock wave at a convex corner*, Acta Astronautica, 6 (1979), pp. 401–412.

[4] T.V. BAZHENOVA, L.G. GVOZDEVA AND M.A. NETTLETON, *Unsteady interactions of shock waves*, Prog. Aero. Sci., 21 (1984), pp. 249–331.

[5] G. BEN-DOR, *Regions and transitions of nonstationary oblique shock-wave diffractions in perfect and imperfect gases*, UTIAS Rep. No. 232, (1978).

[6] G. BEN-DOR, *Steady, pseudo-steady and unsteady shock wave reflections*, Prog. Aero. Sci., 25 (1988), pp. 329–412.

[7] G. BEN-DOR AND I.I. GLASS, *Domains and boundaries of non-stationary oblique shock-wave reflexions. 1. Diatomic gas*, J. Fluid Mech., 92 (1979), pp. 459–496.

[8] G. BEN-DOR AND I.I. GLASS, *Domains and boundaries of non-stationary oblique shock-wave reflexions. 2. Monatomic gas*, J. Fluid Mech., 96 (1980), pp. 735–756.

[9] G. BEN-DOR, K. TAKAYAMA, AND T. KAWAUCHI, *The transition from regular to Mach reflexion and from Mach to regular reflexion in truly non-stationary flows*, J. Fluid Mech., 100 (1980), pp. 147–160.

[10] G. BEN-DOR AND K. TAKAYAMA, *Analytical prediction of the transition from Mach to regular reflection over cylindrical concave wedges*, J. Fluid Mech., 158 (1985), pp. 365–380.

[11] M. BERGER AND P. COLELLA, *Local adaptive mesh refinement for shock hydrodynamics*, J. Comp. Phys., 82 (1989), pp. 64–84.

[12] P. COLELLA, *Multidimensional upwind methods for hyperbolic conservation laws*, Lawrence Berkeley Laboratory Rep. LBL-17023 (1984).

[13] P. COLELLA AND H.M. GLAZ, *Efficient solution algorithms for the Riemann problem for real gases*, J. Comp. Phys., 59 (1985), pp. 264–289.

[14] P. COLELLA AND L.F. HENDERSON, *The von Neumann paradox for the diffraction of weak shock waves*, Lawrence Livermore National Laboratory Rep. UCRL-100285 (1988).

[15] A.YU. DEM'YANOV AND A.V. PANASENKO, *Numerical solution to the problem of the diffraction of a plane shock wave by a convex corner*, Fluid Dynamics, 16 (1981), pp. 720–725. Translated from the original Russian.

[16] R.L. DESCHAMBAULT AND I.I. GLASS, *An update on non-stationary oblique shock-wave reflections: actual isopycnics and numerical experiments*, J. Fluid Mech., 131 (1983), pp. 27–57.

[17] J.M. DEWEY AND D.J. MCMILLIN, *Observation and analysis of the Mach reflection of weak uniform plane shock waves. Part 1. Observations and Part 2. Analysis*, J. Fluid Mech., 152 (1985), pp. 49–81.

[18] I.I. GLASS, *Some aspects of shock-wave research*, AIAA J., 25 (1987), pp. 214–229. See also AIAA Report AIAA - 86 - 0306 with the same title and author.

[19] H.M. GLAZ, *Numerical computations in gas dynamics with high resolution schemes*, in Shock Tubes and Waves, Proc. Sixteenth Intl. Symp. on Shock Tubes and Waves, H. Grönig, editor, VCH Publishers (1988), 1988, pp. 75–88.

[20] H.M. GLAZ, P. COLELLA, I.I.GLASS AND R.L.DESCHAMBAULT, *A numerical study of oblique shock-wave reflections with experimental comparisons*, Proc. R. Soc. Lond., A398 (1985), pp. 117–140.

[21] H.M. GLAZ, P. COLELLA, I.I.GLASS AND R.L.DESCHAMBAULT, *A detailed numerical, graphical, and experimental study of oblique shock wave reflections*, Lawrence Berkeley Laboratory Rep. LBL-20033 (1985).

[22] H.M. GLAZ, P.A. WALTER, I.I. GLASS AND T.C.J. HU, *Oblique shock wave reflections in SF_6: A comparison of calculation and experiment*, AIAA J. Prog. in Astr. and Aero., 106 (1986), pp. 359–387.

[23] H.M. GLAZ, P. COLELLA, J.P. COLLINS AND R.E. FERGUSON, *Nonequilibrium effects in oblique shock-wave reflection*, AIAA J., 26 (1988), pp. 698–705.

[24] L.G. GVOZDEVA, T.V. BAZHENOVA, O.A. PREDVODITELEVA AND V.P. FOKEEV, *Mach reflection of shock waves in real gases*, Astronautica Acta, 14 (1969), pp. 503–508.

[25] L.F. HENDERSON, *Regions and boundaries for diffracting shock wave systems*, Z. angew.
      Math. Mech., 67 (1987), pp. 73–86.

[26] L.F. HENDERSON AND A. LOZZI, *Experiments on transition of Mach reflexion*, J. Fluid Mech.,
      68 (1975), pp. 139–155.

[27] L.F. HENDERSON AND A. LOZZI, *Further experiments on transition to Mach reflexion*, J.
      Fluid Mech., 94 (1979), pp. 541–559.

[28] R.G. HINDMAN, P. KUTLER, AND D. ANDERSON, *A two-dimensional unsteady Euler-equation
      solver for flow regions with arbitrary boundaries*, AIAA Rep. 79-1465 (1979).

[29] H. HORNUNG, *Regular and Mach reflection of shock waves*, Ann. Rev. Fluid Mech., 18 (1985),
      pp. 33–58.

[30] H.G. HORNUNG AND J.R. TAYLOR, *Transition from regular to Mach reflection of shock
      waves. Part 1. The effect of viscosity in the pseudosteady case*, J. Fluid Mech., 123 (1982),
      pp. 143–153.

[31] T.C.J. HU AND I.I. GLASS, *Pseudostationary oblique shock-wave reflections in sulphur
      hexafluoride(SF_6):interferometric and numerical results*, Proc. R. Soc. Lond., A408 (1986),
      pp. 321–344.

[32] M. IIMURA, H. MAEKAWA AND H. HONMA, *Oblique reflection of weak shock waves in carbon
      dioxide*, in *Proceedings of the 1988 National Symposium on Shock Wave Phenomena*, K.
      Takayama, editor, Tohoku University, Japan, 1989, pp. 1–10.

[33] T. IKUI, K. MATSUO, T. AOKI, AND N. KONDOH, *Mach reflection of a shock wave from an
      inclined wall*, Memoirs of the Faculty of Engineering, Kyushu University, 41 (1981), pp.
      361–380.

[34] V.P. KOROBEINIKOV, ED., *Nonstationary interactions of shock and detonation waves in gases*,
      Nauka, Moscow, USSR, 1986. In Russian.

[35] P. KUTLER AND V. SHANKAR, *Diffraction of a shock wave by a compression corner:Part I -
      regular reflection*, AIAA J., 15 (1977), pp. 197–203.

[36] C.K. LAW AND I.I. GLASS, *Diffraction of strong shock waves by a sharp compressive corner*,
      C.A.S.I. Trans., 4 (1971), pp. 2–12.

[37] J.-H. LEE AND I.I. GLASS, *Pseudo-stationary oblique-shock-wave reflections in frozen and
      equilibrium air*, Prog. in Aero. Sciences, 21 (1984), pp. 33–80.

[38] V.N. LYAKHOV, V.V. PODLUBNY, AND V.V. TITARENKO, *Influence of Shock Waves and Jets
      on Elements of Structures*, Mashinostroenie, Moscow, 1989. In Russian.

[39] A. MAJDA, *Compressible Fluid Flow and Systems of Conservation Laws in Several Space
      Variables*, Springer - Verlag, 1984.

[40] K. MATSUO, T. AOKI, N. KONDOH, AND S. NAKANO, *An experiment on double Mach reflec-
      tion of a shock wave using sulfur hexafluoride*, in *Proceedings of the 1988 National Sym-
      posium on Shock Wave Phenomena*, K. Takayama, editor, Tohoku University, Japan (ISSN
      0915 - 4884), 1989, pp. 11–20.

[41] H. MIRELS, *Mach reflection flowfields associated with strong shocks*, AIAA J., 23 (1984),
      pp. 522–529.

[42] D.C. PACK, *The reflexion and diffraction of shock waves*, J. Fluid Mech., 18 (1964), pp.
      549–576.

[43] A. SAKURAI, L.F. HENDERSON, K. TAKAYAMA, AND P. COLELLA, *On the von Neumann
      paradox of weak reflection*, Fluid Dynamics Research, 4 (1989), pp. 333–345.

[44] G.P. SCHNEYER, *Numerical simulation of regular and Mach reflections*, Physics of Fluids,
      18 (1975), pp. 1119–1124.

[45] A.N. SEMENOV, M.P. SYSHCHIKOVA, AND M.K. BEREZKINA, *Experimental investigation of
      Mach reflection in a shock tube*, Soviet Physics - Technical Physics, 15 (1970), pp. 795–803.

[46] V. SHANKAR, P. KUTLER, AND D. ANDERSON, *Diffraction of a shock wave by a compression
      corner: Part II - single Mach reflection*, AIAA J., 16 (1978), pp. 4–5.

[47] M. SHIROUZO AND I.I. GLASS, *Evaluation of assumptionss and criteria in pseudostationary
      oblique shock-wave reflections*, Proc. R. Soc. Lond., A406 (1986), pp. 75–92.

[48] J.T. URBANOWICZ, *Pseudo-stationary oblique-shock-wave reflections in low gamma gases -
      isobutane and sulphur hexafluoride*, UTIAS Tech. Note No. 267,1988.

[49] J.T. URBANOWICZ AND I.I. GLASS, *Oblique-shock-wave reflections in low gamma gases -
      sulfurhexafluoride (SF_6) and isobutane [$CH(CH_3)_3$]*, preprint, 1989.

[50] J.M. WHEELER, *An interferometric investigation of the regular to Mach reflection transition
      boundary in pseudostationary flow in air*, UTIAS Tech. Note No. 256,1986.

NONLINEAR WAVES: OVERVIEW AND PROBLEMS

JAMES GLIMM*†‡

Abstract. The subject of nonlinear hyperbolic waves is surveyed, with an emphasis on the discussion of a number of open problems.

Key words. Conservation Laws, Riemann Problems.

AMS(MOS) subject classifications. 76N15, 65M99, 35L65

1. Introduction. The questions which modern applied science asks of the area of nonlinear hyperbolic equations concern an analysis of the equations, a search for effective numerical methods and an understanding of the solutions.

The analysis of equations involves mathematical questions of existence, uniqueness and regularity. It is the special features of nonlinear hyperbolic equations which give these standard mathematical concerns a broader scientific relevance. The basic conservation laws of physics are hyperbolic, and fit into the discussion here. They do not in general have regular solutions. An exact classification of the allowed singularities (i.e. the nonlinear waves) is an open problem in the presence of realistic or complex physics, chemistry, etc. and/or higher spatial dimensions. Existence and uniqueness of solutions is not a consequence of the fact that the equations "come from physics" and thus "must be O.K." The equations come from degenerate simplifications of physics in which all length scales have been eliminated. Serious mathematical work remains to determine the formulations of these equations which have satisfactory mathematical properties, and thus provide a suitable starting place for effective numerical computation and for scientific understanding.

We next discuss the modification of equations. Simplified versions of complex equations capture the essential difficulties in a form which can be analyzed conveniently and understood. Equations are also rendered more complex through the inclusion of additional physical phenomena. Such steps are important for experimental validation and for applications. Often the complication involves additional terms or equations containing a small parameter, and the limit as the parameter tends to zero is of great interest. This interest in small parameters can be traced back to the physics, where events on very different length and time scales arise in a single problem. Another approach to this hierarchy of equations, physics, and length and time scales is to analyze asymptotic properties (including intermediate time scales) of the solution as $t \to \infty$.

Effective computational measures address basically the same difficulties and issues, but with different tools and approaches. The presence of widely varying

---

*Department of Applied Mathematics and Statistics, SUNY at Stony Brook, Stony Brook, NY, 11794-3600.

†Supported in part by the Applied Mathematical Sciences Program of the DOE, grant DE-FG02-88ER25053

‡Supported in part by the NSF Grant DMS-8619856

length and time scales and in some cases of underspecified physics can degrade the accuracy and the resolution of numerical methods. Three dimensional and especially complex or chaotic solutions are typically underresolved computationally.

The search for effective numerical methods can be broken into two main themes: concerns driven by computer hardware and concerns driven by features of the solution.

Effective numerical methods which are driven by solution features, i.e. by physics, depend on the study of nonlinear waves. Most modern numerical methods for the solution of nonlinear conservation laws employ Riemann solvers, i.e. the exact or approximate solution of nonlinear wave interactions, as part of the numerical algorithm. Equally important is the use of limiters to avoid Gibbs phenomena overshoots and oscillations associated with the discrete approximation to discontinuous solutions.

Of the many possible issues associated with the analysis of solutions, we focus on chaos. By chaos, we mean a situation in which the microscopically correct equations are ill posed or ill conditioned (through sensitive dependence on initial conditions) for the time periods of interest and must be interpreted stochastically. The stochastic interpretation leads to new equations, useful on larger length and time scales.

Further background on the topics discussed here can be found in recent general articles of the author and in references cited there [21-23, 26].

2. Nonlinear Wave Structures. The nonlinear waves which arise in many of the examples of complex physics (elastic plastic deformation, magneto hydrodynamics, chemically reactive fluids, oil reservoir models, granular material, etc.) are currently being explored. Striking, novel and complex mathematical phenomena have recently been discovered in these examples, including crossing shocks, bifurcation loci, shock admissibility dependence on viscosity, and inadmissible Lax shocks. The class of solved Riemann problems continues to increase, as a result of examining these physically motivated examples in detail. The general theories in which these new phenomena are imbedded are to a large extent a subject for future research. Perhaps the most pressing question in this circumstance is, having found the trees, to discover the forest. Wave interactions are technically related to the subject of ordinary differential equations in the large. This fact suggests an approach to the construction of general theories for Riemann solutions.

We also ask whether these novel wave structures have counterparts in experimental science. We believe the answer will be positive, and to the extent that this is the case, mathematical theory is ahead of experiment, in making predictions about nature. The motivating example of three phase flow in oil reservoirs is not a promising place to resolve this question. Three phase flow experiments are difficult, inconclusive and seldom performed. The equations themselves are not known definitively, and for this reason, the topological argument of Shearer [46] that an umbilic point must occur on topological grounds for any plausible three phase flow equation is significant. Ting has argued that umbilic points also occur in elastic plastic flow [47]. Gariazar has examined common constitutive laws for common metals [15] and found that the umbilic point will occur in uniaxial compression, at compressions within the plastic region [16]. It remains to be determined whether these umbilic points are an artifact of a consititutive law or whether they reflect a true property of nature.

In view of the considerable progress which has been made with non strictly hyperbolic conservation laws at the level of wave interactions and Riemann solutions, it is a very interesting question of examine these same equations from the point of view of the general theory. This means considering general Cauchy data, not just Riemann (scale invariant) data. We mention two recent results of this type. A general existence theorem for one of the conservation law systems with quadratic flux and an isolated umbilic point was proved [34] using the method of compensated compactness. At the umbilic point, the entropy functions required by this method have singularities. It was shown that for a restricted subclass of entropies, the singularity was missing, and that the proof could be completed using this restricted class of entropies. For another system in this class, the stability to perturbations of finite amplitude of a viscous shock wave was demonstrated [38]. On the basis of these examples and the related work of others, it appears that the general theory of conservation laws will admit extensions to allow a loss of e.g. strict hyperbolicity or genuine nonlinearity. Since many conservation laws arising in science appear to have such features, such extensions would be of considerable interest.

On the basis of the above discussion, answers to the following questions might help to define the overall structure of nonlinear hyperbolic wave interactions:

1. A complete study of bifurcations. A classification of generic unfoldings of the underresolved physics of Riemann problems would be both interesting and useful. A set of bifurcation loci for Riemann problem left states was proved to be *full* (for left and mid states in the complementary set, there is no bifurcation resulting from variation of the left state) in [13, 14]. Remaining problems concern bifurcation as the conservation law or its viscous terms (used even in the case of zero viscosity to define admissibility) are varied. Moreover, bifurcation for wave curves passing through the resonant (umbilic) set has yet to be addressed. A classification of the bifurcation unfoldings which result from left states (or conservation laws, etc.) located on the bifurcation loci has not been given.

2. The local theory of multiple resonant eigenvalues (higher order umbilic points). For a higher dimensional state space, the resonant (umbilic) set is a manifold with singularities. The local behavior of Riemann solutions near the resonant set will depend on the dimension and the codimension of the resonant set, and on its local singularity structure as a subset of the state space. Beyond this, there will be some number of "bifurcation parameters" which partition the Riemann solutions into invariance classes.

3. A topological characterization of removable vs. nonremovable resonance. For what class of problems is a resonance required? Is there an example where it can be observed experimentally?

4. Asymptotics, large systems with small parameters, rate limiting subsystems

and "stiff" Riemann problems.

5. A resolution of entropy conditions. The physical entropy principle is not only that entropy will increase across a shock, but that the admissible solution will be the one which *maximizes* the rate of entropy production. Entropy is defined in many physical circumstances. For example the equations of two fluid Buckley-Leverett flow in porous media are described by a single nonconvex conservation law. Entropy, in the sense of thermodynamics can be understood in this context [1], and yields the well known entropy condition of Oleinik in this example. However, to obtain the Oleinik entropy condition, the above strong form of the physical entropy condition is needed. The equations for three phase flow in porous media inspired much of the recent work on Riemann solutions, in which it was realized that a number of mathematically motivated entropy conditions were inadequate. Thus it would be worthwhile to return to the first principles of physics and to formulate a physically based entropy condition for three phase flow and for the quadratic flux Riemann problems.

6. Nonuniqueness and nonexistence of Riemann solutions. Symmetry breaking (non scale invariant or higher dimensional) solutions for Riemann data. There is enough evidence that this phenomena will occur, but we have neither enough examples nor enough theory to predict under what circumstances it should be expected.

7. Discretized flux functions. Both the qualitative (wave structure) and the quantitative (convergence rates) aspects of convergence are of interest. How should flux discretization be performed in order to preserve some specific aspect of the Riemann solution wave structure?

8. Special classes of subsystems containing important examples. E.g. mechanical systems, with a state space given as a tensor product of configuration space and a momentum space, or more generally as a cotangent bundle over a configuration space manifold.

9. The use of known Riemann solutions as a test for numerical methods.

3. Relaxation Phenomena. The internal structure of a discontinuity refers to any modification of the equation and the underlying physics which replaces a discontinuous solution by a continuous one (having a large gradient). The internal structure is of interest partly as a test of admissibility of the discontinuity and partly because of the more refined level of resolution and physics which is described from this approach. The conservation law is scale invariant, and thus has no length scales in it, while the internal structure necessarily has at least one length scale (its width) and may have more. For example consider chemically reactive fluid dymanics. With even moderately complex chemistry, there will be multiple reactions, and reaction zones, each with individual length scales (the width of an individual reaction). In the conservation law, the relative speeds of the interactions are lost, as all times and lengths have been set to zero. It is in this way that the physics described by the conservation law becomes underspecified.

There are two approaches to the internal structure of a discontinuity. Either new equations are added to enlarge the system or new terms are added to the original system, without a change in the number of dependent variables. There are other and more complex possibilities, such as the fluid equations giving way to the Boltzmann equation, in which an infinite number of new variables are used and the old variables are not a subsystem of the new, but are only recovered through an asymptotic limit. Such situations, while they do occur, are outside the scope of the present discussion.

Common examples of approximate discontinuities with internal structure are shock waves, chemical reaction fronts, phase transitions, and plastic shear bands. The internal structure involves concepts from nonequilibrium thermodynamics. The use of higher order terms in the equations is the simplest and most familiar way to introduce internal structure into a discontinuity. The coefficients (viscosity, heat conduction, reaction rates etc.) in these terms necessarily have a dimension. The coefficients are known as transport coefficients; they are defined in principle from nonequilibrium thermodynamics. Once the coefficients are known, equilibrium thermodynamics is used exclusively.

The other approach, which in many examples is more fundamental, is to enlarge the system. We regard the nonequilibrium variables and reactions as divided into fast and slow. This division is relative to the region internal to the discontinuity; even the slow variables could be fast relative to typical fluid processes. Then the fast variables are set to their instantaneous equilibrium, relative to the specified values of the slow variables. This describes an approximation in which the ratio of the fast to slow time scales becomes infinite. Another description would be to say that the fast variables are at thermodynamic equilibrium, relative to constraints set by the values taken on by the slow variables. The slow variables are governed by differential equations derived from nonequilibrium thermodynamics applied to this limiting situation. A typical equation for the slow variable is an ordinary differential equation, i.e. a Lagrangian time derivative set equal to a reaction or relaxation rate source term. The lower order source terms have a dimension and introduce the length scale which characterizes the internal structure of the discontinuity. The equations for chemically reacting fluids have exactly this form, and can be regarded as a completely worked out example of the point of view proposed here.

A comparison of these two approaches has been worked out by T.-P. Liu [37], and is summarized in his lectures in this volume. Liu considers the lowest order nonequilibrium contribution to the internal energy of a (diatomic) gas, namely the vibrational energy in the lowest energy state of the molecule. Thus there are now two contributions to the internal energy, this one vibrational mode and all remaining internal energy contributions. The vibrational energy has a preferred value as a function of the other thermodynamic variables, namely its equilibrium value. There is also a relaxation rate, defined in principle from statistical physics, but in practice determined by mearurement, for return of the vibrational energy to this preferred value. The result is an enlarged system, with vibrational energy as the new dependent variable. Liu's asymptotic analysis, as the relaxation rate goes to infinity, leads to the smaller system, augmented with a higher order viscosity term, and a computation of the viscosity coefficient in terms of the nonequilibrium relaxation process. For a quantitatively correct description of rarefied gas dynamics, this model is too simple, and the full chemistry of N_2, O_2, CO_2, H_2O, etc., including free radicals, dissociation and partially ionized atoms, is needed. Realistic models of chemistry for rarefied gas dynamics and internal shock wave structure can involve up to 100 variables. Such systems are typically very stiff, are still approximate, and depend on rate laws which are not known precisely.

Liu's analysis assumes that the original system of conservation laws is strictly hyperbolic and genuinely nonlinear. In a neighborhood of a phase transition and especially along a phase boundary, genuine nonlinearity typically fails for the conservation laws describing gas dynamics [40]. Presumably dissociation and ionization have similar effects on the convexity of the rarefaction and shock Hugoniot wave curves, and hence on the structure of the Riemann solutions. The metastable treatment of dissociation assumes that species concentrations are dependent variables, and that their evolution is governed by rate laws. In the equilibrium description, all reactions have been driven to completion and all concentrations are at equilibrium values. Thus these two descriptions differ in the number of dependent variables employed. It would be of interest to extend Liu's analysis to a wider range of cases, and to remove the restrictive hypotheses in it.

Caginalp [5, 7] considers phase transitions on the level of the heat equation alone. The simple system describes the Stefan problem, and the augmented system includes a Landau-Ginsberg equation. Caginalp discusses a number of asymptotic limits of the augmented system [8, 9], and gives physical interpretations of the assumptions on which these limits are based. Anisotropy is important in this context, as it provides the symmetry breaking which initiates the dendritic growth of fingers [6]. Rabie, Fowles and Fickett [43] replace compressible fluid dyamics by Berger's equation and their augmented system then has two equations. They examine the wave structure, and compare it to detonation waves, a point of view carried further by Menikoff [41].

Efforts to describe metastable phase transitions by the addition of higher order terms in the compressible (equilibrium thermodynamics) fluid equations have led to solutions in qualitative disagreement with experiment, as well as with physical principles. It should be recalled that for common materials and for most of the phase transition parameter space (excepting a region near critical points), the influence of a phase boundary is felt for a distance of only a few atoms from the phase boundary location. On the length scale of these few atoms, the continuum description of matter does not make a lot of sense. Thus the view, sometimes expressed, that on philosophical grounds, there should be a gradual transition between the phases, is valid, if at all, only within the context of quantum mechanics and statistical physics. In this case the continuous variable is the fraction of intermolecular bonds in the lattice, or the quantum mechanical probability for the location of the bonding electrons, etc. Correlation functions for particle density are studied in this approach.

The mathematical structure associated with metastability is further clouded by the occurrence of elliptic regions in some formulations of the equations. According to a linearized analysis, the equations are then unstable, and presumably unphysical. They are at least ill posed in the sense of Hadamard. Detailed mathematical analyses of a Riemann problem with an elliptic region [31, 32, 35] did not reveal pathology which would disqualify these equations for use in physical models. The theory in these examples is not complete, and especially the questions of admissibility of shock waves and the stability of wave curves should be addressed. Examples of computational solutions for equations with elliptic regions are known to have solutions without obvious pathology as well [4, 18]. There are examples, such as Stone's model for three phase flow in porous media, where the elliptic region appears to have a very small influence on the overall solution. In most cases, the elliptic regions result from the elimination of some variable in a larger system. In this sense they are not fundamentally correct. Whether they are acceptable as an approximation in a specific case seems to depend on the details of the situation.

A correct theory should predict a number which can be verified by experiment. For conservation laws, the wave speed is a basic quantity to be predicted. In the case of metastable phase transitions, this task is complicated, for some parameter values, by the occurrence of interface instabilities, which lead to fingers (dendrites) and which produce a mixed phase mushy zone. The propagation speed of this dynamic mushy zone is not contained in a one dimensional analysis using microscopically correct thermodynamics and rate laws. In other words, there are no physcially admissible Riemann solutions to the one dimensional conservation laws in such cases. The equations of chemically reactive fluids may also fail to have physically admissible one dimensional Riemann solutions. For some parameter ranges, the wave front may lose its planar symmetry and become krinkled or become fully three dimensional, through the interaction with chaotically distributed hot spot reaction centers. Recent progress on this issue has been obtained [39]; older literature can be traced from this reference as well. The question of complex, or chaotic internal interface structure suggests the following point of view. In such cases, the question of physical admissibility is a modeling question, i.e. a judgement to be made on the basis of the level of detail desired in the model equations. The admissible solutions for microscopic physics and for macroscopic physics need not be the same. A change in admissibility rules is really a change in the meaning of the equations, i.e. a change in the equations themselves. This point of view can be taken further, and of course we realize that there is no need for the equations of microscopic and macroscopic physics to coincide, even when they are both continuum theories. The relation between these two solutions (or equations) is the topic of the next section.

Specific questions posed by the above discussion include:

1. In various physical examples of relaxation phenomena, it would be desirable to determine correct equations, mathematical properties of the solutions, including the structure of the nonlinear waves, and asymptotic limits giving relations between various distinct descriptions of the phenomena.

2. Which properties of a larger system lead to elliptic regions in an asymptotically limiting subsystem?

3. Is there a principle, similar to the Maxwell construction, which will replace

a system of conservation laws having an elliptic region with a system having an umbilic line, or surface of codimension one in state space? Does this construction depend on additional physical information, such as the specification of the pairs of states on the opposite sides of the elliptic region joined by tie lines, as in the case of a phase transition? Is there a physical basis for introducing tie lines in the case of the elliptic region which arises in Stone's model?

4. What is the proper test for dimensional symmetry breaking of a one dimensional Riemann solution? Symmetry breaking should be added to the admissibility criteria, and when the criteria fails, there would be no admissible (one dimensional) Riemann solution. The same comments apply to the breaking of scale invariance symmetry.

5. The mathematical theory of elliptic regions needs to be examined more fully, especially to determine the importance of viscous profiles and conditions for uniqueness of solutions.

4. Surface Instabilities. The nonlinear waves considered in the two previous sections are one dimensional, and in three dimensions, they define surfaces. Sometimes the surfaces are unstable, and when this occurs, a spatially organized chaos results. Examples are the vortices which result from the roll up (Kelvin-Helmholtz instability) of a slip surface, and the fingers which result from a number of contexts: the Taylor-Saffman instability in the displacement of fluids of different viscosity in porous media, the Rayleigh-Taylor and Richtmyer-Meshkov instabilities resulting from the acceleration of an interface between fluids of different densities, the evolution of a metastable phase boundary giving rise to the formation of dendrites and a multiphase transitional mushy zone between the two pure phases. Instabilities in chemically reactive fronts were referred to in the previous section.

Surface instabilities give rise to a chaotic mixing region, which can be thought of as an internal layer between two distinct phases, fluids, or states of the conservation law. In the case of vortices, the mixing occurs first of all in the momentum equation, and for this reason is modeled at the simplest level by a diffusion term in this equation. The coefficient of the diffusion term is viscosity, and the required viscosity to model the turbulent mixing layer is larger than the microscopically defined viscosity; it is called eddy viscosity to distinguish it from the latter. Similarly the simplest model of fingering induced mixing is a diffusion term in the conservation of mass equation. Again it has a much larger coefficient than the mass diffusion terms of microscopic physics. We call these simple mixing theories the effective diffusion approximation. In the language of physics, they provide a renormalization, in which bare, or microscopically meaningful parameters are replaced by effective or macroscopically meaningful ones.

For many purposes the effective diffusion approximation does not give a sufficiently accurate description of the mixing layer. The effective diffusion approximation gives a smeared out boundary in contrast to the often observed sharp boundary to the mixing region. The theories which set the effective diffusion parameter (the eddy viscosity, etc.) are phenomenological and tend to be very context dependent.

For this reason, the key parameter in this theory is known with assurance only if it has been experimentally determined. Of even greater importance, the effective diffusion approximation contains no length scales beyond the total width of the mixing region. It represents an approximation in which all mixing occurs at a microscopic scale. The internal structure of the mixing layer is more complicated. It is less well mixed and somewhat lumpy, as we now explain.

The initial distribution of unstable modes (vortices or fingers) on the unstable interface is governed by the theory of the most unstable mode. The pure conservation laws are unstable on all length scales, with the shortest length scales having the most rapid growth. For this reason, these equations must be modified by the inclusion of length dependent terms (interface width, surface tension, (microscopic) viscosity, curvature dependent melting points, etc.) which stabilize all but a finite number of wave lengths. Of the remaining unstable modes, the one with the fastest growth rate is called the most unstable. That mode (or that range of wave lengths) is presumed to provide the initial disturbance to the interface, in the absence of some explicit initialization containing other length scales.

However initialized, the modes grow and interact. There is a significant tendency for merger and growth of wave lengths. Presumably this is due to our picture of the mixing region as a thin layer or thickened interface, and to the well known tendency in two dimensional turbulence for length scales to increase. In any case the merger of modes and the growth of length scales produces a dynamic renormalization in the dimensionality of the equation, and a change in the algebraic growth rate of the interface thickness. The distribution of length scales in the mixing layer can be thought of as a random variable. It is time dependent, and ranges from the minimum size of the most unstable wave length (or what is typically nearly the same, the smallest unstable wave length) up to a possible maximum value of the current interface thickness. The distribution of length scales is also typically spatially dependent, and is a function of the distance through the mixing layer. Thus the mixing layer need not be homogeneous, but may contain distinct sublayers, with different statistical distributions of length scales within each layer. This statistical distribution of spatially and temporally dependent length scales is completely missing in the effective diffusion approximation.

We now specialize the discussion to the Rayleigh-Taylor instability and we consider only one aspect of the spatially dependent length scale distribution, namely the interface width as a function of time. According to experiment [44], the interface thickness, or height, $h(t)$, has the form $h = \alpha g A t^2$, where g is the accelerational force on the interface, t is time, and $A = \frac{\rho_1 - \rho_2}{\rho_1 + \rho_2}$ is the Atwood number characterizing the density contrast at the interface, with ρ_i, $i = 1, 2$ denoting the densities in the two fluids. The first computations of the unstable interface which show quantitative agreement with the experimental value of α for a time regime up to and beyond bubble merger are reported in [25]. Control of numerical diffusion through a front tracking algorithm appears to be the essential requirement in obtaining this quantitative agreement with experiment. See also the paper of Zhang in this volume, where the computations are discussed in more detail and also see the related computations of Zufiria [49], who also obtains agreement with the experimental value of α, for a more limited time and parameter range. Because of the sensitivity of the unstable interfaces to modifications of the physics, the computations are no doubt also sensitive to details in the numerical methods. For this reason it is very desirable to present not only carefully controlled analyses of each method, but also of carefully controlled comparisons between the methods.

The outer edge of the Rayleigh-Taylor mixing region adjacent to the undisturbed heavy fluid is dominated by bubbles; for this reason we refer to it as the bubble envelope. Now we adapt the language of multiphase flow and consider the transition from bubbly to slug flow. These two bubbly and slug flow regimes have distinct equations, or constitutive laws, but are both derived in principle from the same underlying physics, namely the Euler or Navier-Stokes equations. In this sense the regimes can be thought of as phases in the statistical description of the flow in terms of bubbles and droplets. Taking this point of view, the transition between the regimes is a phase transition. The order paramater of this transition is the void fraction; for small void fraction, bubbly flow is stable and for large void fraction, slug flow is stable. The metastable process for the bubble to slug flow transition to take place is bubble merger, which is exactly the dominant process at the Rayleigh-Taylor bubble envelope. From this point of view, the role and importance of statistical models for the bubble merger process [17, 24, 25, 45, 50] becomes clear. These models have as their goal to yield rate laws and constitutive relations for the metastable transition regions. In particular they should yield the internal structure of the Rayleigh-Taylor bubble envelope.

Major questions concerning the theory of unstable interfaces and chaotic mixing layers are open.

1. The importance of microscopic length scales, viscosity, surface tension, interfacial thickness, (mass) diffusion or compressibility, has been an area of active research. However, the exact role of these features in regularizing the equations to the extent that the solutions are well defined for all time has not been established. Without regularization, the solutions are known or suspected of containing essential singularities in the form of cusps, which appear to preclude existence beyond a limited time.

2. Does the mixing zone have a constant width or grow as some power ot t? Usually the power of t is known with some level of assurance, but the coefficient in front of the power and the dimensionless groups of variables it depends upon may not be known, depending on the specific situation.

3. Mode splitting, coupling, merging and stretching are the important ingredients of the dynamics of mixing layer chaos. Theories for the rates governing these events are needed.

4. Distributions of length scales within the mixing zone are needed. Stable statistical measures of quantities which are reproducible, both experimentally and computationally are needed. Point measures of solution variables are not useful in the description of chaos, while statistical correlation functions have proven useful in the study of turbulence, for example.

5. Does the idea of fractal dimension, or of a renormalization group fixed point have a value in this context?

6. Chaotic mixing layers are very sensitive to numerical error and difficult to compute. An analysis of the accuracy of numerical methods for these problems would be very useful. For the same reason, comparison to experiment is important.

5. Stochastic Phenomena. The issues to be discussed here are similar to those raised in §4. The main difference is that stochastic phenomena do not always concern mixing and whether or not it is concerned with mixing, they do not have to have to be concentrated in or caused by instabilities of thin layers. To illustrate this point, as well as to introduce the next section, we refer to the problem of determining constitutive relations and properties of real materials. It is well known that the atomic contribution to material strength will give properties of pure crystals, which are very different from (normal) real materials. Common materials are not pure crystals, but have defects in their crystal lattice structure, impurities, voids, microfractures and domain walls, each of which can be modeled on a statistical basis, in terms of a density.

Similarly, the heterogeneities in a petroleum reservoir occur on many length scales. Some heterogeneities are not accessible for measurement, and can be infered on a statistical basis. Others, such as the vertical behavior in the vicinity of a well bore, can be measured at a very fine scale, but it is not practical to use this detail in a computation, so again a statistical treatment is called for. Weather forecasting data also illustrates the point that the available data may be too fine grained to be usable in a practical sense, and averaged data, including the statistical variability of averaged data may be a more useful level of description of the problem.

6. Equations of State. The equation of state problem extends beyond the fluid equilibrium thermodynamic equations of state, to elastic moduli, constitutive relations, yield surfaces, rate laws, reaction rates and other material dependent descriptions of matter needed to complete the definition of conservation laws. It is the portion of the conservation law which is not specified from the first principles of physics on the basis of conservation of mass, momentum, etc. The comments of this section apply as well to the transport coefficients, which are the coefficients of the higher order terms which are added to the conservation laws, such as the coefficients of viscosity, diffusion, thermal conductivity, etc.

There are two aspects to this problem. The first is: given the equation of state, to determine its consequences for the solution of the conservation laws, the nonlinear wave structure, and the numerical algorithms. This problem is the topic of §2.

The second problem is to determine the equation of state itself. With the increasing accuracy of continuum computations, we may be reaching a point where errors in the equation of state could be the dominant factor in limiting the overall validity of a computation. Equations of state originate in the microphysics of subcontinuum length scales, and their specification draws on subjects such as statistical physics, many body theory and quantum mechanics at a fundamental level.

Thus an explanation is needed to justify the inclusion of this question in an article oriented towards a continuum mathematics audience. Although the equations of state originate in the subcontinuum length scales, for many purposes the problems do not stay there. In many cases, there are important intermediate structures which have a profound influence on the equation of state, and which are totally continuum in nature. This is exactly the point of the two previous sections, which we are now repeating using different language. Thus, for example, one could use a continuum theory to study cracks in an elastic body, and then, in the spirit of statistical physics, combine the theories of individual cracks or groups of them in interaction, to give an effective theory for the strength of a material with a given state of micro-crack formation. In other words, important aspects, and in a number of cases, the most important aspects, of the determination of the equation of state are problems of continuum science.

To further illustrate the point being made, consider the example of petroleum reservoir simulation. Here the relative permeability and the porosity are basic material response functions, in the sense of the equation of state as discussed above. Measurements can be made on well core samples, typically about six inches long. This defines the length scale of the microscopic physics for this problem. (We do not enter into the program of predicting core sample response functions from physics and rock properties at the scale of rock pores, i.e. the *truly microscopic* physics of the problem.) The next measurable length scale is the inter-well spacing, about one quarter mile. However, information is needed on intermediate scales by the computations. On the basis of statistics and geology, one reconstructs plausible patterns of heterogeneity for the intermediate scales. This is then used to correct the measured relative permeability functions. The modified relative permeability functions are known as pseudo-functions, and they are supposed to contain composite information concerning both the intermediate heterogeneities and the permeabilities as measured from core samples. This range of questions concerning the scale up of predictions and measurements from the microscopic to the macroscopic levels is of basic importance to petroleum reservoir engineering and is an area of considerable current activity.

7. Two Dimensional Wave Structures. The wave interaction problem is a scattering problem [20]. The data for a Riemann problem is by definition scale invariant, and thus defines the origin as a scattering center. At positive times, elementary waves (defined by the intersection of two or more one dimensional wave fronts) propagate away from the scattering center. The elementary waves are joined by the one dimensional wave fronts which, through their intersections, define these elementary waves. At large distance from the scattering center, the solution is determined from the solution of one dimensional Riemann problems. Going to reduced variables, $\eta = \frac{x}{t}$, $\zeta = \frac{y}{t}$, the time derivatives are eliminated from the equations, and in the new variables, the system is hyperbolic at least for large radii, with the radially inward direction being timelike. It has known Cauchy data (at large radii). However, in general there are elliptic regions at smaller radii, when the solution is considered in the reduced variables, or partially elliptic regions, where some but not all of the characteristics are complex.

Analysis of any but the simplest problems of this type in two dimensions will require the type of functional analysis estimates and convergence studies which are needed for the analysis of general data in one space dimension. The study of a single elementary wave uses ideas similar to those found in the study of one dimensional Riemann problems, with the distinction that here the nonlinearities and state space complications tend to be more severe. In this context, the wave curves are known as shock polars, and the analysis of a single elementary wave involves the intersection of various shock polars, one for each one dimensional wave front belong to the elementary wave. The intersections may be nonunique, or may fail to exist, indicating that given wave configurations may exist only over limited regions of parameter space, and that the possibility of non-uniqueness is more of a problem for higher dimensional wave theory than it is in typical one dimensional wave interactions. As in the earlier sections, non-uniqueness, admissibility, entropy conditions and internal structure are closely related topics. Not very much is known about the internal structure of higher dimensional elementary waves, and so we indicate two approaches which might be fruitful.

Characteristic energy methods were developed by Liu [36] and extended by Chern and Liu [12] to study large time asymptotic limits and to develop the theory of diffusion waves in one space dimension. The proof of convergence of the Navier-Stokes equation to the Euler equation [30] also uses characteristic energy methods, as well as an analysis of an initial layer, and the evolution of an initial shock wave discontinuity as Navier-Stokes data. For the purposes of the present discussion, we note that within the initial layer, there are three nonlinear waves, which are geometrically distinct, but still in interaction. The mechanism of their interaction and time scale for the duration of the initial layer is set by the diffusive (parabolic) transport of information between the distinct waves. These initial layer and characteristic energy techniques may be useful in two dimensions for the study of internal structure of two dimensional elementary waves.

The classic approach to internal structure for a single wave in one dimension is through the analysis of ODE trajectories which describe the traveling wave in state space. To apply this method to Riemann problems it is necessary to join such curves, each nearly equal to a trajectory for a single such traveling wave. In the approximation for which each one dimensional wave is exactly a traveling wave or jump discontinuity, the method of intersecting shock polars gives a geometric construction of the solution. The method of formal matched asymptotic expansions have also proved useful for the study of two dimensional wave interactions. Here the matching is used to join the distinct one dimensional waves, while the formal asymptotic expansions describe the single waves in the approximation of zero wave strength (the acoustic limit). This method was recently applied to the study of the kink mode instability in a shear layer discontinuity at high Mach number. The kink mode wave pattern was known from shock polar analysis, see [10, 11, 33] and from computations [48]. The expansions showed the instability of the unperturbed shear flow and thus gave a theory of the initiation of this wave configuration from an unperturbed shear flow state [2]. An extension of this analysis concerned the bifurcation diagram of the shock polars [3]. Matched asymptotic expansions have also been used in a rigorous theoretical analysis for the large time limit in one space dimension [27]. On this basis, we mention asymptotic methods for use in mathematical proofs, in the study of two dimensional wave interaction problems.

The above techniques succeed in joining one dimensional elementary waves in regions where the solution is slowly changing and the waves themselves are widely separated. The problem we pose has waves meeting at a point, so the juncture occurs where the solution is rapidly changing. For this reason one should not rule out the occurrence of new phenomena. A solved problem for the interaction of viscous waves is the shock interaction with a viscous boundary layer [42]. This interaction produces a lambda shock, which is a structure which would not be predicted either from the inviscid theory of a shock interacting with a boundary, or from the viscous theory of a single shock wave. A one dimensional analog problem with similar mathematical difficulties would be to understand the internal structure (viscous shock layers) associated with the crossing point of two shock waves.

The questions discussed in the previous sections will all be important for higher dimensional wave interactions as well. In addition, we pose a few specific questions.

1. Generalize the classification of [19] for two dimensional elementary waves to general equations of state, as formulated by [40].

2. Prove an existence theorem for the oblique reflection of a shock wave by a ramp. The case of regular reflection is easiest and is the proper starting point. Weak waves can be assumed if this is helpful. A more detailed discussion of this problem, including ideas for the construction of an iteration scheme to prove existence are presented in [19]. The major interest in this problem derives from unresolved differences between proposed bifurcation criteria or possible nonuniqueness for the overlap region in which both regular reflection and Mach reflection are possible on the basis of simple shock polar analyses.

3. Determine bifurcation criteria for two dimensional elementary waves. There is a large amount known concerning this problem. See [29] for background information and a deeper discussion of this area. A recent paper of Grove and Menikoff involves bifurcations in non localized wave interactions arising from noncentered rarefaction waves [28], an issue which is part of the general bifurcation problem.

4. What is the role of scale invariance symmetry breaking for higher dimensional elementary waves?

5. The correct function space for a general existence theory for higher dimensional conservation laws depends on the equation of state, because local singularities are allowed, and occur in centered (cylindrical) waves. The order of the allowed singularity, and the L_p space it belongs to is limited by the equation of state. This relation has not been worked out, and so the existence theory and large time asymptotics for radial solutions would be of interest.

References

1. A. Aavatsmark, To Appear, *"Capillary Energy and Entropy Condition for the Buckley-Leverett Equation,"* Contemporary Mathematics.

2. M. Artola and A. Majda, 1987, *"Nonlinear Development of Instabilities in Supersonic Vortex Sheets,"* Physica D **28**, pp. 253-281.

3. M. Artola and A. Majda, 1989, *"Nonlinear Kink Modes for Supersonic Vortex Sheets,"* Phys. Fluids.

4. J. B. Bell, J. A. Trangenstein, and G. R. Shubin, 1986, *"Conservation Laws of Mixed Type Describing Three-Phase Flow in Porous Media,"* SIAM J. Appl. Math. **46**, pp. 1000-1017.

5. G. Caginalp, 1986, *"An Analysis of a Phase Field Model of a Free Boundary,"* Archive for Rational Mechanics and Analysis **92**, pp. 205-245.

6. G. Caginalp, 1986, *"The Role of Microscopic Anisotropy in the Macroscopic Behavior of a Phase Field Boundary,"* Ann. Phys. **172**, pp. 136-146.

7. G. Caginalp, To Appear, Phase Field Models: Some Conjectures on Theorems for their Sharp Interface Limits

8. G. Caginalp, To Appear, Stefan and Hele-Shaw Type Models as Asymptotic Limits of the Phase Field Equations

9. G. Caginalp, To Appear, *"The Dynamics of a Conserved Phase Field System: Stephan-like, Hele-Shaw and Cahn-Hilliard Models as Asymptotic Limits,"* IMA J. Applied Math..

10. Tung Chang and Ling Hsiao, 1988, *The Riemann problem and Interaction of Waves in Gas Dynamics* (John Wiley, New York).

11. Guiqiang Chen, 1987, *"Overtaking of Shocks of the same kind in the Isentorpic Steady Supersonic Plane Flow,"* Acta Math. Sinica **7**, pp. 311-327.

12. I-Liang Chern and T.-P. Liu, 1987, *"Convergence to Diffusion Waves of Solutions for Viscous Conservation Laws,"* Comm. in Math. Phys. **110** , pp. 503-517.

13. F. Furtado, 1989, *"Stability of Nonlinear Waves for Conservation Laws,"* New York University Thesis.

14. F. Furtado, Eli Isaacson, D. Marchesin, and B. Plohr, To Appear, Stability of Riemann Solutions in the Large

15. X. Garaizar, 1989, *"The Small Anisotropy Formulation of Elastic Deformation,"* Acta Applicandae Mathematica **14**, pp. 259-268.

16. X. Garaizar, 1989, Private Communication

17. C. Gardner, J. Glimm, O. McBryan, R. Menikoff, D. H. Sharp, and Q. Zhang, 1988, "*The Dynamics of Bubble Growth for Rayleigh-Taylor Unstable Interfaces,*" Phys. of Fluids **31**, pp. 447-465.

18. H. Gilquin, 1989, "*Glimm's scheme and conservation laws of mixed type,*" SIAM Jour. Sci. Stat. Computing **10**, pp. 133-153.

19. J. Glimm, C. Klingenberg, O. McBryan, B. Plohr, D. Sharp, and S. Yaniv, 1985, "*Front Tracking and Two Dimensional Riemann Problems,*" Advances in Appl. Math. **6**, pp. 259-290.

20. J. Glimm and D. H. Sharp, 1986, "*An S Matrix Theory for Classical Nonlinear Physics,*" Foundations of Physics **16**, pp. 125-141.

21. J. Glimm and David H. Sharp, 1987, "*Numerical Analysis and the Scientific Method,*" IBM J. Research and Development **31**, pp. 169-177.

22. J. Glimm, 1988, "*The Interactions of Nonlinear Hyperbolic Waves,*" Comm. Pure Appl. Math. **41**, pp. 569-590.

23. J. Glimm, Jan 1988, "*The Continuous Structure of Discontinuities,*" in *Proceedings of Nice Conference*.

24. J. Glimm and X.L. Li, 1988, "*On the Validation of the Sharp-Wheeler Bubble Merger Model from Experimental and Computational Data,*" Phys. of Fluids **31**, pp. 2077-2085.

25. J. Glimm, X. L. Li, R. Menikoff, D. H. Sharp, and Q. Zhang, To appear, A Numerical Study of Bubble Interactions in Rayleigh-Taylor Instability for Compressible Fluids

26. J. Glimm, To appear, "*Scientific Computing: von Neumann's vision, today's realities and the promise of the future,*" in *The Legacy of John von Neumann*, ed. J. Impagliazzo (Amer. Math. Soc., Providence).

27. J. Goodman and X. Xin, To Appear, Viscous Limits for Piecewise Smooth Solutions to Systems of Conservation Laws

28. J. W. Grove and R. Menikoff, 1988, "*The Anomalous Reflection of a Shock Wave through a Material Interface,*" in preparation.

29. L. F. Henderson, 1988, "*On the Refraction of Longitudinal Waves in Compressible Media,*" LLNL Report UCRL-53853.

30. D. Hoff and T.-P. Liu, To Appear, "*The Inviscid Limit for the Navier-Stokes equations of Compressible, Isentropic flow with shock data,*" Indiana J. Math..

31. H. Holden, 1987, "*On the Riemann Problem for a Prototype of a Mixed Type Conservation Law,*" Comm. Pure Appl. Math. **40**, pp. 229-264.

32. H. Holden and L. Holden, To Appear, "*On the Riemann problem for a Prototype of a Mixed Type Conservation Law II,*" Contemporary Mathematics.

33. Ling Hsiao and Tung Chang, 1980 Acta Appl. Math. Sinica **4**, pp. 343-375.

34. P.-T. Kan, 1989, "*On the Cauchy Problem of a* 2 × 2 *System of Nonstrictly Hyperbolic Conservation Laws,*" NYU Thesis.

35. B. Keyfitz, To Appear, "*Criterion for Certain Wave Structures in Systems that Change Type,*" Contemporary Mathematics.

36. T.-P. Liu, 1985, "*Nonlinear stability of shock waves for viscous conservation laws,*" Memoir, AMS:328, pp. 1-108.

37. T.-P. Liu, 1987, "*Hyperbolic Conservation Laws with Relaxation,*" Comm Math Phys **108**, pp. 153-175.

38. T.-P. Liu and X. Xin, To Appear, Stability of Viscous Shock Wave Asociated with a System of Nonstrictly Hyperbolic Conservation Laws

39. A. Majda and V. Roytburd, To Appear, "*Numerical Study of the Mechanisms for Initiation of Reacting Shock Waves,*" Siam J. Sci Stat Comp.

40. R. Menikoff and B. Plohr, 1989, "*Riemann Problem for Fluid Flow of Real Materials,*" Rev. Mod. Phys. **61**, pp. 75-130.

41. R. Menikoff, 1989, Private Communication

42. R. von Mises, 1958, *Mathematical Theory of Compressible Fluid Flow* (Academic Press, New York).

43. R. L. Rabie, G. R. Fowles, and W. Fickett, 1979, "*The Polymorphic Detonation,*" Phys. of Fluids **22**, pp. 422-435.

44. K. I. Read, 1984, "*Experimental Investigation of Turbulent Mixing by Rayleigh-Taylor Instability,*" Physica 12D, pp. 45-48.

45. D. H. Sharp and J. A. Wheeler, 1961, "*Late Stage of Rayleigh-Taylor Instability,*" Institute for Defense Analyses.

46. M. Shearer, 1987, "*Loss of Strict Hyperbolicity in the Buckley-Leverett Equations of Three Phase Flow in a Porous Medium.,*" in *Numerical Simulation in Oil Recovery,* ed. M. Wheeler (Springer Verlag, New York).

47. Z. Tang and T. C. T. Ting, 1987, "*Wave Curves for the Riemann Problem of Plane Waves in Simple Isotropic Elastic Solids,*" Int. J. Eng. Science **25**, pp. 1343-1381.

48. P. Woodward, 1985, "*Simulation of the Kelvin-Helmholtz Instability of a Supersonic Slipsuface with a Piecewise Parabolic Method,*" Proc. INRIA Workshop on Numerical Methods for Euler Equations, p. 114.

49. J. A. Zufiria, , "*Vortex-in-Cell Simulation of Bubble Competition in Rayleigh-Taylor Instability*," Preprint, 1988.

50. J. A. Zufiria, 1988, "*Bubble Competition in Rayleigh-Taylor Instability*," Phys. of Fluids **31**, pp. 440-446.

THE GROWTH AND INTERACTION OF BUBBLES
IN RAYLEIGH-TAYLOR UNSTABLE INTERFACES

JAMES GLIMM[a,b,c], XIAO LIN LI[c,d], RALPH MENIKOFF[e,f],
DAVID H. SHARP[e,f] AND QIANG ZHANG[c,g]

Abstract. The dynamic behavior of Rayleigh-Taylor unstable interfaces may be simplified in terms of dynamics of fundamental modes and the interaction between these modes. A dynamic equation is proposed to capture the dominant behavior of single bubbles and spikes in the linear, free fall and terminal velocity stages. The interaction between bubbles, characterized by the process of bubble merger, is studied by investigating the motion of the outer envelope of the bubbles. The front tracking method is used for simulating the motion of two compressible fluids of different density under the influence of gravity.

Key words. Bubble, Rayleigh-Taylor Instability, Chaotic Flow.

AMS(MOS) subject classifications. 76-04, 76N10, 76T05, 76E30

1. Introduction. The Rayleigh-Taylor instability is a fingering instability between two fluids with different density. Although the system is in equilibrium when the light fluid supports the heavy fluid by a flat interface with its normal direction parallel to the direction of gravity or external forces, such equilibrium is unstable under the influence of these forces. Any small perturbation will drive the system out of this unstable equilibrium state. Then an instability develops and bubbles and spikes are formed. A bubble is a portion of the light fluid penetrating into the heavy fluid and a spike is a portion of the heavy fluid penetrating into the light fluid. At a later stage of the instability, spikes may pinch off to form droplets.

The problem of mixing of two fluids under the influence of gravity was first investigated by Rayleigh [1] and later by Taylor [2]. Since then, various methods have been used to study this classical problem, such as nonlinear integral equations [3,4], boundary integral techniques [5,6], conformal mapping [7], modeling [8,9], vortex-in-cell methods [10,11], high order Godunov methods [12], front tracking [13,14,15] etc. Most of this work has been carried out in the limit of incompressible fluids or in the limit of single component systems. (The other component is a vacuum.) For a review of Rayleigh-Taylor instability and its applications to science and engineering, see reference [16]. We present here the results of our study on the development of single mode Rayleigh-Taylor instability, i.e. the development of spikes and bubbles, and on the interactions between the bubbles in compressible fluids.

[a] Department of Applied Mathematics and Statistics, SUNY at Stony Brook, Stony Brook, NY, 11794-3600.
[b] Supported in part by the Applied Mathematical Sciences Program of the DOE, grant DE-FG02-88ER25053
[c] Supported in part by the NSF Grant DMS-8619856
[d] Department of Applied Mathematics, New Jersey Institute of Technology, Newark, NJ 07102
[e] Theoretical Division, Los Alamos National Laboratory, Los Alamos, NM 87545
[f] Supported by the U.S.Department of Energy.
[g] Courant Institute of Mathematical Sciences, New York University, New York, NY 10012

For two dimensional compressible, inviscid fluids, the motion of the fluids is governed by two dimensional Euler equations,

$$\frac{\partial \rho}{\partial t} + \frac{\partial \rho u}{\partial x} + \frac{\partial \rho v}{\partial z} = 0,$$

$$\frac{\partial \rho u}{\partial t} + \frac{\partial (\rho u^2 + P)}{\partial x} + \frac{\partial \rho u v}{\partial z} = 0,$$

$$\frac{\partial \rho v}{\partial t} + \frac{\partial \rho u v}{\partial x} + \frac{\partial (\rho v^2 + P)}{\partial z} = 0,$$

$$\frac{\partial \rho e}{\partial t} + \frac{\partial [\rho u(e + PV)]}{\partial x} + \frac{\partial [\rho v(e + PV)]}{\partial z} = \rho v g,$$

where u is the x-component of the velocity, v is the z-component of the velocity, e is the specific total energy, P is pressure, V is specific volume and g is gravity. Here we have assumed that the gravity points in the positive z direction. Our systems are characterized by two dimensionless quantities, the Atwood number $A = \frac{\rho_h - \rho_l}{\rho_h + \rho_l}$ and the compressibility $M^2 = \frac{\lambda g}{c_h^2}$ and by the equation of state. Here ρ_h is the density of heavy fluid, ρ_l is the density of light fluid, λ is the wavelength of the perturbation and c_h is the speed of sound in heavy fluid. We used the polytropic equation of state $e = \frac{(\gamma-1)P}{\rho}$ with $\gamma = 1.4$ in our simulations. A range of density ratios and compressibility were studied. The numerical data on single mode systems were analyzed by using an ODE which models the entire motion of the bubble or spike. The results on single mode systems provide a basis for the study of the interaction between bubbles of different modes. We observed that, in chaotic flow, the magnitude of the terminal velocity of a large bubble exceeds the value for the corresponding single mode system due to the interaction between the bubbles. A superposition hypothesis is proposed to capture the leading order correction to the bubble velocity. Our simulations show agreement between the superposition hypothesis and numerical results. The agreement is better in with high density ratio and low compressibility than in systems with low density ratio or high compressibility. The cause of such phenomena will be discussed.

We use the front tracking method to study the motion of a periodic array of bubbles and spikes (i.e. single mode system) and to study the interactions between bubbles of different modes. The front tracking method contains a one dimensional moving grid embedded in the two dimensional computational grid. It preserves sharp discontinuities and provides high resolution around the areas of interest, i.e. nearby and on the interface between two mixing fluids.

2. Motion of single mode bubbles and spikes. When two fluids are separated by a flat interface with its normal vector parallel to the direction of gravity or external forces, the solution of the Euler equations is an exponentially stratified distribution of density and pressure along the direction of gravity or external forces. For systems with small deviations from such a flat interface, the Euler equations can be linearized in terms of the amplitude of the perturbation [14,17]. When Fourier analysis is applied to the perturbation, the Fourier modes do not couple with each other in the linearized equations. An analytic solution exists for the linearized Euler equations. In our simulation, we use the solution of the linearized Euler equations to initialize our system and the full Euler equations with front tracking to update the evolution of the system. In this section, we consider the single mode system, which is a periodic array of bubbles and spikes, or equivalently a single bubble or spike with periodic boundary conditions. The top and bottom of the computational domain are reflecting boundaries.

When a bubble or a spike emerges from a small sinusoidal perturbation on a flat interface, it follows the stages of linear growth, free fall and terminal velocity. For a single component system, the asymptotic behavior of the spike is free fall. In the linear regime, the dynamics of the system is mainly governed by the linearized Euler equations. The velocity grows exponentially with time. The exponential growth rate σ is determined by a transient equation derived from the linearized Euler equations. In the free fall regime, the acceleration reaches a maximum absolute value, which we call the renormalized gravity g_R. The velocity varies linearly with time in the free fall regime. In the terminal velocity regime, the velocity approaches a limiting value (terminal velocity v_∞) with a decay rate b. A comparison of the numerical results and the asymptotic behavior of a spike in each regime is given in Fig. 1. Here we use a dimensionless acceleration $\frac{a}{g}$, a dimensionless velocity $\frac{v}{c_h}$, a dimensionless length $\frac{gh}{c_h^2}$ and a dimensionless time $\frac{gt}{c_h}$.

The entire motion of bubble and spike may be described by an ODE

$$\frac{dv}{dt} = \frac{\sigma v (1 - \frac{v}{v_\infty})}{\frac{\sigma}{b}\frac{v}{v_\infty} + (1 - \frac{v}{v_\infty}) + [\frac{\sigma v_\infty}{g_R} - (1 + \sqrt{\frac{\sigma}{b}})^2]\frac{v}{v_\infty}(1 - \frac{v}{v_\infty})},$$

which has the solution

$$t - t_0 = \frac{1}{\sigma}ln(\frac{v_t}{v_0}) + [\frac{1}{g_R} - (\frac{1}{\sqrt{\sigma}} + \frac{1}{\sqrt{b}})^2\frac{1}{v_\infty}](v_t - v_0) - \frac{1}{b}ln(\frac{v_\infty - v_t}{v_\infty - v_0}).$$

Each term on the right hand side of the above expression has a clear physical meaning. The first term is the contribution from the linear regime; the second term is that of free fall regime and the third term is the contribution from the asymptotic regime. Extensive validation of this model has been performed for a range of Atwood numbers A and compressibility M. The A and M dependence of the parameters σ, g_R, b and v_∞ have been explored in [18]. In Fig. 2 we show an example of the comparison between the results of numerical simulation of the full two dimensional Euler equations and the results from fitting the solution given above. In Figs. 3 and 4, we show the interface at successive time steps and density and pressure contour plots for systems with $A = \frac{1}{5}, \frac{9}{11}$ and $M^2 = 0.5$. For systems with small Atwood number A, the interface consists of two interpenetrating fluids of similar shape. Secondary instabilities appears along the side of spike. (See Fig. 3.) As $A \to 0$, the pattern of the two fluids will be symmetric with phase difference π. For high density ratio systems, the spike is thinner with less roll up shed off the edge of its tip. (See Fig. 4.) For systems of high compressibility, the velocity of the bubble or spike becomes supersonic relative to the sound speed in the heavy

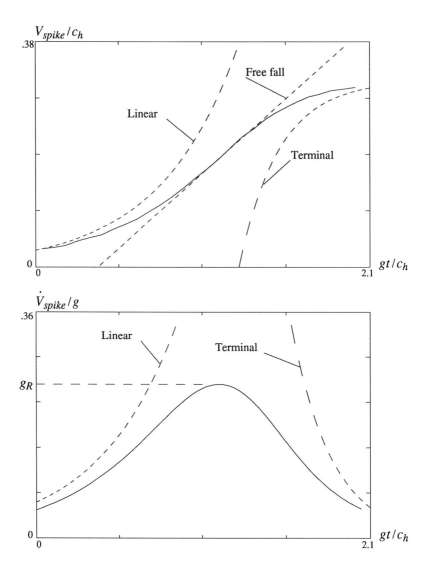

Figure 1. The comparison of the spike velocity and the spike acceleration of the numerical result to its linear and large time asymptotic behavior for $D = 2$, $M^2 = .5$ and $\gamma = 1.4$. The solid lines are the numerical results obtained by using a 80 by 640 grid in a computational domain 1×8.

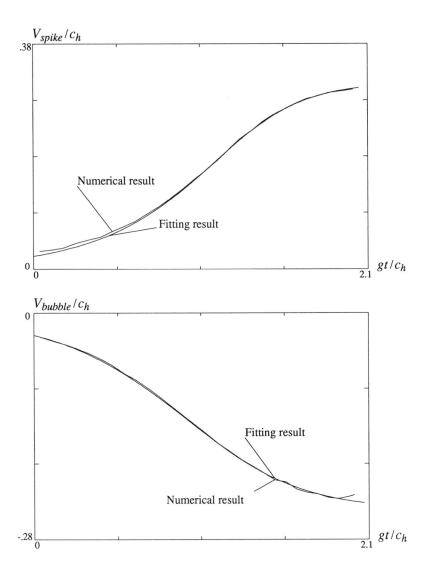

Figure 2. Plots of spike velocity and bubble belocity versus time are compared with the best three parameter fit to the solution of the ODE superimposed, for the values $A = 1/3$, $M^2 = 0.5$, $\gamma = 1.4$. The numerical results are obtained by using a 80 by 640 grid in a computational domain 1×8.

material at the late times, but it remains subsonic relative to the sound speed in the light material. The effects of grid size, the remeshing frequency, the amplitude of perturbation and boundary effects at the top and bottom of the computational domain have been tested and studied. We refer to reference [14] for the details of these studies.

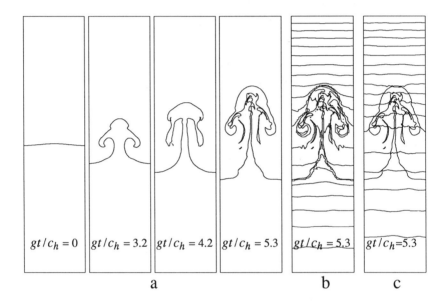

Figure 3: Plots of the interface position, density and pressure contours for $A = 1/5$, $M^2 = 0.5$, $\gamma = 1.4$ in a computation domain 1×6 with a 40 by 240 grid. Only the upper two thirds of the computational region is shown in the plot because nothing of interest occures in the remainder of the computation. (a) The interface position for successive time steps. (b) The density contour plot. (c) The pressure contour plot.

From a dimensional argument, the terminal velocity of bubble should be proportional to $\sqrt{\lambda g}$, i.e.

$$v_\infty = c_1 \sqrt{\lambda g}.$$

Here c_1 is constant of proportionality and it is a function of the dimensionless parameters A, M and γ only. In Fig. 5 we plot c_1 for a range of Atwood number A and compressibility M. It shows that c_1 has a strong dependence on M and for a given value of M^2, the dependence on A is approximately $\sqrt{A}$ in systems with low compressibility. Since we used the same value (1.4) for γ in all of our simulations, the dependence of c_1 on γ is not explored in this study.

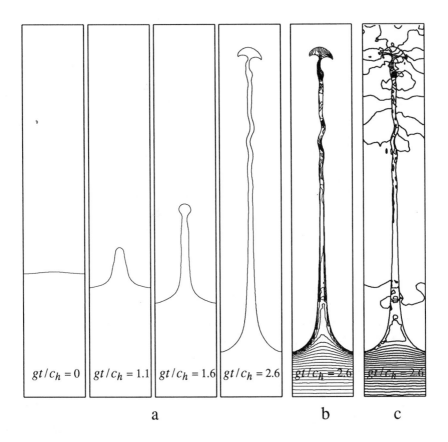

Figure 4: Plots of the interface position, density and pressure contours for $A = 0.01$, $M^2 = .5$, $\gamma = 1.4$ in a computation domain 1×10 with a 20 by 200 grid. Only the upper four fifths of the computational region is shown in the plot because nothing of interest occurs in the remainder of the computation. (a) The interface position for successive time steps. (b) The density contour plot. (c) The pressure contour plot.

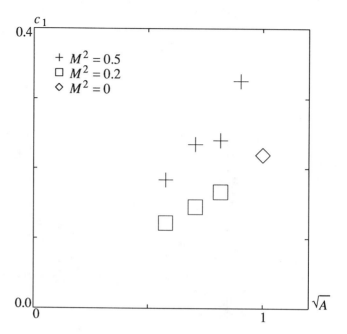

Figure 5: The dependence of c_1 on A and M. c_1 has a strong dependence on M. For a given value of M^2 the dependence on A is approximately $\sqrt{A}$ in systems of low compressibility. The value of c_1 for an incompressible fluid ($M^2 = 0$) is taken from reference [7].

3. Interaction between bubbles. To study the interactions between bubbles, we initialize our systems with an ensemble of bubbles of different wavelength. The bubble of large wavelength has small growth rate σ but large terminal velocity v_∞. Therefore small bubbles run faster than the large bubbles initially. But the large bubble will catch up and run ahead of the small ones eventually and emerge on the outer envelope of the interface between the fluids. In our simulation, we observe that, at a late stage of chaotic flow, the velocity of the large bubble exceeds the terminal velocity for the corresponding single mode bubble. In other words, it exceeds the terminal velocity of single mode bubble of the same wavelength. The increment of large bubble velocity is due to the interaction between neighboring bubbles. During the process of interaction, large bubbles gain velocity from small ones. At the end of the interaction, the small bubble washes downstream quickly. We call such a process bubble merger since the number of bubbles in the outer envelope is reduced as the result of interactions. After n generations of bubble merger, the number of bubble at the outer envelope will be reduced by a factor of 2^n. This phenomenon of bubble merger has been observed in the experiments of Read [19] and in our numerical simulations [15].

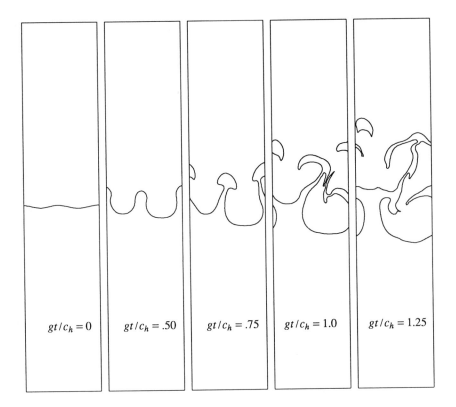

Figure 6: Successive times in a two bubble merger process. The compressibility and density ratio for this case re $M^2 = 0.1$ and Atwood number $A = 2/3$. It can be seen that the large bubble overtakes the smaller one at $gt/c_h = 1.0$. The velocity of the large bubble is accelerated during the merger while the velocity of the small bubble is reversed, see Figure 7.

We propose a superposition hypothesis for the bubble velocity in the chaotic regime. An envelope is constructed by connecting the tips of adjacent bubbles with a sinusoidal function. The hypothesis states that, to leading order, the velocity of each bubble is the sum of the velocity of each individual bubble plus the velocity of the envelope, $v = v_{individual} + v_{envelope}$. At the tip of the more advanced bubble, the bubble and the envelope are in phase. Therefore, the envelope acts as a bubble of longer wavelength at that point. Similarly, the envelope acts as a spike of longer wavelength at the tip of the less advanced bubble, since the bubble and the envelope are out of phase at that point. The velocity of individual bubble and the velocity of the envelope can each be evaluated from the single mode theory. Since the envelope has a longer wavelength than the individual bubbles, the velocity of the individual bubble $v_{individual}$ dominate the the bubble velocity v in the early stage of the merger process. At the later stage of the merger process, the envelope velocity $v_{envelope}$ will be the dominant contribution to the bubble velocity v.

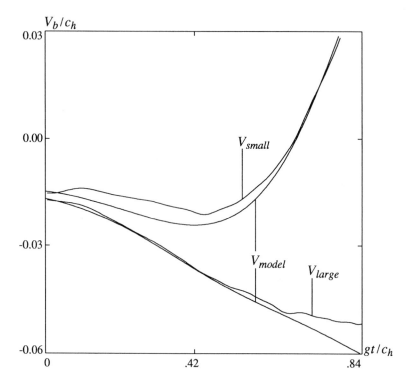

Figure 7: The plot of bubble velocities vs. time for the two bubble merger simulation. The result shows that the small bubble is accelerated at the beginning and is then decelerated after about $gt/c_h = .4.2$. The small bubble is washed downstream after its velocity is reversed. The large bubble is under constant acceleration. The smooth curves represent the bubble motion as predicted by the superposition hypothesis.

The superposition hypothesis has been compared with the experimental data of Read [19] and with the results of our numerical simulations of the full Euler equations. The relative error between superposition theory and the results of numerical simulations or experimental data is less than 20% for systems with $A > \frac{2}{3}$ and $M^2 \leq .1$, and about 30% for systems with small density ratio or large compressibility. In the latter case, the superposition principle is valid only for a finite time interval. This time interval can be understood as resulting from a nonlinearity in the bubble interaction due to density stratification [15].

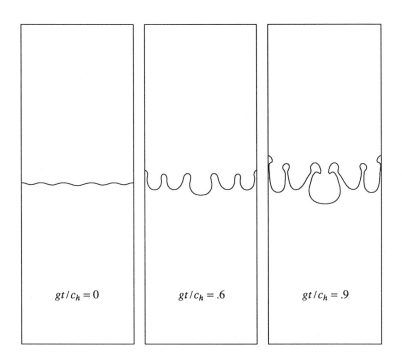

$gt/c_h = 0$ $gt/c_h = .6$ $gt/c_h = .9$

Figure 8: The interface evolution of a five bubble simulation. The compressibility in this case is $M^2 = 0.1$ and the Atwood number is $A = 1/11$. The velocity analysis showed that the superposition model is applicable to the largest bubble within an error of 15%.

In Fig. 6, we show the interface between two fluids at successive times in a two bubble merger process. The comparison of the result of the superposition hypothesis and the numerical result of Fig. 6 is given in Fig. 7. The behavior at small bubble velocities indicates clearly the contribution from the envelope. Initially, the single mode bubble velocity dominates the total velocity since the envelope has small growth rate due to its long wavelength. When contributions from the single bubble and the envelope have the same magnitude but opposite signs, the bubble stops accelerating. After that point, the velocity of the envelope dominates the total velocity. Then the small bubble de-accelerates and washed downstream quickly. Similar behavior shows up in a simulation for a system of five bubbles. (See Figs. 8 and 9.)

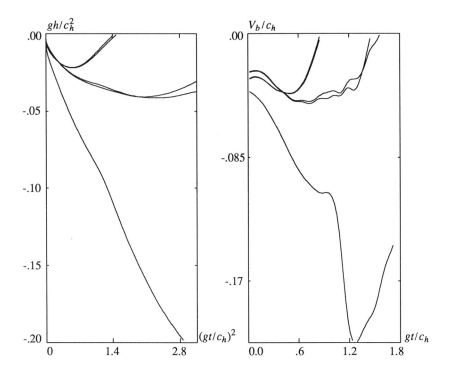

Figure 9: The left plot displays dimensionless bubble heights vs. dimensionless t^2 in a simulation with 5 initial bubbles. The Atwood number in this case is $A = 1/11$, and the compressibility is $M^2 = 0.1$. The right picture shows the dimensionless velocity vs. dimensionless t in the same case. The superposition model of the bubble velocity is valid up to $gt/c_h = 0.9$.

By dimensional arguments, one expects that the position of the bubble will be proportional to time t. However, for chaotic flow, the radius of a large bubble will

Figure 10: Plots of interfaces in the random simulation of Rayleigh-Taylor instability. The density ratio is $A = 1/3$ and the compressibility is $M^2 = 0.1$. The acceleration of the bubble envelope is in good agreement with the experiment of Read for $1\frac{1}{2}$ generations of bubble merger. The acceleration decreases after this time due to the multiphase connectivity, which is different in the exactly two dimensional computation and the approximately two dimensional experiments.

increase due to interactions between the bubbles. Consequently, the terminal velocity of the large bubble increases. By taking this into account, one can show that the position of the bubble is proportional to t^2, i.e. $z = \alpha A g t^2$. Read reported a range of values for α in his experiments [19], with $\alpha = .06$ being a fairly typical value. Values of α in the range of $0.04 \sim 0.05$ and $0.05 \sim 0.06$ were reported respectively by Young [12] and by Zufuria [10] on the basis of their numerical simulations. In our study, we found that α is not a constant. α fall in the range $0.055 \sim 0.065$ at early stages of interaction and in ranges $0.038 \sim 0.044$ at late stage of the simulations [15]. In Fig. 9, the slope of the large bubble curves corresponds to the value of α. We observe that the reduction of α from about .06 to about .04 is due to the multi-connectivity of the interface in the deep chaotic regime. In Young's numerical simulations, the interface between two fluids was not tracked [12]. Therefore effective multi-connectivity occurred during early stages of his simulations. We propose this as a possible explanation for the small values of α which be observed. The discrepancy between the value of α observed at late times in our numerical simulations and the value observed in experiments [19] results from the difference between exact two dimensional numerical simulations and an approximately two dimensional experiment. For example, the ratio of thickness to width is 1:6 in Read's experiments [19]. The computationally isolated segments of fluids in the $x - z$ plane may be connected in the third dimension (y direction) in experiments. Such discrepancies may be resolved in three dimensional calculations which will provide a more realistic approximation to the experimental conditions.

The interface configuration of a random system at the initial and final times of simulation is shown in Fig. 10. We see that the small structures (bubbles) merge into large structures.

Due to the exponential stratification of the density distribution of the unperturbed fluid, the effective Atwood number decreases as the bubble moves into the heavy fluid. The reduction of Atwood number results in a non-monotonicity of the bubble velocity. A turnover of bubble velocity is observed in our numerical simulation. Since such turnover phenomena have not been taken into account in the single mode theory, the superposition theory is not applicable when the effective Atwood number has been reduced substantially. To get a better understanding of the phenomenon of velocity turnover in a single mode system and the failure of the superposition hypothesis in a multi-mode system, we use the initial density distribution of light and heavy fluid to approximate the dynamic effective Atwood number $A_{effective}$. For a flat interface, the density distribution is

$$\rho_i(z) = \rho_i(0)exp(\frac{\gamma g z}{c_i^2}), \ i = l, h.$$

When a bubble reaches the position z, we approximate the effective Atwood number as

$$A_{effective}(z) = \frac{\rho_h(z) - \rho_l(z)}{\rho_h(z) + \rho_l(z)} = \frac{(1 + A)exp(\gamma M^2 \frac{2A}{1+A} \frac{z}{\lambda}) - (1 - A)}{(1 + A)exp(\gamma M^2 \frac{2A}{1+A} \frac{z}{\lambda}) + (1 - A)}.$$

For a single mode system, the turnover phenomenon should occur before the effective Atwood number $A_{effective}$ vanishes. For a multi-mode system, the superposition theory is applicable as long as $A_{effective} \approx A = A_{effective}(z = 0)$. In fig. 11, we plot the approximate effective Atwood number $vs.$ $\frac{z}{\lambda}$. Since $A_{effective}$ decreases more rapidly in a system of small density ratio or large compressibility, the superposition theory fails at small value of $\frac{z}{\lambda}$ in these systems.

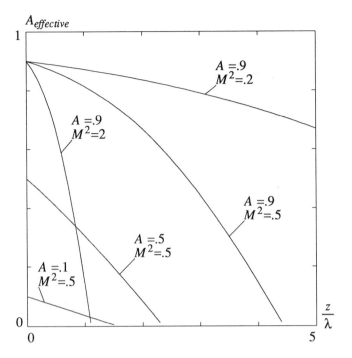

Figure 11: The plot of approximate effective Atwood number as bubble reaches position z. $A_{\text{effective}}$ decreases more rapidly in the system with small initial Atwood number or large compressibility than in the system with large Atwood number and small compressibility. The decreasing of the effective Atwood number is the source the turnover phenomenon in single mode system and the failure of superposition theory in multi-mode system.

One should not confuse the turnover of the bubble velocity in a single mode system with the turnover of the velocity of the small bubble in the multi-mode system . The former is due the stratified density distribution and latter is due to the interactions between bubbles, i.e. the contribution of the envelope velocity to the total velocity of the small bubble.

4. Acknowledgement. We would like to thank the Institute of Mathematics and its Applications for providing us on a CRAY-2 for portions of our study of the single mode problem.

REFERENCES

[1] LORD RAYLEIGH, *Investigation of the Character of the Equilibrium of An Incompressible Heavy Fluid of Variable Density*, Scientific Papers, Vol. II (Cambridge Univ. Press, Cambridge, England, 1900), p200.

[2] G. I. TAYLOR, *The instability of liquid surfaces when accelerated in a direction perpendicular to their planes. I*, Proc. R. Soc. London Ser. A 201, 192 (1950).

[3] G. BIRKHOFF AND D. CARTER, *Rising Plane Bubbles*, J. Math. Mech. 6, 769 (1957).

[4] P. R. GARABEDIAN, *On steady-state bubbles generated by Taylor instability*, Proc. R. Soc. London A 241, 423 (1957).

[5] G.R.BAKER, D.I.MEIRON AND S.A.ORSZAG, *Vortex simulation of the Rayleigh-Taylor instability*, Phys. Fluids 23, 1485 (1980).

[6] D.I.MEIRON AND S.A.ORSZAG, *Nonlinear Effects of Multifrequency Hydrodynamic Instabilities on Ablatively Accelerated thin Shells*, Phys. Fluids 25, 1653 (1982).

[7] R.MENIKOFF AND C.ZEMARK, *Rayleigh-Taylor Instability and the Use of Conformal Maps for Ideal Fluid Flow*, J. Comput. Phys. 51, 28 (1983).

[8] D.H.SHARP AND J.A.WHEELER, *Late Stage of Rayleigh-Taylor Instability"*, Institute for Defense Analyses, (1961).

[9] J.GLIMM AND X.L.LI, *Validation of the Sharp-Wheeler Bubble Merge Model from Experimental and Computational Data*, Phys. of Fluids 31, 2077 (1988).

[10] JUAN ZUFIRIA, *Vortex-in-Cell simulation of Bubble competition in Rayleigh-Taylor instability*, Phys. Fluids 31, 440 (1988).

[11] G. TRYGGVASON, *Numerical Simulation of The Rayleigh-Taylor Instability*, Journal of Computational Physics 75, 253 (1988).

[12] D. L. YOUNGS, *Numerical Simulation of Turbulent Mixing by Rayleigh-Taylor Instability*, Physica D 12, 32 (1984).

[13] J.GLIMM, O.MCBRYAN, R.MENIKOFF AND D.H.SHARP, *Front Tracking applied to Rayleigh-Taylor Instability*, SIAM J. Sci. Stat. Comput. 7, 177 (1987).

[14] C.L.GARDNER, J.GLIMM, O.MCBRYAN, R.MENIKOFF, D.H.SHARP AND Q.ZHANG, *The dynamics of bubble growth for Rayleigh-Taylor unstable interfaces*, Phys. Fluids 31, 447 (1988).

[15] J.GLIMM, R.MENIKOFF, X.L.LI, D.H.SHARP AND Q. ZHANG, *A Numerical Study of Bubble Interactions in Rayleigh-Taylor Instability for Compressible Fluids*, to appear.

[16] D.H.SHARP, *An Overview of Rayleigh-Taylor Instability*, Physica D 12, 32 (1984).

[17] I.B.BERNSTEIN AND D.L.BOOK, *Effect of compressibility on the Rayleigh-Taylor instability*, Phys. Fluids 26, 453 (1983).

[18] QIANG ZHANG, *A Model for the Motion of Spike and Bubble*, to appear.

[19] K.I.READ, *Experimental Investigation of Turbulent Mixing by Rayleigh-Taylor*, Physica D 12, 45 (1984).

FRONT TRACKING, OIL RESERVOIRS, ENGINEERING SCALE PROBLEMS AND MASS CONSERVATION

JAMES GLIMM,* BRENT LINDQUIST† AND QIANG ZHANG‡

Abstract. A critical analysis is given of the mechanisms for mass conservation loss for the front tracking algorithm of the authors and co-workers in the context of two phase incompressible flow in porous media. We describe the resolution to some of the non-conservative aspects of the method, and suggest methods for dealing with the remainder.

Key words. front tracking, mass conservation

AMS(MOS) subject classifications. 76T05, 65M99, 35L65

1. Introduction. Two phase, incompressible flow in porous media is described by a set of PDEs consisting of a subsystem of hyperbolic equations, which describe conservation of the fluid components that thermodynamically combine into the two distinct flowing phases, coupled to a subsystem of equations of elliptic type. The parametric functions in these equations describe the physical properties of the reservoir (petrophysical data) and the physical/thermodynamic properties of the flowing fluids (petrofluid data). Engineering scale problems involve the use of tabulated petrophysical and petrofluid data applicable to real reservoir fields. Such data includes discontinuous rock properties in addition to smooth variations.

We adopt an IMPES type solution method for this set of equations; namely the two subsystems are treated as parametrically coupled, and each subsystem is solved in sequence using highly adapted methods. For the hyperbolic subsystem we use the front tracking algorithm of the authors and co-workers; for the elliptic subsystem we use finite elements. In the original form of the method the solution conserves mass only in the limit of arbitrarily small numerical discretization. We have performed a critical analysis to understand the mechanisms of conservation loss and present here a brief discussion of our conclusions as well as corrections that have been or are in the process of being implemented. Our goal is a front tracking method for flow in porous media that is conservative on all length scales of numerical discretization.

---

*Department of Applied Mathematics and Statistics, SUNY at Stony Brook, Stony Brook, NY, 11794-3600. Supported in part by the Army Research Organization, grant DAAL03-89-K0017; the National Science Foundation, grant DMS-8619856; and the Applied Mathematical Sciences subprogram, U.S. Department of Energy, contract No. DE-FG02-88ER25053.

†Department of Applied Mathematics and Statistics, SUNY at Stony Brook, Stony Brook, NY, 11794-3600. Supported in part by the Applied Mathematical Sciences subprogram, U.S. Department of Energy, contract No. DE-FG02-88ER25053.

‡Courant Institute of Mathematical Sciences, New York University, 251 Mercer St., New York, NY, 10012. Supported in part by the Applied Mathematical Sciences subprogram, U.S. Department of Energy, contract No. DE-FG02-88ER25053.

1.1 The system of equations. Consider as example the flow of two immiscible, incompressible fluid phases. The system of equations which describe the flow of these two phases in a porous medium are

$$(1.1a) \qquad \alpha(x)\phi(x)\frac{\partial s}{\partial t} + \nabla \cdot \alpha(x)\vec{v}(x)f(s) = 0,$$

$$(1.1b) \qquad \begin{cases} \nabla \cdot \alpha(x)\vec{v}(x) = 0, \\ \vec{v}(x) = -\lambda(s,x)\mathbf{K}(x) \cdot \nabla P. \end{cases}$$

Equation (1.1a) is a single equation representing the conservation of volume of the two incompressible phases, phase 1 occupying fraction s of the available pore volume in the rock and phase 2 occupying $1 - s$ of the available pore volume. In general a fraction s_c of phase 1, and a fraction s_r of phase 2 are inextricably bound to the rock, therefore s varies between s_c and $1 - s_r$. A region of the reservoir in which s is constant and equal to one of its limiting values is a region of single phase flow. In a region in which s varies smoothly, and lies strictly within its bounding limits, both phases are flowing. The discontinuity waves that occur in the solution of (1.1a) describe discontinuous transitions in s.

The first of equations (1.1b) expresses the incompressibility of the flowing fluids; the second, Darcy's law, relates the total fluid velocity $\vec{v}$ to the gradient of the pressure field P in the reservoir. In (1.1a), $f(s)\vec{v}(x)$ is the fraction of the total fluid velocity carried by phase 1. For simplicity of presentation, we neglect gravitational terms in (1.1) though they are included in the analysis. For simplicity, we also neglect point source and sink terms which describe point injection and production sites (wells) for the fluid phases, and appear, especially in two dimensional calculations, on the left hand sides of (1.1a) and the first of (1.1b). The effects of other neglected terms such as surface tension, chemical reactions, compressibility, etc. present in more complex flows are not included in our analysis. The other parameters in (1.1) specify the petrophysical data (PPD) and petrofluid data of the reservoir:

$\alpha(x)$ is a geometrical factor accounting for volume affects not specifically accounted for by the independent spatial variables in (1.1);

    1) $\alpha(x)$ is the cross-sectional area of a 1 dimensional reservoir.

    2) $\alpha(x)$ is the thickness in the third dimension for a two dimensional reservoir.

    3) $\alpha(x) = 1$ for a fully three dimensional calculation.

$\phi(x)$ is the porosity (volume fraction of pore space) of the rock medium.

$\mathbf{K}(x)$ is the rock permeability tensor describing the connectedness of the geometrical pathways through the rock pores.

$\lambda(s,x)$ is the total relative fluid transmissibility describing how the presence of phase 1 affects the flow of phase 2 and vice-versa. It has explicit x dependence as its functional form may differ according to the local rock type.

In an engineering scale problem, this data is usually specified in tabulated form and may contain information on the location of sharp transitions across faults, layer structures, and barrier regions.

1.2 Conservation form for hyperbolic equations. Consider the system of conservation laws

(1.2)
$$s_t + F(s)_x = 0.$$

The conservative formulation for a finite difference scheme is based on the integral (weak) formulation of (1.2)

(1.3)
$$\int_t^{t+dt} dt \int_{x_i-1/2}^{x_i+1/2} dx(s_t + F(s)_x) = 0,$$

over a numerical grid block centered at x_i, and can be expressed in the form

(1.4) $\quad S_i^{n+1} = S_i^n + \dfrac{\Delta t}{\Delta x}\{G_{i-1/2}(S_{i-p-1}, ..., S_{i+q}) - G_{i+1/2}(S_{i-p}, ..., S_{i+q+1})\},$

where S is the volume integrated mass in the mesh block centered at x_i. The numerical flux G, defined over a stencil of $p + q + 2$ grid blocks, must satisfy

$$G(S, ..., S) = F(S),$$

and, more trivially, but important for our considerations,

(1.5)
$$G_{i+1/2}(\cdot) = G_{(i+1)-1/2}(\cdot).$$

1.3 Conservation form for elliptic equations.

DEFINITION. *A solution $\vec{v}$ of the elliptic system (1.1b) is conservative with respect to a grid of lines G, if it satisfies $\oint \vec{v} \cdot \hat{n} d\ell = 0$ for every closed path consisting of lines of G.*

DEFINITION. *A solution $\vec{v}$ of the elliptic system (1.1b) is conservative in a region Ω if it satisfies $\oint \vec{v} \cdot \hat{n} d\ell = 0$ for every closed path in Ω.*

1.4 The solution method and mass conservation. The front tracking scheme of the authors and co-workers [7,8] for solving (1.1a) consists of a conservative scheme of the form (1.4) defined on a regular rectangular, two dimensional grid G_H, in conjunction with moving one dimensional grids (curves) to track the evolving discontinuity surfaces. The propagation of the tracked curves (the *front* solution) is achieved via spatially local Riemann problem solutions in the direction normal to the curve. The method also takes into account flow of fluid tangential to the discontinuity curves. The solution away from the tracked curves (the *interior* solution) is obtained using the grid G_H with an upwind scheme of the form (1.4). These *front* and *interior* schemes are coupled, the tracked curves providing 'boundary values' for the interior solution and the Riemann problem data taking into account interior solution values.

The elliptic system (1.1b) is solved by combining the two equations into a single elliptic equation for the pressure field P, and solving by finite elements [7,8,11]. The finite element mesh G_E is a mixture of rectangles and triangles, whose edges match all discontinuity surfaces in the solution and in the PPD. The finite elements are standard; tensor product Lagrangian basis functions on the rectangles, and triangle basis elements. The velocity field is obtained by analytic differentiation of the basis function representation of the pressure field.

Five items have been identified as responsible for loss of mass conservation of the method. These items are

1) the discretization of the medium properties,

2) the physical limits for the solution variable s, $\quad s_c \leq s \leq 1 - s_r$,

3) the implementation of the conservation form (1.4) near physical discontinuities in the medium properties (faults, layers and barriers),

4) the conservative properties of the velocity field $\vec{v}$,

5) the tracking of moving fluid discontinuities.

As these items interconnect, it is difficult to state precisely their relative order of importance; in our test example the first three issues are more crucial for mass conservation achievement than the last two. In the remainder of this paper we consider each of these five items separately.

2. Discretization of the medium properties. It is important in the front tracking method, for both mass conservation and resolution of the phase discontinuity behavior, that the PPD be represented in a smooth (C^0) fashion away from faults rather that in a piecewise constant (block centered) manner. Consider the first order Enquist-Osher scheme for (1.4). It has the form (in one space dimension)

$$(2.1) \qquad G_{i-1/2}(S_{i-1}, S_i, \cdot) = \frac{1}{2} \left[F_{i-1}(\cdot) + F_i(\cdot) - \int_{S_{i-1}}^{S_i} \left| \frac{\partial F(\cdot)}{\partial S} \right| dS \right].$$

The unspecified arguments ($\cdot$) in (2.1) include the explicit x dependence of F required for (1.1a),

$$F(S, x) = \alpha(x)v(x)f(S).$$

Using block centered PPD, a logical choice might be

$$F_{i-1}(\cdot) = F(S_{i-1}, x_{i-1}),$$
$$F_i(\cdot) = F(S_i, x_i).$$

However, with cell centering, the requirement (1.5) of cell boundary continuity on the numerical fluxes implies that, for intervals (S_{i-1}, S_i) over which $\partial F/\partial S$ changes sign, the evaluation of the last term in (2.1) be done as

$$\int_{S_{i-1}}^{S(x_{i-1/2})} \left| \frac{\partial F(S(x), x)}{\partial S} \right| dS + \int_{S(x_{i-1/2})}^{S_i} \left| \frac{\partial F(S(x), x)}{\partial S} \right| dS.$$

This in turn requires a map of the interval (x_{i-1}, x_i) onto (S_{i-1}, S_i) due to the discontinuity in the PPD at $x_{i-1/2}$. While it is possible to devise ways to achieve this, such a choice is equivalent to an ad hoc smoothing of the data.

Further, the use of block centered PPD requires the specification of greater amounts of data in order to perform calculations on refined meshes. While the specification of such additional data can be automated, this again results in an ad hoc smoothing method **that is intimately coupled to the grid used in the solution of (1.1a).** In addition, as not all PPD remains smooth as the numerical discretization lengths $\Delta x \to 0$; these physical discontinuities are an inherent aspect of the PPD data that must be discretized accurately on all length scales.

The use of block centered PPD can also play a role in introducing spurious instabilities for front tracking algorithms. In Figure 1 we show a tracked curve passing through two mesh blocks. If point p_1 on the tracked curve is propagated using the PPD from mesh block B_1, and p_2 is propagated using the PPD from B_2, a kink will develop in the front, which under physically unstable flow regimes will grow.

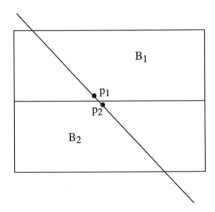

Figure 1 A jump discontinuity in the PPD, as in the specification of piecewise constant data, between mesh blocks B_1 and B_2 can result in numerically based discontinuous propagation behavior of the two points p_1 and p_2 on the tracked curve.

It is relatively easy to provide a representation of the PPD that is continuous in the appropriate regions, and resolves the discontinuous structure as well. We illustrate an automatic discretization method which achieves this. The method has the additional feature that it discretizes the PPD on a grid that is independent of the grids G_H and G_E. This allows a representation of the PPD **that can be held fixed** while mesh refinement studies are done for the solution methods used on (1.1a) and (1.1b). While the idea behind this discretization method is not new [12,14], we reiterate that smooth representation of the PPD is necessary for use in conjunction with the front tracking method.

This discretization method is illustrated in Figure 2. Figure 2a depicts the two dimensional areal plan of an inclined reservoir bed. The reservoir contains two fault lines F_1 and F_2. The two dimensional slice follows the local inclination of the middle

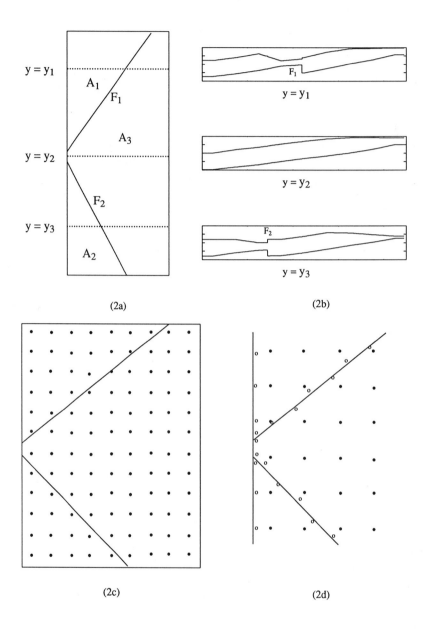

(2a)

(2b)

(2c)

(2d)

Figure 2 a) An areal view (x vs y) of a reservoir field having two fault lines F_1 and F_2. b) Three vertical plans (x vs z) through the reservoir field. c) A demonstration of the placement (dark circles) of given field petrophysical data. d) An enlargement of a region in c) showing a possible choice of points where additional field data is required in order to fit the fault and boundary structure.

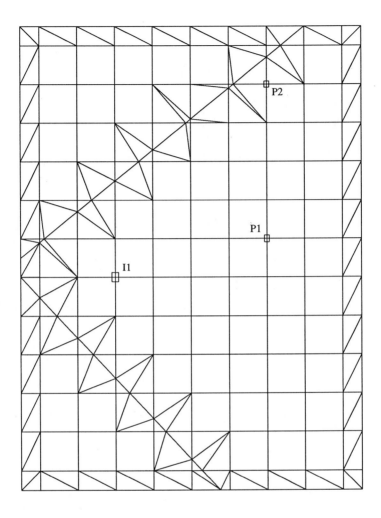

Figure 2e A tessellation of the geophysical structure of the reservoir to produce C^0 continuity of the petrophysical data.

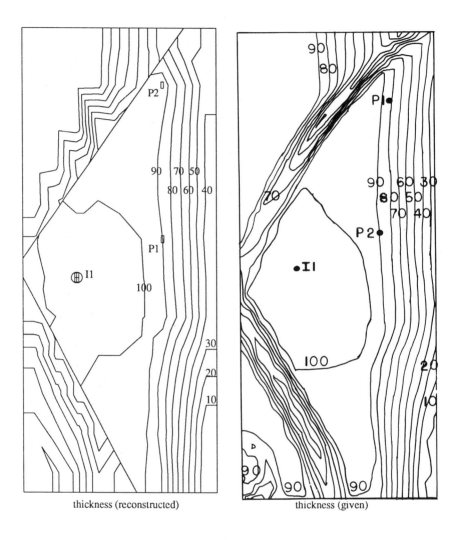

thickness (reconstructed)

thickness (given)

Figure 3 Comparison of algorithmically smoothed reservoir thickness data with field measured data. The numerical smoothing algorithm used field data defined on a rectangular 9 by 12 grid and knowledge of the fault locations to produce a smooth approximation to the reservoir thickness.

of the reservoir bed; three vertical plans of the bed are depicted in Figure 2b. PPD was specified from field readings at the corners (black points in Figure 2c) of a rectangular grid G_R. To obtain the required representation of this data, additional PPD is required along each side of the fault lines, and along the boundaries of the computational domain. The unshaded points in Figure 2d (a close up of a small area of Figure 2c) demonstrate one possible placement for specification of this additional data. A tessellation T of the grid G_R into a mesh of rectangles and triangles is achieved by triangulating those rectangles of G_R that are cut by faults, or lie next to the computational boundary, in such a manner that

- the faults are coincident with triangle sides,
- triangle nodes lie either at the corners of G_R, on the fault lines, or on the boundaries.

Such a tessellation is shown in Figure 2e. C^0 smoothness of the PPD away from the discontinuities is then achieved by employing, for example, linear (bilinear) interpolation on the triangles (rectangles) of T.

The efficacy of such a discretization is demonstrated in Figure 3. Figure 3b shows initially specified contours of $\alpha(x)$ for the reservoir under discussion in Figure 2. PPD were specified on the corners of the rectangular 9 by 12 grid G_R of Figure 2c. Data on each side of the fault lines and at the boundaries of the reservoir were obtained by constant extrapolation from the closest point of G_R. The resultant piecewise continuous discretization of the data is shown as a contour plot in Figure 3a. In spite of the coarseness of the grid G_R, the resultant piecewise continuous discretization of the PPD on T agrees extremely well with the measured data in the large area A_3 and in A_2. The representation in the small triangular region A_1 is not as good. However, in this particular calculation, the active region of the reservoir was constrained (by specification of the rock permeability values) to lie only in A_3, so no effort was made to improve the representation of the data in A_1.

While T is useful for providing interpolation of the PPD, it is inappropriate to compute numerical derivatives of this data directly from the linear/bilinear representation it provides (which would result in piecewise constant/linear derivatives). This is due to the extreme aspect ratios that may develop for some triangles. Rather, derivatives such as those required to compute the local gravitational strength (neglected in (1.1)), can be achieved by usual finite differences. Figure 4 illustrates this. The gradient $d\delta/dx$ of some petrophysical quantity δ at the point p_0 can be obtained by central differencing over a distance of 2h (Figure 4a). The values of δ at p_{-1} and p_1 are obtainable by interpolation on the tessellation T. This finite difference scheme must be modified near faults and boundaries in an appropriate one-sided manner as illustrated in Figures 4b,c and d. In Figure 4b, the centered difference at p_0 is based on an irregular stencil of length $h_1 + h_2$. If p_0 lies exactly on a horizontal fault as in Figure 4c, two derivatives are required, one on each side of the fault. If the fault kinks, it may be necessary to resort to a difference based on a triangle for one of the sides of the fault, as illustrated in Figure 4d.

3. Physical solution limits. For incompressible flow the fluid volume fraction s, and hence the numerical integrated volume fraction S, are bounded above and

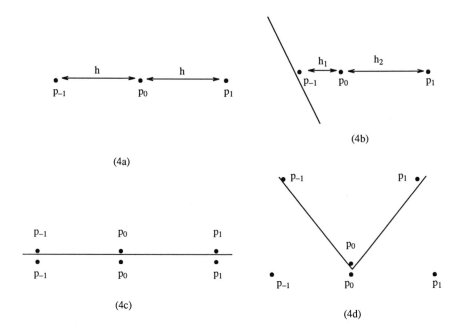

Figure 4 A centered finite difference stencil a) used to compute
derivatives (here we illustrate d/dx) of the petrophysical data must
be modified in appropriate one-sided ways b) c) and d) in the vicinity of discontinuities and the reservoir boundary.

below,

$$s_c \leq s \leq 1 - s_r, \quad S_c \leq S \leq 1 - S_r, \quad S_c = s_c, \quad S_r = s_r.$$

Furthermore, functional evaluations may be defined only for s (S) lying within this
bounded domain. The numerical scheme (1.4), or its two dimensional extension
to irregularly shaped domains, while conservative, provides no guarantee that the
numerical solution will remain within these bounds. In fact (1.4) only guarantees
conservation if the numerical flux function G is definable for **every** numerical value
of S generated. In practice only a finite extension of the domain $[S_c, 1 - S_r]$ is required. Given petrophysical and petrofluid data based on tabulated experimental
data, even limited extension is usually impractical. One is then forced to truncate
the solution whenever it reaches its limiting values. In order to maintain mass conservation, this truncated mass must be reintroduced into the solution in a physically
realistic manner. We discuss a fully two-dimensional, unsplit version of (1.4) which
includes the reintroduction of truncated mass. (Directionally split schemes are less
preferred as truncated mass must be stored each directional sweep, and the excess
masses reintroduced after the final sweep.)

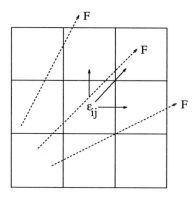

Figure 5 Schematic apportioning of truncated mass ϵ_{ij} into downstream (indicated by flux direction F) mesh blocks.

Consider a rectangular mesh block ij removed from faults or moving fronts. Let $\tilde{S}_{ij}^{n+1}$ denote the solution obtained for ij by applying (1.4) in an unsplit, two dimensional form,

$$(3.1) \quad \tilde{S}_{ij}^{n+1} = S_{ij}^{n} + \frac{\Delta t}{\Delta x}\{G_{i-1/2,j}(\cdot) - G_{i+1/2,j}(\cdot)\} + \frac{\Delta t}{\Delta y}\{G_{i,j-1/2}(\cdot) - G_{i,j+1/2}(\cdot)\}.$$

Define

$$\hat{S}_{ij}^{n+1} = \begin{cases} S_c, & \text{if } \tilde{S}_{ij}^{n+1} < S_c \\ 1 - S_r, & \text{if } \tilde{S}_{ij}^{n+1} > 1 - S_r \\ \tilde{S}_{ij}^{n+1}, & \text{otherwise.} \end{cases}$$

Let $\epsilon_{ij} \equiv \tilde{S}_{ij}^{n+1} - \hat{S}_{ij}^{n+1}$ represent the clipped mass for block ij that must be restored to the solution. For $\epsilon_{ij} > 0$ (< 0), we apportion the clipped mass into appropriate downstream (upstream) blocks in proportion to the carrying capacities of these blocks. This is indicated schematically in Figure 5. If no appropriate downstream (upstream) blocks are available, or they have insufficient carrying capacity, any unallocated clipped mass is accumulated until it can be distributed.

For mesh blocks cut by faults, a clipped mass is calculated for each polygonal area into which the mesh block is cut by the fault. Distribution of the clipped mass again takes place, but now into the appropriate polygon areas.

The algorithm consists of two passes over the mesh. On the first pass the ϵ_{ij} are stored; on the second the clipped mass is distributed. If the number of mesh blocks having $\epsilon \neq 0$ is dense in the mesh, (i.e. on the average, any given mesh block lies downstream from several mesh block containing excess mass) the apportioning of the ϵ_{ij} becomes a constrained optimization problem amongst coupled mesh blocks. If however, the number of mesh blocks in which $\epsilon_{ij} \neq 0$ is sparse, such that downstream mesh blocks receiving mass are in one-to-one correspondence with mesh blocks having excess mass, the distribution of this mass can be done by a direct sweep through the mesh treating one mesh block at a time. Based on our early experience we expect the number of mesh blocks containing truncated mass to be relatively sparse; therefore we have implemented the latter, simpler scheme for distributing the truncated mass.

4. The conservative form in the vicinity of faults. With the PPD smoothly discretized on a fixed grid, we now turn to the solution of (1.1a) in the vicinity of faults. A common implementation of (3.1) in the vicinity of faults is to stair-case the faults to conform to the grid G_H, the stair-casing becoming finer as G_H is refined. However this choice leads to spurious bending of the tracked discontinuity surfaces and the potential for spurious fingering in unstable flow regimes. This is illustrated in Figure 6 where a tracked interface traveling obliquely to a stair-cased fault encounters a series of corners. The front movement around these corners results in curvature.

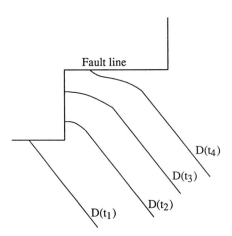

Figure 6 Bending of a tracked propagating discontinuity wave $D(t)$, $(t_i > t_j$ for $i > j)$, traveling obliquely to a fault line represented in a stair-case fashion.

One is then constrained to exact representation of the faults (as in Figure 7) which is achieved in the front tracking scheme by representing them as tracked, unmoving waves. The conservative scheme (3.1) must be modified to handle those mesh blocks of G_H cut by such faults. An appropriately volume averaged solution value must be stored for each of the separate regions produced (Figure 7). (3.1) is modified in the obvious way as a sum of fluxes flowing normally through each of the sides of the irregularly shaped polygons thus formed. The problem one is now forced to solve is the restriction due to the CFL condition which reduces the allowed maximum timestep by the ratio of the smallest area of all such polygons

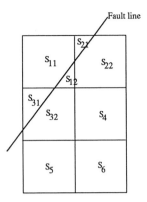

Figure 7 Modification of grid and solution representation required for exact representation of fault lines.

formed to that of the regular rectangles on G_H,

$$\frac{\min \{A_{\text{polgon}}\}}{A_{\text{regular rectangle}}}.$$

Several authors (Leveque [9,10], Chern and Colella [5], Berger and Leveque [1]) have treated this problem for the Euler equations of compressible fluid dynamics in order to overcome the CFL restrictions. It is a common feature of such conservative interface methods to allow excess entropy production, resulting in shock wall heating, slip line heating, or fluid mixing and entrainment in compressible fluid flow. The approach we take here is similar to that of Chern and Colella, in that conservation is restored by placement of mass in adjacent cells when CFL limits are encountered. See §3.

5. Conservative properties of $\vec{v}$. As the subsystem (1.1a) contains the non-hyperbolic variable $\vec{v}$, it is necessary that this velocity field $\vec{v}$ be conservative with respect to the grid G_H to ensure that an algorithm of the form (3.1) remains conservative when applied to (1.1a). However, in order to avoid an undesirable coupling between the grids G_H and G_E, it is then desirable that $\vec{v}$ be conservative in the complete computational domain, i.e. $\vec{v}$ must be divergence free everywhere.

The finite element method currently in use for front tracking calculations [7,8,11] does not have this conservative property. The velocity field it produces is, in general, not conservative in the region of computation, and can have spurious source/sink regions especially near corners and boundaries. Raviart and Thomas [13] have developed a mixed finite element method for solving

(6.1a) $$\nabla \cdot \vec{v} = f, \qquad f \in L^2,$$

(6.1b) $$\vec{v} = -\nabla u.$$

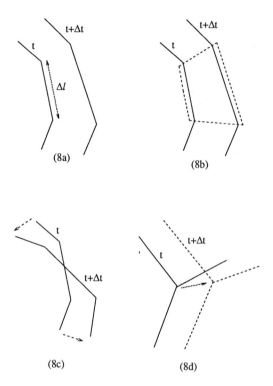

Figure 8 Schematic illustration of issues to be dealt with in deriving a conservative formulation for propagating tracked waves.

Solutions for $\vec{v}$ and u are developed in two separate spaces, V_h and U_h, of polynomial elements. Through judicious choice of the properties of the basis functions in these two spaces the numerical solution $\vec{v}$ solves (6.1a) exactly. Chavant, Jaffré et. al. [2,3,4] and Douglas, Ewing and Wheeler [6] have adapted this mixed finite element approach to two phase incompressible flow in two dimensions. This body of work is characterized by also solving (1.1a) by a Galerkin procedure. We are in the process of implementing this mixed finite element method for the solution of (1.1b) in combination with front tracking for the solution of (1.1a).

6. Tracking of moving fronts. The tracking curves are propagated [7,8] in a non-conservative manner. As mentioned in §1.4, these one dimensional grids are composed of piecewise linear bonds. The movement of the tracked curve is achieved by propagating the end points of each bond via information from Riemann solutions. Figure 8a illustrates the propagation of a bond of length $\Delta\ell$. The movement of the bond's two end points results in the movement of an amount of mass along the entire bond. This method of propagating the front will conserve mass only in the limit $\Delta \ell \to 0$.

We are in the process of investigating different approaches for achieving a conservative front propagation algorithm. The most likely approach would be one consistent with the integral formulation (1.3). One such bond oriented version is indicated by the integration path (dashed line) shown in Figure 8b. However, the front propagation is not usually as straightforward as suggested by Figure 8b. One possible complication is depicted in Figure 8c. Further complications exist at points where two or more tracked fronts join (Figure 8d), or when separate tracked curves interact. In addition the details of the coupling of a conservative approach for the front to the method (3.1) used in obtaining the solution away from the front remain to be worked out. Preferably, any proposed conservative scheme for the fronts should be extendible to systems and compressible flow.

7. Example calculations. Figure 9 shows the results of a calculation for the areal plan of the reservoir field shown in Figure 2. The PPD (top of formation, α, ϕ, rock permeability K) were discretized according to §2, based upon a 9 by 12 rectangular grid G_R. Local gravitational strengths were calculated using finite differences as discussed in §2. The hyperbolic equation (1.1a) was solved using front tracking. The faults F_1 and F_2 were represented as tracked, unmoving waves. Water was injected at constant rate into well I1, and fluid pumped at constant rates from wells P1 and P2. The interface between the resultant two phase, water swept region and the single phase, undisturbed oil region was tracked. The solution in the region away from the tracked discontinuities was calculated using the first order Enquist-Osher scheme on a 9 by 12 regular rectangular grid. Since the Raviart-Thomas based mixed finite element method has not as yet been completely installed, (1.1b) was solved with the original finite element method described in §1.4. Linear/bilinear basis functions were used on the mesh G_E. This mesh adapts to the moving interfaces, hence it changes each timestep. There is no correlation between the grids G_E and G_H.

Figure 9 shows the tracked phase discontinuity interface at selected times during the first 33 years of the calculation. Figure 10 shows the percentage mass balance error for the water component as a function of time. The mass balance error is defined as

$$(7.1) \qquad \mathcal{E}_M \equiv \frac{M_{\text{present}} - M_{\text{initial}} - M_{\text{injected}} + M_{\text{produced}}}{M_{\text{injected}}}$$

where M represents water mass. Note that $\mathcal{E}_M$ is a 'forgiving' dimensionless measure of mass balance error since, as M_{injected} typically increases in time, $\mathcal{E}_M$ can decrease in spite of an increase in the absolute magnitude of the numerator of (7.1). The calculation was performed using the mass conservation corrections discussed in §§2, 3 and 4. The velocity field obtained was not conservative (§4) over the entire domain of the calculation, and no correction was applied for the front movement (§6).

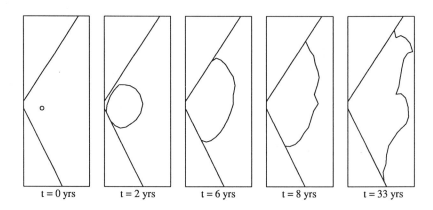

t = 0 yrs   t = 2 yrs   t = 6 yrs   t = 8 yrs   t = 33 yrs

Figure 9 Calculation for the reservoir field described in figures 2d and 3. The phase discontinuity delineating the two phase region swept by the injected water from the single phase, undisplaced oil region is shown at selected times.

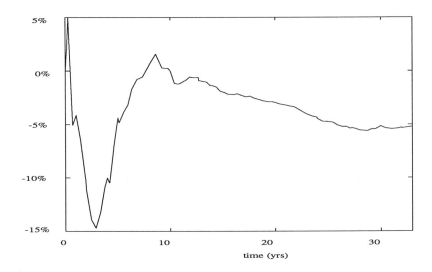

Figure 10 Water mass balance error $\mathcal{E}_M$ for the calculation of Figure 9.

The resultant mass balance error after 33 years of simulation time is $\approx -6\%$. The mass balance error is indeterminate at $t = 0$ as the denominator of (7.1) goes to zero. The initial mass balance errors are dominated by three things: 1) the non-conservative front propagation scheme (§6), 2) the lack of correction of mass balance errors for the scheme (1.4) in mesh blocks cut by the initial phase discontinuity interface (which encloses an area smaller than the size of a grid block for a range of initial times), 3) the inability of the finite element method implementation to resolve the velocity field around the point source. This last cause is the most critical and has long been of problematic concern in reservoir simulation (see for example the treatments in [4] and [6]). An analytic treatment of the velocity divergence in the vicinity of wells has been included in this calculation, but match-up with the finite element solution is problematic. The initial error is also amplified by the smallness of the denominator. At late times, the error is primarily due to the nonconservative use of the velocity field.

Acknowledgements. The authors wish to thank Statoil, Norway for supplying the realistic petrophysical and petrofluid field data used in this study, and for their support of the development of front tracking for reservoir calculations. We also gratefully acknowledge the continuing support of the Institute for Energy Technology, Norway.

REFERENCES

[1]   M. BERGER AND R. J. LEVEQUE, *An adaptive cartesian mesh algorithm for the Euler equations in arbitrary geometry.*, AIAA 89-1930, 9th Computational Fluid Dynamics Conference, Buffalo, NY, June 1989.

[2]   G. CHAVENT AND J. JAFFRÉ, *Mathematical Models and Finite Elements for Reservoir Simulation*, North Holland, Amsterdam, 1986.

[3]   G. CHAVENT, G. COHEN, J. JAFFRÉ, M. DUPUY, AND I. RIBÉRA, *Simulation of two dimensional waterflooding by using mixed finite elements*, Soc. Pet. Eng. J., 24 (1984), pp. 382-389.

[4]   G. CHAVENT, J. JAFFRÉ, R. EYMARD, D. GUERILLOT, AND L. WEILL, *Discontinuous and mixed finite elements for two-phase incompressible flow*, SPE 16018, 9th SPE Symposium on Reservoir Simulation, San Antonio.

[5]   I-L. CHERN AND P. COLELLA, *A conservative front tracking method for hyperbolic conservation laws*, J. Computational Physics (to appear).

[6]   J. DOUGLAS, JR., R. E. EWING, AND M. F. WHEELER, *The approximation of the pressure by a mixed method in the simulation of miscible displacement*, R.A.I.R.O. Analyse numérique, 17 (1983), pp. 17–33.

[7]   J. GLIMM, E. ISAACSON, D. MARCHESIN, AND O. MCBRYAN, *Front tracking for hyperbolic systems*, Adv. Appl. Math., 2 (1981), pp. 91–119.

[8]   J. GLIMM, W. B. LINDQUIST, O. MCBRYAN, AND L PADMANABHAN, *A front tracking reservoir simulator, five-spot validation studies and the water coning problem*, SIAM Frontiers in Appl. Math., 1 (1983), pp. 107–135.

[9]   R. J. LEVEQUE, *Large time step shock-capturing techniques for scalar conservation laws*, SIAM J. Numer. Anal., 19 (1982), pp. 1091–1109.

[10]   ─────────── , *A large time step generalization of Godunov's method for systems of conservation laws*, SIAM J. Numer. Anal., 22 (1985), pp. 1051–1073.

[11]   O. MCBRYAN, *Elliptic and hyperbolic interface refinement*, in *Boundary Layers and Interior Layers – Computational and Asymptotic Methods*, J. Miller (ed.), Boole Press, Dublin, 1980.

[12]   L PADMANABHAN. Chevron Oil Field Research, private communication.

[13]   P. A. RAVIART AND J. M. THOMAS, *A mixed finite element method for second order elliptic problems*, in *Mathematical Aspects of Finite Element Methods*, Lecture Notes in Mathematics 606, Springer-Verlag, New York, 1977, pp. 292–315.

[14]   Y. SHARMA. Cray Research Inc., private communication.

COLLISIONLESS SOLUTIONS TO THE FOUR VELOCITY BROADWELL EQUATIONS

J. M. GREENBERG* AND CLEVE MOLER†

Introduction. In this note we shall examine special collisionless solutions to the four velocity Broadwell equations. These solutions are new and seem to have gone unnoticed by other investigators who have worked in this area[1]. These solutions are apparently stable; that is in numerical simulations they appear as the asymptotic state of the evolving system.

The basic quantities of interest are particle densities $\tilde{r}$, $\tilde{l}$, $\tilde{u}$, and $\tilde{d}$. Specifically $\tilde{r}(x, y, t)$ represents the number of particles per unit area at $(\tilde{x}, \tilde{y})$ at time $\tilde{t}$ travelling with velocity $w\mathbf{e}_1$. The densities $\tilde{l}$, $\tilde{u}$, and $\tilde{d}$ have a similar interpretation except that the particles travel with velocities $-w\mathbf{e}_1$, $w\mathbf{e}_2$, and $-w\mathbf{e}_2$ respectively. The evolution equations for the densities are

(1.1)
$$\left.\begin{array}{l} \tilde{r}_{\tilde{t}} + w\tilde{r}_{\tilde{x}} = \mathcal{C} \\ \tilde{l}_{\tilde{t}} - w\tilde{l}_{\tilde{x}} = \mathcal{C} \\ \tilde{u}_{\tilde{t}} + w\tilde{u}_{\tilde{y}} = -\mathcal{C} \\ \tilde{d}_{\tilde{t}} - w\tilde{d}_{\tilde{y}} = -\mathcal{C} \end{array}\right\}$$

where the collision term $\mathcal{C}$ is given by

(1.2)
$$\mathcal{C} = K(\tilde{u}\tilde{d} - \tilde{r}\tilde{l}).$$

Dimensional consistency implies that

(1.3)
$$\dim(K) = \frac{\text{Area}}{(\# \text{ of particles}) \times \text{Time}}.$$

Most of the authors dealing with (1.1), or one of its generalizations, treat K as a constant. In what follows we let

(1.4)
$$\tilde{\rho}(\tilde{x}, \tilde{y}, \tilde{t}) = (\tilde{r} + \tilde{l} + \tilde{u} + \tilde{d})(\tilde{x}, \tilde{y}, \tilde{t})$$

be the number of particles per unit area at $(\tilde{x}, \tilde{y})$ at time $\tilde{t}$ and model K by

(1.5)
$$K = \frac{1}{\epsilon\tilde{\rho}(\tilde{x}, \tilde{y}, \tilde{t})}$$

---

*University of Maryland Baltimore County, Department of Mathematics & Statistics, Catonsville, MD 21228

†Ardent Computer Corporation, Sunnyvale, CA

[1] For an excellent survey of the literature on this and similar models the reader should see Platkowski and Illner [1] and the references therein.

where ϵ is a fixed relaxation time.

Under the change of variables

(1.6) $\qquad x = \dfrac{\tilde{x}}{w\epsilon}, \quad y = \dfrac{\tilde{y}}{w\epsilon}, \quad t = \dfrac{\tilde{t}}{\epsilon}, \text{ and } (r, l, u, d) = (w\epsilon)^2 (\tilde{r}, \tilde{l}, \tilde{u}, \tilde{d})$

the system (1.1), (1.2), (1.4) and (1.5) scales to

(1.7) $\qquad \left. \begin{aligned} r_t + r_x &= (ud - rl)/\rho \\ l_t - l_x &= (ud - rl)/\rho \\ u_t + u_y &= (rl - ud)/\rho \\ d_t - d_y &= (rl - ud)/\rho \end{aligned} \right\}$

where

(1.8) $\qquad\qquad\qquad\qquad \rho = r + l + u + d$

is the number density of particles at (x, y) at time t. In a recent paper, [2], we examined solutions to (1.7) and (1.8) which were independent of y and thus satisfied

(1.9) $\qquad \left. \begin{aligned} r_t + r_x &= (ud - rl)/\rho \\ l_t - l_x &= (ud - rl)/\rho \\ u_t &= (rl - ud)/\rho \\ d_t &= (rl - ud)/\rho \end{aligned} \right\}$

where again ρ is given by (1.8). Our results dealt with solutions to (1.8) and (1.9) generated by bounded, compactly supported initial data. If $-a < x < a$ was the support interval of the data, the solutions obtained in [2] had the additional property that

(1.10) $r \equiv u \equiv d \equiv 0, \ x < -a$ and $l_t - l_x = 0, \quad -(a + t) < x < -a$ and $t > 0$

and

(1.11) $\quad l \equiv u \equiv d \equiv 0, \ x > a$ and $r_t + r_x = 0, \quad a < x < (a + t)$ and $t > 0$

and thus the interaction region was the strip $-a < x < a$ and $t > 0$. The principal result of [2] was that in the interval $-a < x < a$:

(1.12) $\qquad\qquad\qquad \lim_{t \to \infty} (r(x, t), l(x, t), (ud)(x, t)) = (0, 0, 0),$

(1.13) $\qquad\qquad\qquad \lim_{t \to \infty} u(x, t) = \max(u(x, 0) - d(x, 0), 0),$

and

(1.14)
$$\lim_{t \to \infty} d(x,t) = \max(0, d(x,0) - u(x,0)).$$

Additional exponential decay estimates were obtained for the more customary form of the Broadwell equations

(1.15)
$$\left.\begin{array}{l} r_t + r_x = (c^2 - rl)/\rho \\ l_t - l_x = (c^2 - rl)/\rho \\ c_t = (rl - c^2)/\rho \end{array}\right\}$$

where now

(1.16)
$$\rho = r + 2c + l.$$

The equations (1.8) and (1.9) reduce to this system on the manifold $u \equiv d \overset{\text{def}}{=} c$.

In this note we shall confine our attention to special collisionless solutions of the full two-dimensional system (1.7) and (1.8). That such solutions were possible was suggested by computations we performed on (1.7) and (1.8) for a variety of initial and boundary conditions. These computations suggested that the long time behavior of the system was characterized by such solutions and motivated our trying to establish that the system did in fact support such solutions.

The collisionless solutions are nonconstant, positive solutions to (1.7) and (1.8) which satisfy the additional identity that $ud - rl = 0$. In section 2 we demonstrate that (1.7) and (1.8) does indeed support such solutions and in section 3 we shall demonstrate how these solutions emerge and characterize the long time behavior of the system. It should be noted that these collisionless solutions are valid for both the collision term considered in our simulations, namely $\mathcal{C} = (ud - rl)/\rho$, and the more customary collision term $\mathcal{C} = (ud - rl)$.

2. Collisionless Solutions. In this section we shall exhibit a class of nonconstant, positive solutions to (1.7) and (1.8) which satisfy the additional constraint:

(2.1)
$$ud - rl \equiv 0.$$

Such solutions must be of the form

(2.2)
$$\left.\begin{array}{l} r(x,y,t) = R_1(x - t, y) \\ l(x,y,t) = L_1(x + t, y) \\ u(x,y,t) = U_1(x, y - t) \\ d(x,y,t) = D_1(x, y + t) \end{array}\right\}$$

or equivalently

(2.3)
$$
\left.
\begin{aligned}
r(x,y,t) &= R_2(x+y-t, x-y-t) \\
l(x,y,t) &= L_2(x+y+t, x-y+t) \\
u(x,y,t) &= U_2(x+y-t, x-y+t) \\
d(x,y,t) &= D_2(x+y+t, x-y-t)
\end{aligned}
\right\} .
$$

The fact that (2.1) must hold implies that R_2, L_2, U_2, and D_2 must satisfy

(2.4)
$$
\begin{aligned}
&R_2(x+y-t, x-y-t)L_2(x+y+t, x-y+t) \\
&\quad = U_2(x+y-t, x-y+t)D_2(x+y+t, x-y-t).
\end{aligned}
$$

If we let

(2.5)
$$
\left.
\begin{aligned}
R_3 &= \ln R_2 \\
L_3 &= \ln L_2 \\
U_3 &= \ln U_2 \\
D_3 &= \ln D_2
\end{aligned}
\right\} ,
$$

then (2.4) is equivalent to

(2.6)
$$
\begin{aligned}
&R_3(x+y-t, x-y-t) + L_3(x+y+t, x-y+t) \\
&\quad = U_3(x+y-t, x-y+t) + D_3(x+y+t, x-y-t).
\end{aligned}
$$

Moreover, if we let

(2.7) $\quad \theta_1 = x+y-t, \theta_2 = x-y-t, \theta_3 = x+y+t, \quad \text{and} \quad \theta_4 = x-y+t,$

and insist that R_3, L_3, U_3, and D_3 satisfy

(2.8) $\qquad R_3(\theta_1, \theta_2) + L_3(\theta_3, \theta_4) = U_3(\theta_1, \theta_4) + D_3(\theta_3, \theta_2)$

for all $\theta_1, \theta_2, \theta_3$, and θ_4, then the functions R_3, L_3, U_3, and D_3 will also satisfy (2.6). The last relation implies that if R_3, L_3, U_3, and D_3 are C^2, then

(2.9)
$$
\frac{\partial^2 R_3}{\partial\theta_1 \partial\theta_2} = \frac{\partial^2 L_3}{\partial\theta_3 \partial\theta_4} = \frac{\partial^2 U_3}{\partial\theta_1 \partial\theta_4} = \frac{\partial^2 D_3}{\partial\theta_2 \partial\theta_3} = 0,
$$

or equivalently that

$$
(2.10) \qquad
\left.
\begin{array}{l}
R_3 = f_1(\theta_1) + f_2(\theta_2) \\
L_3 = f_3(\theta_3) + f_4(\theta_4) \\
U_3 = f_5(\theta_1) + f_6(\theta_4) \\
D_3 = f_7(\theta_3) + f_8(\theta_2)
\end{array}
\right\} .
$$

In order that the functions $f_1 - f_8$ satisfy (2.8), we must also have

$$
(2.11) \qquad f_5 = f_1, \quad f_6 = f_4 \ , f_7 = f_3 \ , \text{and } f_8 = f_2.
$$

If we now let

$$
(2.12) \qquad F_1 = \exp(f_1), \quad F_2 = \exp(f_2), \quad F_3 = \exp(f_3), \text{ and } F_4 = \exp(f_4) \ ,
$$

then the collisionless solution, (2.3), reduces to

$$
(2.13) \qquad
\left.
\begin{array}{l}
r(x,y,t) = F_1(x+y-t)F_2(x-y-t) \\
l(x,y,t) = F_3(x+y+t)F_4(x-y+t) \\
u(x,y,t) = F_1(x+y-t)F_4(x-y+t) \\
d(x,y,t) = F_3(x+y+t)F_2(x-y-t)
\end{array}
\right\}
$$

and the density, $\rho = r + l + u + d$, is given by

$$
(2.14) \qquad \rho = (F_1(x+y-t) + F_3(x+y+t))(F_2(x-y-t) + F_4(x-y+t))
$$

where $F_1 - F_4$ are arbitrary positive functions.

The solutions given by (2.13) satisfy no obvious boundary conditions. Of somewhat more interest are collisionless solutions which satisfy

$$
(2.15) \qquad r(0,y,t) = l(0,y,t) \quad \text{and} \quad r(1,y,t) = l(1,y,t) \ , 0 < y < 1,
$$

and

$$
(2.16) \qquad u(x,0,t) = d(x,0,t) \quad \text{and} \quad u(x,1,t) = d(x,1,t) \ , 0 < x < 1.
$$

These conditions obtain if the flow takes place in the unit square $0 < x < 1$ and $0 < y < 1$ and specular reflection occurs at the walls $x = 0$ and 1 and $y = 0$ and 1.

The boundary conditions (2.15) imply that $F_1 - F_4$ must satisfy the additional constraints

$$
\left.
\begin{array}{c}
F_1(y-t)F_2(-y-t) = F_3(y+t)F_4(-y+t) , \\
\text{and} \\
F_1(1+y-t)F_2(1-y-t) = F_3(1+y+t)F_4(1-y+t)
\end{array}
\right\},
$$
(2.17)

while (2.16) implies that

$$
\left.
\begin{array}{c}
F_1(x-t)F_4(x+t) = F_3(x+t)F_4(x-t) , \\
\text{and} \\
F_1(1+x-t)F_4(x-1+t) = F_3(1+x+t)F_4(-1+x-t)
\end{array}
\right\}.
$$
(2.18)

These last conditions are met if $s \to F_1(s)$ is a positive, 2-periodic function and if

$$
\left.
\begin{array}{c}
F_2(s) = F_1(s) \\
F_3(s) = F_1(-s) \\
F_4(s) = F_1(-s)
\end{array}
\right\}.
$$
(2.19)

With this choice

$$
\left.
\begin{array}{l}
r(x,y,t) = F_1(x+y-t)F_1(x-y-t) \\
l(x,y,t) = F_1(-x-y-t)F_1(-x+y-t) \\
u(x,y,t) = F_1(x+y-t)F_1(-x+y-t) \\
d(x,y,t) = F_1(-x-y-t)F_1(x-y-t)
\end{array}
\right\},
$$
(2.20)

and

$$
r(0,0,t) = l(0,0,t) = u(0,0,t) = d(0,0,t) = F_1^2(-t).
$$
(2.21)

3. Numerical Results. In this section we look at numerical simulations for the system (1.7) and (1.8). These are solved subject the specular reflection conditions:

$$
r(0,y,t) = l(0,y,t) \text{ and } r(1,y,t) = l(1,y,t) , 0 < y < 1
$$
(3.1)

and

$$
u(x,0,t) = d(x,0,t) \text{ and } u(x,1,t) = d(x,1,t) , 0 < x < 1.
$$
(3.2)

We use a fractional step procedure to integrate (1.7) and (1.8). We assume the approximate solution $(r^n, l^n, u^n, d^n)(x, y), (x, y) \in (0, 1) \times (0, 1)$, is known at time $t = n\delta$. Over the first half step we approximately solve

(3.3)
$$r_t = l_t = (ud - rl)/\rho \quad \text{and} \quad u_t = d_t = (rl - ud)/\rho.$$

The results of this step are

(3.4)
$$r^{n+\frac{1}{2}}(x, y) = \frac{(2 - \delta)r^n(x, y) + 2\delta \frac{(r^n(x,y)+u^n(x,y))(r^n(x,y)+d^n(x,y))}{\rho^n(x,y)}}{(2 + \delta)}$$

$$l^{n+\frac{1}{2}}(x, y) = \frac{(2 - \delta)l^n(x, y) + 2\delta \frac{(l^n(x,y)+u^n(x,y))(l^n(x,y)+d^n(x,y))}{\rho^n(x,y)}}{(2 + \delta)}$$

$$u^{n+\frac{1}{2}}(x, y) = \frac{(2 - \delta)u^n(x, y) + 2\delta \frac{(u^n(x,y)+l^n(x,y))(u^n(x,y)+r^n(x,y))}{\rho^n(x,y)}}{(2 + \delta)}$$

$$d^{n+\frac{1}{2}}(x, y) = \frac{(2 - \delta)d^n(x, y) + 2\delta \frac{(d^n(x,y)+l^n(x,y))(d^n(x,y)+r^n(x,y))}{\rho^n(x,y)}}{(2 + \delta)}$$

where

(3.5)
$$\rho^n(x, y) = r^n(x, y) + l^n(x, y) + u^n(x, y) + d^n(x, y).$$

Over the second half step we let r, l, u, and d passively advect. and the results of this step are:

(3.6)
$$r^{n+1}(x, y) = \begin{cases} l^{n+1/2}(\delta - x, y) & , 0 < x < \delta \text{ and } 0 < y < 1 \\ r^{n+1/2}(x - \delta, y) & , \delta < x < 1 \text{ and } 0 < y < 1 \end{cases},$$

(3.7)
$$l^{n+1}(x, y) = \begin{cases} l^{n+1/2}(x + \delta, y) & , 0 < x < 1 - \delta \text{ and } 0 < y < 1 \\ r^{n+1/2}(2 - x - \delta, y) & , 1 - \delta < x < 1 \text{ and } 0 < y < 1 \end{cases},$$

(3.8)
$$u^{n+1}(x, y) = \begin{cases} d^{n+1/2}(x, \delta - y) & , 0 < x < 1 \text{ and } 0 < y < \delta \\ u^{n+1/2}(x, y - \delta) & , 0 < x < 1 \text{ and } \delta < y < 1 \end{cases},$$

and

(3.9)
$$d^{n+1}(x, y) = \begin{cases} d^{n+1/2}(x, y + \delta) & , 0 < x < 1 \text{ and } 0 < y < 1 - \delta \\ u^{n+1/2}(x, 2 - y - \delta) & , 0 < x < 1 \text{ and } 1 - \delta < y < 1 \end{cases}.$$

All of the simulations were run with the discontinuous initial data described below. We let

$$(3.10) \qquad X(y, \lambda) = \frac{1}{2} + \lambda y(y - 1/2)(y - 1) \ , 0 \le y \le 1 \text{ and } 0 \le \lambda,$$

and

$$(3.11) \quad (r_0, l_0, u_0, d_0)(x, y, t) = \begin{cases} (1, 1, u_1, 1/u) \ , 0 < x < X(y, \lambda) \text{ and } 0 < y < 1 \\ (1, 1, u_2, 1/u_2) \ , X(y, \lambda) < x < 1 \text{ and } 0 < y < 1 \end{cases}.$$

The data is incompatible with the boundary conditions (3.2) and thus introduces strong discontinuities which propagate in the y direction.

In simulation 1 we take $u_1 = 5$, $u_2 = 2$, and $\lambda = 3$ while in simulation 2 we take $u_1 = 5$, $u_2 = 0.2$, and $\lambda = 3$. For simulation 2 the density is initially constant and equal to 7.2. For each simulation we show the initial stages of the motion at times t from 0 to 1.75 in increments of .25 and the latter stages at times t from 46 to 49.75 in increments of .25. We also show for each simulation four summary diagnostics graphs which demonstrate that the solutions being computed converge to collisionless solutions of the type described in Section 2 (see (2.20)).

Each summary graph has the following layout. In the upper left hand corner is a graph of $r(0, 0, t)$ versus time over the interval $46 \le t \le 50$; in the upper right hand corner is a graph of $\rho(0, 0, t)/4 = (r(0, 0, t) + l(0, 0, t) + u(0, 0, t) + d(0, 0, t))/4$ versus time over the interval $46 \le t \le 50$; in the bottom left hand corner we show

$$\text{error}(t) \stackrel{\text{def}}{=} \max[|r(0, 0, t) - l(0, 0, t)|, |r(0, 0, t)|, -u(0, 0, t)|, |r(0, 0, t) - d(0, 0, t)|]$$

versus time over $46 \le t \le 50$, and finally in the bottom right hand corner we show $\text{maxcollision}(t) = \max_{(x, y)} |(ud - rl)(x, y, t)|$ versus time over $46 \le t \le 50$. It is the structure of the last graph which demonstrates that our solutions have converged to the collisionless waves described in (2.20).

REFERENCES

[1]   T. PLATKOWSKI AND R. ILLNER, *Discrete Velocity Models of the Boltzmann Equation: A Survey of the Mathematical Aspects of the Theory*, SIAM Review, 30, (1988), pp. 213–255.

[2]   J. M. GREENBERG AND L. L. AIST, *Decay Theorems for the Four Velocity Broadwell Equations*, submitted to Arch. Rat. Mech. and Anal.

148

SIMULATION I

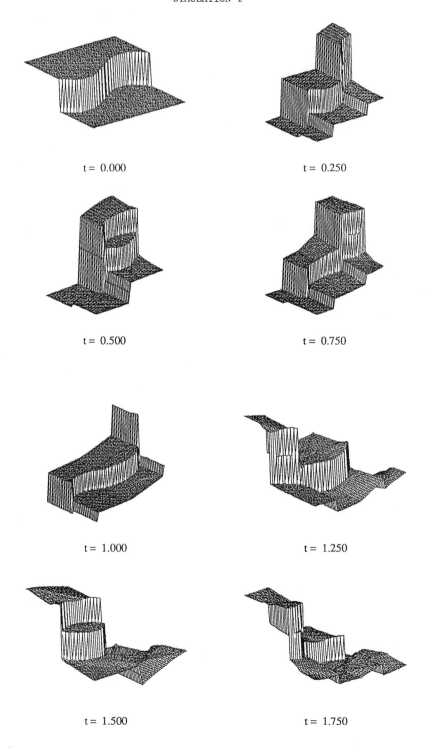

t = 0.000   t = 0.250

t = 0.500   t = 0.750

t = 1.000   t = 1.250

t = 1.500   t = 1.750

SIMULATION I

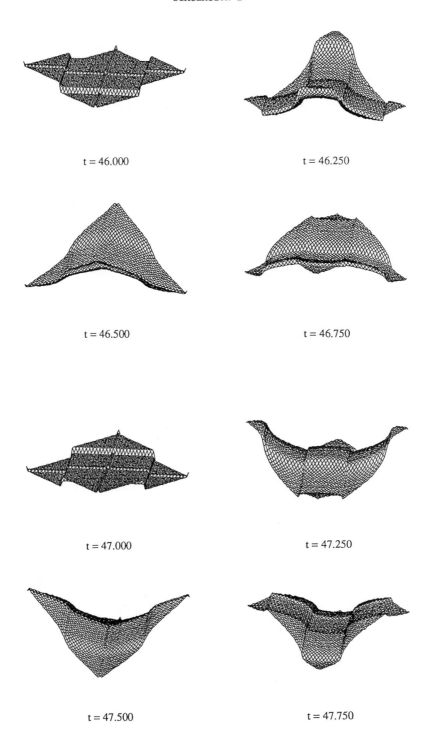

t = 46.000

t = 46.250

t = 46.500

t = 46.750

t = 47.000

t = 47.250

t = 47.500

t = 47.750

SIMULATION I

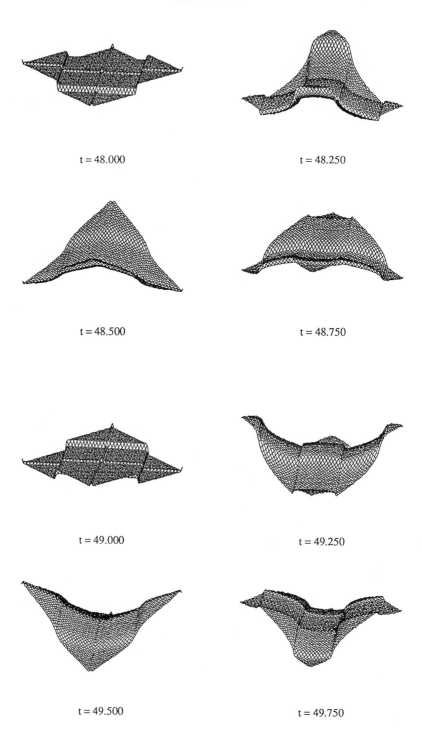

t = 48.000

t = 48.250

t = 48.500

t = 48.750

t = 49.000

t = 49.250

t = 49.500

t = 49.750

SIMULATION I

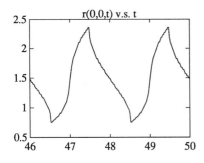

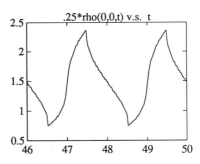

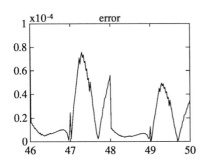

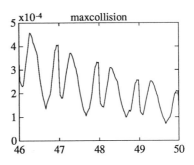

SIMULATION II

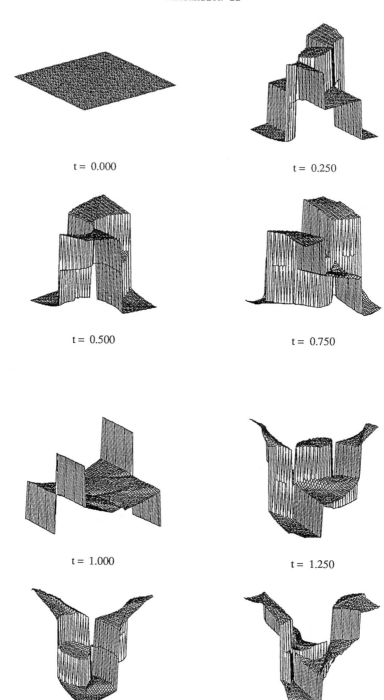

t = 0.000

t = 0.250

t = 0.500

t = 0.750

t = 1.000

t = 1.250

t = 1.500

t = 1.750

SIMULATION II

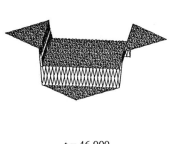

t = 46.000

t = 46.250

t = 46.500

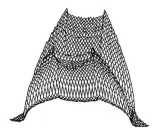

t = 46.750

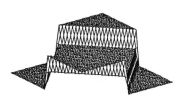

t = 47.000

t = 47.250

t = 47.500

t = 47.750

SIMULATION II

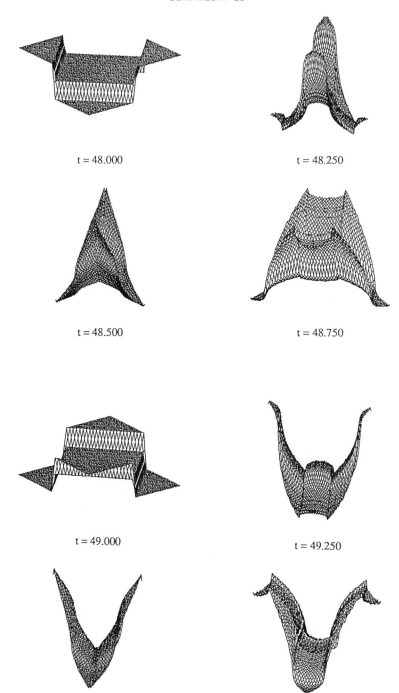

t = 48.000

t = 48.250

t = 48.500

t = 48.750

t = 49.000

t = 49.250

t = 49.500

t = 49.750

SIMULATION II

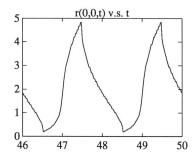

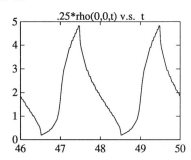

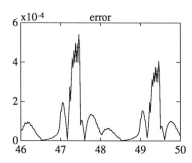

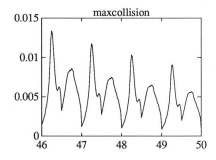

ANOMALOUS REFLECTION OF A SHOCK WAVE AT A FLUID INTERFACE*

JOHN W. GROVE† AND RALPH MENIKOFF‡

Abstract. Several wave patterns can be produced by the interaction of a shock wave with a fluid interface. We focus on the case when the shock passes from a medium of high to low acoustic impedance. Curvature of either the shock front or contact causes the flow to bifurcate from a locally self-similar quasi-stationary shock diffraction, to an unsteady anomalous reflection. This process is analogous to the transition from a regular to a Mach reflection when the reflected wave is a rarefaction instead of a shock. These bifurcations have been incorporated into a front tracking code that provides an accurate description of wave interactions. Numerical results for two illustrative cases are described; a planar shock passing over a bubble, and an expanding shock impacting a planar contact.

Key words. anomalous reflection, front tracking

AMS(MOS) subject classifications. 76-06 76L05

1. Introduction.

The collision of a shock wave with a fluid interface produces a variety of complicated wave diffractions [1,2,12]. In the simplest case these consist of pseudo-stationary self-similar waves that can be described by solutions to Riemann Problems for the supersonic steady-state Euler equations. In more complicated cases and in particular when one or both of the colliding waves is curved, these regular diffraction patterns can bifurcate into complex composites of individual wave interactions between the scattered waves.

The purpose of this analysis is to understand the particular bifurcation behavior of the collision of a shock in a dense fluid with an interface between the dense fluid and a much lighter fluid. Two basic cases will be considered. The collision of a shock in water with a bubble of air, and the diffraction of a cylindrically expanding underwater shock wave with the water's surface. It will be seen that initially these interactions produce regular shock diffractions with reflected Prandtl–Meyer waves. Subsequently these regular waves bifurcated to form anomalous waves that are analogous to non-centered Mach reflections whose reflected waves are rarefactions. We will describe a method to include this analysis into a front tracking numerical method that allows enhanced resolution computations of these interactions.

*This article is a condensed version of reference [9] which will appear elsewhere.

†Department of Applied Mathematics and Statistics, State University of New York at Stony Brook, Stony Brook, NY 11794. Supported in part by the U. S. Army Research Office, grant no. DAAL03-89-K-0017.

‡Theoretical Division, Los Alamos National Laboratory, Los Alamos, NM 87545. Supported by the U. S. Department of Energy.

2. The Equations of Motion.

In the absence of heat conduction and viscosity, the fluid flow is governed by the Euler equations that describe the laws of conservation of mass, momentum and energy respectively.

(2.1a)
$$\partial_t \rho + \nabla \cdot (\rho \mathbf{q}) = 0,$$

(2.1b)
$$\partial_t (\rho \mathbf{q}) + \nabla \cdot (\rho \mathbf{q} \otimes \mathbf{q}) + \nabla P = \rho \mathbf{g},$$

(2.1c)
$$\partial_t (\rho \mathcal{E}) + \nabla \cdot \rho \mathbf{q} (\mathcal{E} + VP) = \rho \mathbf{q} \cdot \mathbf{g}.$$

Here, ρ is the mass density, $\mathbf{q}$ is the particle velocity, $\mathbf{g}$ is the gravitational acceleration, $\mathcal{E} = \frac{1}{2}|\mathbf{q}|^2 + E$ is the total specific energy, E is the specific internal energy, and P is the pressure. Gravity will be neglected since the interactions considered here occur on short time-scales. The equilibrium thermodynamic pressure $P(V, E)$, where $V = 1/\rho$ is the specific volume, is referred to as the equation of state and describes the fluid properties.

It is well known that system (2.1) is hyperbolic, and the characteristic modes correspond to the propagation of sound waves and fluid particles through the medium. The sound waves propagate in all directions from their source with a velocity c with respect to the fluid, where the sound speed c satisfies $c^2 = \partial P / \partial \rho$ at constant entropy. Another important measure of sound propagation is the Lagrangian sound speed or acoustic impedance given by ρc.

3. Elementary Wave Nodes and the Supersonic Steady State Riemann Problem.

An elementary wave node is a point of interaction between two waves that is both stationary and self-similar [7]. It can be shown [6, 13 pp. 405–409] that there are four basic elementary nodes. These are the crossing of two shocks moving in opposite directions (cross node), the overtaking of one shock by another moving in the same direction (overtake node), the splitting of a shock wave due to interaction with other waves or boundaries to produce a Mach reflection (Mach node), and the collision of a shock with a fluid interface (diffraction node). All of these waves are characterized by the solution of a Riemann problem for a steady state flow, where the data is provided by the states behind the interacting waves. We will primarily be concerned with the last of these interactions, but bifurcations in this node will lead to the production of all of the other elementary nodes.

For a stationary planar flow, system (2.1) reduces to a 4×4 system that is hyperbolic in the restricted variables provided the Mach number $M = |\mathbf{q}|/c$ is greater than one, i.e., the flow is supersonic. The streamlines or particle trajectories define the time-like direction. The hyperbolic modes in this case are associated with two families of sound waves, and a doubly linearly degenerate characteristic family. If θ and q are the polar coordinates of the particle velocity $\mathbf{q}$, then the sonic waves have characteristic directions with polar angles $\theta \pm A$, where A is the Mach angle, $\sin A = M^{-1}$. Waves of these families are either stationary shock waves or steady state centered rarefaction waves also called Prandtl–Meyer waves. Waves of the degenerate family are a combination of a contact discontinuity and a vortex sheet across which the pressure and flow direction θ are continuous while the other variables may experience jumps.

Following the general analysis of systems of hyperbolic conservation laws [14], we see that the wave curve for a sonic wave family consists of two branches corresponding to either a shock or a simple wave. The shock branch is commonly called a shock polar [4, pp. 294–317] and actually forms a closed and bounded loop where the two sonic families meet at the point where the stationary shock is normal to the incoming flow. If we let the state ahead of the wave be denoted by the subscript 0, a straightforward derivation of the Rankine-Hugoniot equations for the system (2.1) shows that the thermodynamics of the states on either side of the shock are related by the Hugoniot equation

$$(3.1) \qquad E = E_0 + \frac{P + P_0}{2}(V_0 - V).$$

A similar derivation applied to the steady state Euler equations shows that the flow velocities on either side of a stationary oblique shock satisfy

$$(3.2) \qquad \tfrac{1}{2}q^2 + H = \tfrac{1}{2}q_0^2 + H_0 \ ,$$

where $H = E + PV$ is the specific enthalpy. The jump in the flow direction is given by

$$(3.3) \qquad \tan(\theta - \theta_0) = \pm \left[\frac{P - P_0}{\rho_0 q_0^2 - (P - P_0)} \right] \cot \beta \ .$$

Here β is the angle between the incoming streamline and the shock wave, and is given by $\sin \beta = \sigma/q_0$, where $\sigma = V_0 m$ is the wave speed of the shock wave with respect to the fluid ahead and m is the mass flux across the shock, $m^2 = -\Delta P/\Delta V$. The difference between the flow direction on either side of the shock is called the turning angle of the wave.

The same analysis when applied to the simple wave curves shows that the entropy is constant inside a Prandtl–Meyer wave. The flow speed and flow direction are related by (3.2) where $H = H(P, S_0)$ and

$$(3.4) \qquad \theta = \theta_0 \mp \int_{P_0}^{P} dP \left. \frac{\cos A}{\rho c q} \right|_S \ .$$

In analogy to the shock polar defined by (3.1)–(3.3) we will call this locus of states a rarefaction polar.

It is easily checked that the two branchs of (3.4) are respectively associated with the $\theta \pm A$ characteristic directions in the sense of Lax. Similarly it can be shown [8] that for most equations of state, the two branches of (3.3) are also associated with the $\theta \pm A$ characteristics in the sense of Lax provided the state downstream from the shock is supersonic. Since θ and P are constant across waves of the degenerate middle family, the Riemann problem for a stationary two-dimensional flow can be solved by finding the intersection of the projections of the wave curves in the $\theta - P$ phase plane.

The are two major differences between the solution to the Riemann problem for a stationary flow and that of a one-dimensional unsteady flow. The Mach number behind the shock wave is given by

$$(3.5) \qquad M = \frac{m}{\rho c}\left(1 + \frac{\rho^2}{\rho_0^2}\cot^2 \beta\right)^{1/2}.$$

For most equations of state [15] $m < \rho c$ and is a monotone function of the pressure along the shock Hugoniot. Thus if β is sufficiently close to $\frac{\pi}{2}$ the flow behind the shock will be subsonic and the steady Euler equations ceases to be hyperbolic. The second reason is that for an normal angle of incidence, the turning angle through the shock is zero. This means that the two branches of the shock polar meet at this point forming a closed and bounded loop. These two issues together imply a loss of existence and uniqueness for the solution to the two dimensional stationary Riemann problem. The resolution is that a bifurcation occurs from a stationary solution to a time dependent solution of the full two dimensional Euler equations.

The actual shape and properties of the shock and rarefaction polars depends on the equation of state. We will make no use of a specific choice of equation of state in our analysis, but we will need to assume that the equation of state satisfies appropriate conditions to guarantee that the shock polar has a unique point at which the state behind the shock becomes sonic, and a unique local extremum in the turning angle. These conditions are satisfied by most ordinary equations of state, and in particular by the polytropic and stiffened polytropic equations of state used in the numerical examples.

4. Anomalous Reflection.

As was mentioned in the introduction, the simpliest case of shock diffraction is that in which the flow near a point of diffraction is scale invariant and pseudo-stationary. This will be the case provided the flow is sufficiently supersonic when measured in a frame that moves with the point [8]. Then the data behind the incoming waves define Riemann data for the downstream scattering of the interacting waves. A representative shock polar diagram for a regular shock diffraction producing a reflected Prandtl–Meyer wave is shown in Fig. 1.

Diffractions of these types have been studied experimentally by several investigators [1,2,11,12], as well as numerically [3,8]. Longer time simulations of the resulting surface instabilities in the fluid interface (called the Richtmyer–Meshkov instability [16,19]) are found in [8,17,20]. One of the interferogrames, Fig. 14 of [12] shows an irregular wave pattern that corresponds to what we call an anomalous reflection. In this wave the angle between the incident shock and the material interface is such that the state behind the shock has become subsonic.

We consider the perturbation of a regular shock diffraction that produces a reflected Prandtl–Meyer wave. Suppose that initially the state behind the incident shock is close to but slightly below the sonic point on the incident shock polar.

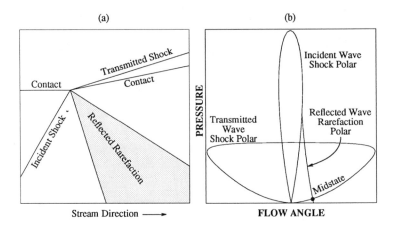

FIG. 1. A sketch of the wave pattern and polar diagrams for a regular shock-contact diffraction that produces a reflected rarefaction wave.

We allow the incident angle to increase while keeping the other variables constant so that the state behind the incident shock passes above the sonic point. Such a situation might occur as a shock diffracts through a bubble as illustrated in Fig. 2. When this happens, the solution can no longer be self-similar since a Prandtl–Meyer wave can only occur in a supersonic flow. Instead the reflected wave begins to overtake and interact with the incident shock, Fig. 2c. This interaction dampens and curves the incident shock near its base on the fluid interface allowing the flow immediately behind the node to return to a supersonic condition. The single point of interaction bifurcates into a degenerate overtake node where the leading edge of the reflected rarefaction overtakes the incident shock, and a sonic diffraction node at the fluid interface. This interaction is a two-dimensional version of the one-dimensional overtaking of a shock by a rarefaction. The composite configuration is in many ways analogous to a regular Mach reflection. In this case the reflected wave is a Prandlt-Meyer wave and instead of a single point of Mach reflection the interaction is spread over the region where the rarefaction interacts with the incident shock. The "Mach" stem an be regarded as the entire region from the point where the incident shock is overtaken by the rarefaction to its base on the fluid interface.

If we allow the incident angle to increase further we will eventually see a second bifurcation in the solution, Fig. 2d. As the material interface continues to diverge from the incident shock, the Mach number near the trailing edge of the reflected rarefaction continues to decrease. The characteristics behind the incident shock are almost parallel to the shock interface near the base of the anomalous reflection. The flow there becomes nearly one-dimension and the rarefaction wave eventually overtakes the incident shock. If there is a great difference in the acoustic impedance between the two materials as in the numerical cases studied here, this second bifurcation will occur as the strength of the incident shock at the fluid interface reduces to zero. The now non-centered rarefaction breaks loose from the fluid interface and begins to propagate away. This second configuration is also analogous to a Mach

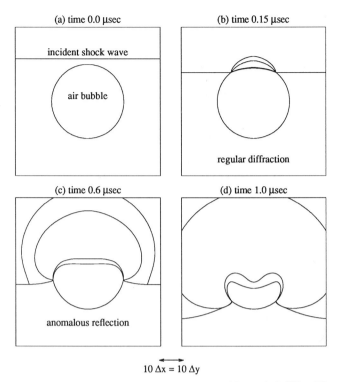

(a) time 0.0 µsec

incident shock wave

air bubble

(b) time 0.15 µsec

regular diffraction

(c) time 0.6 µsec

anomalous reflection

(d) time 1.0 µsec

$10 \, \Delta x = 10 \, \Delta y$

FIG. 2. The collision of a shock wave in water with an air bubble. The fluids ahead of the shock are at normal conditions of 1 atm. pressure, with the density of water 1 g/cc and air 0.0012 g/cc. The pressure behind the incident shock is 10 Kbar with a shocked water density of 1.195 g/cc. The grid is 60×60.

reflection. Here the Mach node corresponds to the interaction region between the rarefaction and incident shock, while the Mach stem is the degenerate wave portion from the trailing edge of the rarefaction to the fluid interface.

5. The Tracking of the Anomalous Reflection Wave.

The qualitative discussion of the anomalous reflection in the previous section can be incorporated into a front tracking code to give an enhanced resolution of the interaction.

The tracking of a regular shock diffraction was described in [8]. The first step in the propagation is the computation of the velocity of the diffraction node with respect to the computational (lab) reference frame. Suppose at time t the node is located at point p_{00}. The node position at time $t + dt$ is found by computing the intersection between the two propagated segments of the incident waves. If this new node position is p_0, then the node velocity is given by $(p_0 - p_{00})/dt$. This velocity defines the Galilean transformation into a frame where the node is at rest. When the state behind the incident shock is supersonic in this frame, it together with the state on the opposite side of the fluid interface provide data for a supersonic steady state Riemann problem whose solution determines the outgoing waves. The outgoing tracked waves are then modified to incorporate this solution.

A bifurcation will occur if the calculated node velocity is such that the state behind the incident shock is subsonic in the frame of the node. If the reflected wave is a Prandtl–Meyer wave this will result in an anomalous reflection. The front tracking implementation of this bifurcation is a straightforward application of the analysis described in the previous section.

First the leading edge of the reflected rarefaction is allowed to break loose from the diffraction node. The intersection p_1 between the propagated rarefaction leading edge and the incident shock is computed and a new overtake node is installed at p_1 by disconnecting the rarefaction leading edge from the diffraction node and connecting it to p_1.

If this reflected rarefaction edge is untracked, then p_1 is found by calculating the characteristic through the old node position corresponding to the state behind the incident shock and computing the intersection of its propagated position with the propagated incident shock. This characteristic makes the Mach angle A with the streamline through the node. Since the bifurcation occurs between times t and $t + dt$, $M \geq 1$ at time t and A is real. This wave moves with sound speed in its normal direction. In this case no new overtake node is tracked.

We are now ready to compute the states and position of the point of shock diffraction after the bifurcation. As was mention previously, the rarefaction expands onto the incident shock causing it weaken. This in turn slows down the node causing the incident shock to curve into the fluid interface. The diffraction node will slow down to the point where the state immediately behind the node becomes sonic. After this the configuration near the node can be computed using the regular case analysis.

The adjusted propagated node position is computed as follows, see Fig. 3. For each number s sufficiently small, let $p(s)$ be the point on the propagated material interface that is located a distance s from p_0 when measured along the curve, the positive direction being oriented away from the node into the region ahead of the incident shock. Let $\beta(s)$ be the angle between the tangent vector to the material interface at $p(s)$ and the directed line segment between the points $p(s)$ and p_1. Let $\mathbf{v}(s)$ be the node velocity found by moving the diffraction node to position $p(s)$, and let $\mathbf{q}(s)$ be the velocity of the flow ahead of the incident shock in the frame that moves with velocity $\mathbf{v}(s)$ with respect to the lab frame. The mass flux across this shock is given by

$$(5.1) \qquad m(s) = \rho_0 |\mathbf{q}(s)| \sin \beta(s) .$$

Given $m(s)$ and the state ahead of the incident shock, the state behind the shock and hence its Mach number $M(s)$ can be found. The new node position is given by $p(s^*)$, where s^* is the root of the equation $M(s^*) = 1$. Finally, the state behind the incident shock with mass flux $m(s^*)$ together with the state on the opposite side of the contact are used as data for a steady state Riemann problem whose solution supplies the states and angles of the transmitted shock, the trailing edge of the reflected rarefaction, and the downstream material interface.

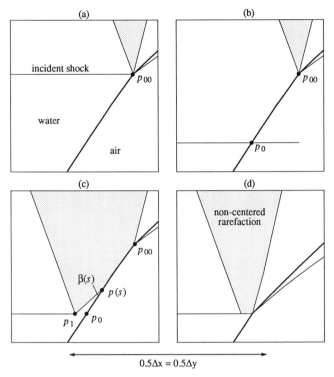

FIG. 3. A diffraction node initially at p_{00} bifurcates into an anomalous reflection. The predicted new node position at p_0 yields a Mach number of 0.984 behind the incident shock. The leading edge of the reflected Prandtl-Meyer wave breaks away from the diffraction node to form an overtake node at p_1. The propagated position of the diffraction node is adjusted to return the flow to sonic behind the node.

The subsequent propagation of the anomalous reflection node is performed in the same way. The bifurcation repeats itself as more of the reflected rarefaction propagates up the incident shock. The leading edge of the reflected rarefaction wave that connects to the diffraction node is not tracked after the first bifurcation.

The secondary bifurcations that occur when the trailing edge of the rarefaction overtakes the incident shock are detected in a couple of ways. If the incident shock is sufficiently weak, i.e., the normal shock Mach number is close to 1, then it is possible for the numerically calculated upstream Mach number to be less than one. This is a purely numerical effect since physically the upstream state is always supersonic. However in nearly sonic cases such numerical undershoot can occur. If such a situation is detected the trailing edge of the reflected rarefaction wave is disengaged from the anomalous reflection node and installed at a new overtake node on the incident shock. The residual shock strength for the portion of the incident shock behind the rarefaction wave is small and the diffraction node at the material interface reduces to the degenerate case of a sonic signal diffracting through a material interface.

The second way in which the secondary bifurcation is detected occurs when the trailing edge of the rarefaction overtakes the shock. Here a new intersection between the incident shock and the trailing edge characteristic is produced. As before the tracked characteristic is disengaged from the diffraction node and a new overtake node is installed at the point of intersection. The residual shock strength at the node is non-zero so the diffraction at the material interface produces an additional expansion wave behind the original one. This new expansion wave is not tracked.

It is possible to make a few remarks about the amount of tracking required for these problems. Since the front tracking method is coupled to a finite difference method for the solution away from the tracked interface (the interior solver), there is always an option between tracking a wave or allowing it to be captured. Of course capturing can result in a considerable loss in resolution in the waves as compared to tracking [5], but it will also simplify the resolution of the interactions. The secondary bifurcations described above are only tracked when the trailing edge of the reflected Prandtl–Meyer wave is tracked. The current algorithm is structured so that at a minimum the two interacting incoming waves are tracked. At this extreme none of the outgoing waves are tracked and no explicit bifurcations in the tracked interface occur. More commonly, the material interface separates different fluids and so must be tracked on both sides of the interaction.

Also, instabilities in the finite difference approximation can affect the accuracy of the solution near the node, especially for stiff materials such as water. Tracking the additional waves seems to considerably reduce these problems. Tracking also allows the use of a much coarser grid, which is important when the diffraction occurs in a small but important zone of a larger simulation. It allows the entire region of diffraction to extend only over only a fraction of a grid block. These remarks show that the amount of tracking is problem dependent, and a compromise can be made between the increased accuracy and stability of front tracking, and the simplicity of a capturing algorithm.

6. Numerical Examples.

Fig. 4 shows a series of frames documenting the collision of a 10 Kbar shock wave with a bubble of air in water. Note in this case the trailing edge of the reflected Prandtl–Meyer wave is not tracked. The states ahead of the incident shock are at one atmosphere pressure and standard temperature. Under these conditions, water is about a thousand times as dense as air. During the initial stage of the interaction regular diffraction patterns are produced.

In less than half of a microsecond an anomalous reflection has formed, and by one microsecond the trailing edge of the rarefaction has also overtaken the incident shock. It is interesting to note that this interaction causes the bubble to collapse into itself. Long time simulations are expected to show the initial bubble split, and the resulting bubbles going into oscillation as they are overcompressed and then expand. This process is important in the transfer of energy as a shock passes through a bubbly fluid. The first diffraction considerably dampens the shock, and much of this energy will eventually be returned to the shock wave in the form of compression waves generated by the expanding bubbles.

165

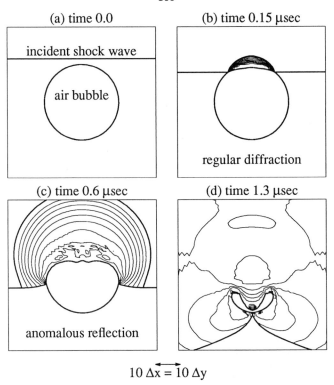

FIG. 4. Log(1 + pressure) contours for the collision of a shock wave in water with an air bubble. The fluids ahead of the shock are at normal conditions of 1 atm. pressure, with the density of water 1 g/cc and air 0.0012 g/cc. The pressure behind the incident shock is 10 Kbar with a shocked water density of 1.195 g/cc. The tracked interface is shown in a dark line. The grid is 60 × 60.

Fig. 5 shows the diffraction of an expanding underwater shock wave through the water's surface. Initially a ten Kbar cylindrically expanding shock wave with a radius of one meter is placed two meters below the water's surface. The interior of the shock wave contains a bubble of hot dense gas. The states exterior to the shock are ambient at one atmosphere pressure and normal temperature. A gravitational acceleration of one g has been added in this case, but due to the rapid time scale on which the diffractions occur the effect of gravity is negligible. Here the entire reflected Prandtl–Meyer wave is captured rather than tracked. The pressure contour plots show that by six milliseconds an anomalous reflection has developed as indicated in the blowup of Fig. 5b shown in Fig. 6. Another interesting feature of this problem is the acceleration of the bubble inside the shock wave by the reflected rarefaction wave. This causes the bubble to rise much faster than it would under just gravity. When the bubble reaches the surface it expands into the atmosphere leading to the formation of a kink in the transmitted shock wave between the region ahead of the surfacing bubble, and the rest of the wave. This kink is an untracked example of the elementary wave called the cross node where two oblique shocks collide.

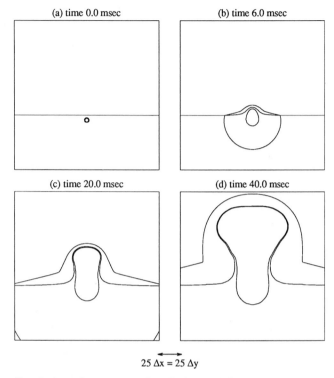

(a) time 0.0 msec   (b) time 6.0 msec

(c) time 20.0 msec   (d) time 40.0 msec

$25\ \Delta x = 25\ \Delta y$

FIG. 5. An underwater expanding shock wave diffracting through the water's surface. An expanding shock wave with an internal pressure of 10 Kbars and initial radius of 1 meter is installed at a depth of 2 meters below the water's surface. The external conditions are ambient at one atmosphere pressure and normal densities for the air and water. The boundary conditions are constant Dirichlet at the initial ambient values. The grid is 150 × 150.

The water in the simulations described above is modeled by what is called a stiffened polytropic equation of state [10,18] where the pressure is given by

$$(6.1) \qquad P(V, E) = \Gamma_0 \rho (E - E_\infty) - (\Gamma_0 + 1) P_\infty \ ,$$

with $\Gamma_0 = 6$, $E_\infty = 0$, and $P_\infty = 3000$ atm. The air is treated as a polytropic gas

$$(6.2) \qquad P(V, E) = (\gamma_0 - 1) \rho E \ ,$$

with $\gamma_0 = 1.4$.

7. Summary.

Compressible fluid flow is characterized by the production and interaction of shock, rarefaction and contact waves. We have studied the diffraction of a shock through a material interface from a medium of high to low acoustic impedance. The bifurcations that occur during the diffraction were analyzed in terms of polar diagrams for steady supersonic flow. This analysis was incorporated into a front

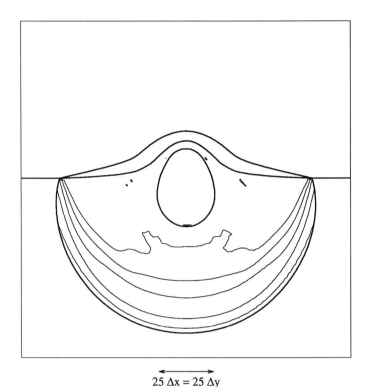

25 Δx = 25 Δy

FIG. 6. A blowup of Fig. 5.1b showing pressure contours scaled from 0.001 – 10 Kbars. The tracked interface is shown superimposed in a dark line over the pressure contours.

tracking code to allow enhanced resolution computations of the interactions. The particular simulations studied were the diffraction of a planar shock in water through an air bubble, and the diffraction of an expanding shock in water through the water's surface. In both cases the anomalous reflection bifurcation plays an important role in correctly computing the flow.

REFERENCES

[1] A. M. ABD-EL-FATTAH AND L. F. HENDERSON, Shock Waves at a Fast-Slow Gas Interface, J. Fluid Mech., 86, pp. 15–32.

[2] A. M. ABD-EL-FATTAH AND L. F. HENDERSON, Shock Waves at a Slow-Fast Gas Interface, J. Fluid Mech., 89, pp. 79–95.

[3] P. COLELLA, L. F. HENDERSON AND E. G. PUCKETT, LLNL Preprint UCRL-100260.

[4] R. COURANT AND K. O. FRIEDRICHS, Supersonic Flow and Shock Waves, Springer Verlag, New York, pp. 294–317.

[5] J. GLIMM, J. W. GROVE, AND X. L. LI, Three Remarks on the Front Tracking Method, Proceedings of the conference in Taormina Sicily.

[6] J. GLIMM, C. KLINGENBERG, O. MCBRYAN, B. PLOHR, D. H. SHARP, AND S. YANIV, Front Tracking and Two Dimensional Riemann Problems, Adv. in Appl. Math., 6, pp. 259–290.

[7] J. GLIMM AND D. H. SHARP, An S-matrix Theory for Classical Nonlinear Physics, Foundations of Physics, 16, pp. 125–141.

[8] J. W. GROVE, The Interaction of Shock Waves with Fluid Interfaces, Adv. Appl. Math., 10, pp. 201–227.

[9] J. W. GROVE AND R. MENIKOFF, The Anomalous Reflection of a Shock Wave through a Material Interface, LNL Preprint LA-UR-89-778.

[10] F. H. HARLOW AND A. A. AMSDEN, Fluid Dynamics, Available from National Technical Information Service U.S. Dept of Commerce, Los Alamos National Laboratory, Los Alamos.

[11] L. F. HENDERSON, On the Refraction of Longitudinal Waves in Compressible Media, LLNL Preprint UCRL-53853.

[12] R. G. JAHN, The Refraction of Shock Waves at a Gaseous Interface, J. Fluid Mech., 1, pp. 457–489.

[13] L. LANDAU AND E. LIFSHITZ, Fluid Mechanics, Addison-Wesley, Reading, Mass..

[14] P. LAX, Hyperbolic Systems of Conservation Laws II, Comm. Pure Appl. Math., 10, pp. 537–556.

[15] R. MENIKOFF AND B. PLOHR, Riemann Problem for Fluid Flow of Real Materials, Rev. Mod. Phys., 61, pp. 75–130.

[16] E. E. MESHKOV, Izv. Akad. Nauk SSSR, Mekh. Zhidk. Gaz., 5, p. 151.

[17] KARNIG O. MIKAELIAN, Simulation of the Richtmyer–Meshkov Instability and Turbulent Mixing in Shock–Tube Experiments, LLNL Preprint UCID-21328.

[18] B. PLOHR, Shockless Acceleration of Thin Plates Modeled by a Tracked Random Choice Method, AIAA J., 26, pp. 470–478.

[19] R. D. RICHTMYER, Taylor Instability in Shock Acceleration of Compressible Fluids, Comm. Pure and Appl. Math., 13, pp. 297–319.

[20] DAVID L. YOUNGS, Numerical Simulation of Turbulent Mixing by Rayleigh-Taylor Instability, Physica, 12D, pp. 32–34.

AN APPLICATION OF CONNECTION MATRIX TO MAGNETOHYDRODYNAMIC SHOCK PROFILES

HARUMI HATTORI† AND KONSTANTIN MISCHAIKOW‡

1. Introduction. In this note we shall summarize how to use the connection matrix to show the existence of a viscous profile to the magnetohydrodynamic (MHD) equations. Although Freistuhler [1] has obtained various intermediate shock profiles in a different parameter regime, our technique is different and interesting to know. This is a technique based on algebraic topology and an extension of Conley's index. As the detailed version will be appearing elsewhere [2], in this note we concentrate on how to use the connection matrix.

2. System. Following the notations of Conley and Smoller [3,4], we write the system as

$$
\begin{aligned}
(a) \quad & B_0 \begin{pmatrix} x_1 \\ x_2 \end{pmatrix} = A \begin{pmatrix} x_1 \\ x_2 \end{pmatrix} + \begin{pmatrix} 0 \\ \varepsilon \end{pmatrix} \\
(b) \quad & B_0 \begin{pmatrix} y_1 \\ y_2 \end{pmatrix} = A \begin{pmatrix} y_1 \\ y_2 \end{pmatrix} \\
(c) \quad & \mu_1 \dot{V} = \frac{1}{2}(x_2^2 + y_2^2) + V - J + p(V, T), \\
(d) \quad & \nu \dot{T} = -Q - f(V, T) + e(V, T).
\end{aligned}
$$
(2.1)

$$
B_0 = \begin{pmatrix} \mu & o \\ o & \nu \end{pmatrix}, \qquad A = \begin{pmatrix} 1 & -\delta \\ -\delta & v \end{pmatrix},
$$

$$
Q = \frac{1}{2}(x, Ax) + \frac{1}{2}(y, Ay) + \varepsilon x_2 + \frac{1}{2}v^2 - JV + E - f(V, T),
$$

where $u = (x_1, x_2, y_1, y_2, V, T)$ are the variables, $\lambda = (\mu, \nu, \mu_1, \nu)$ are the viscosity parameters which are positive, and $\varepsilon, \delta, J,$ and E are constants. The variables V and T correspond to specific volume and temperature, x_1 and y_1 are velocities, and x_2 and y_2 are magnetic field intensities. The functions $p, f,$ and e are pressure, the Helmheltz free energy, and internal energy. System (2.1) is a gradient like system and if we set $p = T^{-1}Q$, it can be written as

$$
(2.2) \qquad\qquad\qquad B\dot{u} = \nabla P(u),
$$

where $B = T^{-1} \operatorname{diag} \{\mu, \nu, \mu, \nu, \mu_1, T^{-1}k\}$. Under the appropriate assumptions [2,3], there are four rest points denoted by u_0, u_1, u_2, u_3 and the values of P satisfy

$$
\begin{aligned}
(a) \quad & P(u_0) > P(u_1) > P(u_2) > P(u_3), \qquad \varepsilon \neq 0, \\
(b) \quad & P(u_0) > P(u_1) = P(u_2) > P(u_3), \qquad \varepsilon = 0.
\end{aligned}
$$
(2.3)

†Department of Mathematics, West Virginia University, supported by the Army Grant DAAL 03-89-G-0088.

‡Department of Mathematics, Michigan State University, supported in part by the NSF.

It is well known that there are the "fast" and "slow" shocks which connect u_0 to u_1 and u_2 to u_3, and they lie on $y = 0$. Noting that $y = 0$ is invariant, in what follows we consider the 4×4 system (2.1 a, c, d) and establish $u_1 \rightarrow u_2$ connection. In other words, we consider the flow on $y = 0$. It should be noted that the result holds for (2.1) because $y = 0$ is invariant. The connections $u_0 \rightarrow u_2, u_0 \rightarrow u_3, u_1 \rightarrow u_2$, and $u_1 \rightarrow u_3$ are called intermediate shocks.

3. Results of Conley - Smoller and Hesaaraki. In this section we summarize the results of Conley-Smoller [3,4] and Hesaaraki [5]. The results of Conley-Smoller are the following.

(C-S1) For each λ, ε, and E, there exists a compact invariant set $S(\lambda, E, \varepsilon)$ containing all bounded solutions. This set is an isolated invariant set and all $S(\lambda, E, \varepsilon)$ are related by continuation.

(C-S2) The homotopy index of $S(\lambda, E, \varepsilon)$ is $\bar{0}$.

(C-S3) No intermediate shocks exist for $\xi(= \mu_1 / \max(\mu, \nu)) >> 0$ and $\varepsilon \neq 0$.

On the other hand, Hesaaraki has shown the following result when $\varepsilon = 0$.

(H-1) The flow is symmetric in the sense that if $x_1 = x_1(t), x_2 = x_2(t), V = V(t)$, and $T = T(t)$ is a solution to (2.1 a, c, d), then $x_1 = -x_1(t), x_2 = -x_2(t)$, and $T = T(t)$ is also a solution.

(H-2) The homotopy indicies of the rest points u_0, u_1, u_2, and u_3 are $\sum^4, \sum^3, \sum^3$, and $\sum^2$, respectively.

4. Connection matrix. We describe the properties of connection matrix necessary for this argument. For details concerning the connection matrix the reader should consult Franzosa [8] and Mischaikow [6].

(4-1) The connection matrix is a linear mapping (denoted by $\triangle$) from the index of a Morse set to the index of a Morse set. In our case it is a mapping from the index of a rest point to the index of a rest point due to the fact the system is a gradient system.

(4-2) The elements of the matrix are zero or one ($\mathbf{Z}_2$ field) and the entry from a rest point with the index $\sum^k$ to a rest point with the index $\sum^{k-1}$ is possibly one (boundary -1 map). If it is one there is a connecting orbit connecting $\sum^k$ to $\sum^{k-1}$.

(4-3) It is strictly upper triangular with respect to the ordering defined by P.

(4-4) $\triangle^2 = 0$.

(4-5) $\text{Ker}\,(\triangle)/Im(\triangle) = H_*(h(S(\lambda, \in, \varepsilon)))$.

Since it may not be easy to figure out what the connection matrix looks like, we construct the connection matrix for the case when $\xi >> 0$ and $\varepsilon \neq 0$ which is denoted as $\triangle_{cs}$. From (4-2) and (4-3), it already looks like

$$
(4.1) \qquad \triangle_{cs} = \begin{array}{c} \\ u_3 \\ u_2 \\ u_1 \\ u_0 \end{array} \begin{array}{cccc} u_3 & u_2 & u_1 & u_0 \\ \begin{pmatrix} 0 & \alpha & \beta & 0 \\ 0 & 0 & 0 & \gamma \\ 0 & 0 & 0 & \delta \\ 0 & 0 & 0 & 0 \end{pmatrix} \end{array}.
$$

where α, β, γ, and δ are unknown elements (0 or 1). Here, for example if $\alpha = 1$, there is a connecting orbit from u_2 to u_3 ($\alpha = 0$ does not mean that there is no connecting orbit). Now using (4-4), we see $\alpha = \delta, \beta = \gamma$. It is interesting to know that the properties of connection matrix determines most of the elements in the connection matrix. In the case when $\xi >> 0$ and $\varepsilon \neq 0$, using (C-S3), we have that $\beta = \gamma = 0$. To determine α (or δ) we use (4-5). The homology is a graded vector space and from the homotopy index we can determine the corresponding homology. For example, if the index of a rest point u_α is $\sum^k$, then the homology of the rest point becomes

$$
(4.2) \qquad H_n(h(u_\alpha)) = \begin{cases} Z_2 & \text{if } n = k \\ 0 & \text{if } n \neq k, \end{cases} \qquad n = 0, 1, 2, \cdots
$$

or equivalently

$$
H_*(h(u_\alpha)) = \quad (\quad 0, \qquad 0, \qquad \cdots, \qquad Z_2, \qquad 0, \qquad 0, \qquad \cdots).
$$
$$
(4.3) \qquad\qquad\qquad\quad 0 - th \quad\quad 1 - st \qquad\qquad\quad k - th \qquad\qquad\; n - th
$$

From (4.3) we say that $H_*(h(u_\alpha))$ has a unique non-trivial 1-dimensional vector space.

In our case the homotopy index of $S(\lambda, E, \varepsilon)$ is $\bar{0}$ and this implies $H_*(h(S(\lambda, E, \varepsilon))) = (0, 0, \cdots)$, and therefore from the right hand side of (4-5) we have zero. Also, as we use a vector space, the left hand side of (4-5) implies Ker $(\triangle)$ − Rank $(\triangle)$. Combining the above observation, we have

$$
(4.4) \qquad\qquad\qquad \text{Ker } (\triangle) - \text{ Rank } (\triangle) = 0.
$$

From this we see α must be one.

5. Existence of $u_1 \to u_2$ connection. In order to find the intermediate shock profile $u_1 \to u_2$, we introduce a flow which changes the parameter. For example,

$$
(5.1) \qquad\qquad\qquad \dot{\varepsilon} = -\xi \varepsilon (\varepsilon - \varepsilon_0)
$$

where $0 < \varepsilon \ll 1$, will do. In this case we should choose ε_0 and ξ so that we have the connection matrix of the previous section at ε_0. Consider the connection matrix for the flow (2.1 a, c, d, and 5.1). Denote the rest points at $\varepsilon = 0$ and $\varepsilon = \varepsilon_0$ by

$\bar{u}_0, \bar{u}_1, \bar{u}_2, \bar{u}_3$, and u_0, u_1, u_2, u_3, respectively. Since we added one unstable manifold at $\varepsilon = 0$ and one stable manifold at $\varepsilon = \varepsilon_0$, the connection matrix looks like

$$(5.2) \qquad \Delta = \begin{array}{c} \\ u_3 \\ u_2 \\ u_1 \\ u_0 \\ \\ \bar{u}_3 \\ \bar{u}_2 \\ \bar{u}_1 \\ \bar{u}_0 \end{array} \begin{array}{cccccccc} u_3 & u_2 & u_1 & u_0 & \bar{u}_3 & \bar{u}_2 & \bar{u}_1 & \bar{u}_0 \\ \left(\begin{array}{cccc|cccc} 0 & 1 & 0 & 0 & 1 & 0 & 0 & 0 \\ 0 & 0 & 0 & 0 & 0 & 1 & \gamma & 0 \\ 0 & 0 & 0 & 1 & 0 & \delta & 1 & 0 \\ 0 & 0 & 0 & 0 & 0 & 0 & 0 & 0 \\ \hline 0 & 0 & 0 & 0 & 0 & \alpha & \beta & 0 \\ 0 & 0 & 0 & 0 & 0 & 0 & 0 & \beta \\ 0 & 0 & 0 & 0 & 0 & 0 & 0 & \alpha \\ 0 & 0 & 0 & 0 & 0 & 0 & 0 & 0 \end{array}\right) \end{array}$$

The 4×4 matrix on the upper left correspond to Δ_{cs} and the 4×4 matrix on the lower right will correspond to the connection matrix for the Hesaaraki's result (denoted by Δ_H) which will be discussed now. The 4×4 matrix on the upper right is called transition matrix (denoted by T). It's diagonal elements are one because there are connecting orbits from $\bar{u}_i$ to u_i for each i due to the flow (5.1). Since we have a totally ordered Morse decomposition which continues for all ε, the transition matrix becomes upper triangular [6]. Therefore, $\delta = 0$. Now computing $\Delta^2 = 0$, we have $\alpha = 1$. Then using $(H-2)$, we see $\beta = 1$. This forces γ to be one. Now we can use the result by Reineck [7] which says that if a nondiagonal element $T(i,j)$, of the transition matrix is one there is a connecting orbit from u_i to u_j for some value of ε between 0 and ε_0.

REFERENCES

[1] FREISTÜHLER, H., *On shocks associated with rotational modes*, in this proceeding.

[2] MISCHAIKOW, K., AND H. HATTORI, *On the existence of intermediate magnetohydrodynamic shock waves*, to appear in Journal of Dynamics and Differential Equations.

[3] CONLEY, C.C., AND J. SMOLLER, *On the structure of magnetohydrodynamic shock waves*, Comm. Pure Appl. Math., 27 (1974), pp. 367-375.

[4] CONLEY, C.C., AND J. SMOLLER, *On the structure of magnetohydrodynamic shock waves II*, J. Math. Pure et Appl., 54 (1975), pp. 429-444.

[5] HESAARAKI, M., *The structure of shock waves in magnetohydrodynamics*, Memoirs of AMS, # 302 (1984).

[6] MISCHAIKOW, K., *Transition systems*, in Proceedings of the Royal Society of Edinburgh, 112A (1989), pp. 155-175.

[7] REINECK, J., *Connecting orbits in one-parameter families of flows*, Ergod. Th. & Dynam. Sys., 8 (1988), pp. 359-374.

[8] FRANZOSA, R., *The connection matrix theory for Morse decompositions*, Trans. AMS vol 311 #2 Feb. (1989), pp. 781-803.

CONVECTION OF DISCONTINUITIES IN SOLUTIONS
OF THE NAVIER-STOKES EQUATIONS
FOR COMPRESSIBLE FLOW*

DAVID HOFF†

We report here on results concerning the existence, uniqueness, and continuous dependence on initial data of discontinuous solutions of the Navier-Stokes equations for one-dimensional compressible fluid flow:

$$(1) \quad \begin{cases} v_t - u_x = 0 \\ u_t + p(v,e)_x = \left(\dfrac{\epsilon u_x}{v}\right)_x \\ \left(\dfrac{u^2}{2} + e\right)_t + (up(v,e))_x = \left(\dfrac{\epsilon u u_x + \lambda T(v,e)_x}{v}\right)_x \end{cases}$$

with Cauchy data

$$(2) \quad \begin{bmatrix} v \\ u \\ e \end{bmatrix}(x,0) = \begin{bmatrix} v_0 \\ u_0 \\ e_0 \end{bmatrix}(x).$$

Here v, u, e, p, and T represent respectively the specific volume, velocity, specific internal energy, pressure, and temperature in a fluid; x is the Lagrangian coordinate, so that $x = $ constant corresponds to a particle path; and ϵ and λ are fixed positive viscosity parameters.

Discontinuous solutions are well-known to be fundamental in the mathematical theory of inviscid flows. Our goal here is therefore to study discontinuous solutions in the presence of viscosity and heat conduction effects. We show that discontinuities in u_0 and $T_0 = T(v_0, e_0)$ become smoothed out in positive time, but that discontinuities in v, e, u_x, and T_x persist for all time and convect along particle trajectories. In addition, for ideal or near-ideal gases, we show that the strengths of these discontinuities decay exponentially in time, more rapidly for smaller viscosity ϵ. These observations are fundamental in our construction and analysis of solutions, theorems 1–3 below.

We begin by giving a heuristic derivation of the essential facts concerning the evolution of jump discontinuities in solutions of (1). Thus suppose that $W = \begin{bmatrix} v \\ u \\ e \end{bmatrix}$

---
*Research supported in part by the NSF under Grant No. DMS–8700071 and by the AFOSR through a sabbatical leave supplement
†Department of Mathematics, Indiana University, Bloomington, IN 47405

is a solution of (1) which is smooth except possibly across a curve of discontinuity $\mathcal{C} = \{x(t), t)\}$. Applying the Rankine-Hugoniot conditions to (1), we find that

$$\dot{x}[v] = [u]$$

$$\dot{x}[u] = \left[p - \frac{\epsilon u_x}{v}\right]$$

$$\dot{x}\left[\frac{u^2}{2} + e\right] = \left[up - \frac{\epsilon u u_x + \lambda T_x}{v}\right] ,$$

where $[\cdot]$ denotes a jump across $\mathcal{C}$. Now, the absence of a smoothing mechanism in the first equation in (1) suggests that $[v] \neq 0$, whereas a reasonable interpretation of the terms $\left(\frac{u_x}{v}\right)_x$ and $\left(\frac{T_x}{v}\right)_x$ as distributions would require that $[u] = [T] = 0$. (See Hoff-Smoller [4] for a more rigorous argument in a special case of (1).) Granting then that $[v] \neq 0 = [u] = [T]$, we conclude first that $\dot{x} = 0$, so that discontinuities convect along particle paths $x = $ constant, and second that, along these particle paths,

(3)
$$[p] = \left[\frac{\epsilon u_x}{v}\right] , \qquad \left[\frac{T_x}{v}\right] = 0.$$

We shall derive from (3) a simple linear ode satisfied by the quantity $[L]$, where $L = \log v$. We make use of the standard divided difference notation

$$g(y_1, y_2) = \begin{cases} \dfrac{g(y_2) - g(y_1)}{y_2 - y_1} , & y_2 \neq y_1 \\ g'(y_1) , & y_2 = y_1 \end{cases}$$

for functions $g = g(y)$; and for functions $f = f(L, e)$ we define the partial functions f^e and f^L by

$$f^e(L) = f^L(e) = f(L, e).$$

Now suppose that (3) holds along the particle path $x = 0$. We let $e_\pm = e(0\pm, t)$, etc., and, abusing notation slightly, we regard p and T as functions of L and e. (3) and (1) then show that

$$\epsilon[L]_t = \left[\frac{\epsilon v_t}{v}\right] = \left[\frac{\epsilon u_x}{v}\right] = [p]$$

$$= p^{e_-}_{L_-, L_+}[L] + p^{L_+}_{e_-, e_+}[e].$$

However, the fact that $[T] = 0$ implies that

(4)
$$[e] = \beta[L]$$

where

$$\beta = -T^{e_-}_{L_-, L_+} / T^{L_+}_{e_-, e_+} .$$

Substituting, we thus find that

(5)
$$[L]_t = \frac{\alpha}{\epsilon}[L]$$

where

(6)
$$\alpha = p^{e_-}_{L_-,L_+} + \beta p^{L_+}_{e_-,e_+} .$$

We thus conclude that

(7)
$$[L](t) = [L](0) \exp\left(\epsilon^{-1} \int_0^t \alpha(s)ds\right) .$$

(7) shows that $[L]$ can grow at most exponentially in time, provided that α is bounded above. The same growth condition evidently holds as well for $[e]$, $[u_x]$, and $[T_x]$ by (3) and (4). Moreover, in the typical case of a near-ideal gas, $p_L < 0$ and $|T_L| \ll 1$, so that β is small and α is negative. In this case (7) shows that these jump discontinuities *decay* exponentially in time, more rapidly for smaller viscosity ϵ. Observe also that, for an ideal gas, T is an invertible function of e, so that $\beta = 0$ and e and T have identical regularity properties.

We now give a precise formulation of our results. First, in the generality desired here, we find that the solution operator for (1) does not enforce an upper bound on e or T. We therefore fix a set $K = [\underline{v}, \overline{v}] \times [\underline{e}, \infty)$ with $0 < \underline{v} < \overline{v}$ and $0 < \underline{e}$, and require that, in K,

(8)
$$\begin{cases} C^{-1}T \le e \le CT, \\ C^{-1} \le T_e \\ |T_e, T_v, p_e| \le C \\ 0 < p \le CT \\ -CT \le p_v \le C \end{cases}$$

for some positive constant C. These conditions are easily seen to be satisfied by ideal gases as well as VanderWaals gases. Next, in order to accommodate different states at $x = \pm\infty$, we fix a smooth function $\widetilde{W}(x) = \begin{bmatrix} \tilde{v} \\ \tilde{u} \\ \tilde{e} \end{bmatrix}(x)$ having values $(\tilde{v}, \tilde{e}) \in int(K)$ and satisfying $\widetilde{W}(x) = W_\pm$ for $\pm x \ge 1$. We then have the following local existence result.

THEOREM 1. *Assume that* (8) *holds in* K *and let* $W_0 = \begin{bmatrix} v_0 \\ u_0 \\ e_0 \end{bmatrix}$ *be given initial data satisfying:*

(a) *the initial values* $(v_0(x), e_0(x))$ *are contained in a set* $K' = [\underline{v}', \overline{v}'] \times [\underline{e}', \infty)$ *contained in the interior of* K;

(b) $W_0 - \widetilde{W} \in L^2(\mathbf{R})$ *and* $v_0, u_0 \in BV(\mathbf{R})$.

Then a weak solution $W(x,t)$ *of* (1)–(2) *exists up to some positive time* τ *and satisfies*

(c) $W(\cdot, t) - \widetilde{W} \in L^2(\mathbf{R}); v(\cdot, t), e(\cdot, t) \in BV_{\text{loc}}(\mathbf{R});$ and $u_x(\cdot, t), T_x(\cdot, t) \in L^2(\mathbf{R});$

(d) the jump conditions (3) hold in the sense that the quantities $p - \dfrac{\epsilon u_x}{v}$ and $\dfrac{T_x}{v}$ are locally Hölder continuous for $t > 0$;

(e) if v_0 is also piecewise H^1, then for positive times the quantities v, e, u_x, and T_x have one-sided limits at the points of discontinuity of v_0, and the jump conditions (3), (4), and (7) hold pointwise.

The proof of Theorem 1, as well as more detailed information concerning the regularity of the solution W, may be found in [1]. We remark that, in addition to the information derived heuristically above concerning jump conditions, the analysis establishes precise rates of smoothing for the variables u and T. Indeed, we find that the estimates $\int u_x(x,t)^2 dx \sim t^{-1/2}$ and $\int T_x(x,t)^2 dx \sim t^{-1}$ are crucial. Moreover, these rates are optimal, even for solutions of the heat equation with data in the given regularity classes. Note, however, that, owing to the discontinuities in v, the second and third equations in (1) fail to have differentiable Green's functions. A rather involved argument is therefore required to establish these rates of smoothing.

The local solution of Theorem 1 can be extended to all of $t > 0$ under more restrictive hypotheses on W_0 and the state functions p and T. We have:

THEOREM 2. *Assume that p and T satisfy the conditions of a "near ideal gas"; that is, in addition to conditions (8), p_v should be negative and $|T_v|$ should be sufficiently small that, for values $(v, e) \in K$, the quantity α appearing in (6) is strictly negative. Let $\widetilde{W} = \begin{bmatrix} \tilde{v} \\ \tilde{u} \\ \tilde{e} \end{bmatrix}$ be a constant vector with $(\tilde{v}, \tilde{e}) \in K'$ (K' is as in Theorem 1), and let $W_0 = \begin{bmatrix} v_0 \\ u_0 \\ e_0 \end{bmatrix}$ be Cauchy data satisfying:*

(a) $(v_0(x), e_0(x)) \in K'$ *a.e.;*

(b) $v_0 - \tilde{v} \in L^2 \cap L^1 \cap BV$, $u_0 - \tilde{u} \in L^2 \cap BV$, $e_0 - \tilde{e} \in L^2 \cap L^1$;

(c) *the L^1, L^2, and BV norms indicated in (b) are sufficiently small.*

Then the Cauchy problem (1)–(2) has a weak solution defined for all of $t > 0$.

Theorem 2 is proved by deriving time-independent estimates for the local solution W and its various derivatives, starting from the entropy equality

(9)
$$\int_{\mathbf{R}} -S\big(v(x,t), e(x,t)\big) dx + \int_0^t \int_{\mathbf{R}} \left(\frac{\epsilon u_x^2}{Tv} + \frac{\lambda T_x^2}{T^2 v} \right) dx\,dt$$
$$= \int_{\mathbf{R}} -S\big(v_0(x), e_0(x)\big) dx.$$

Here S is the physical entropy defined by $S_v = \dfrac{p}{T}$, $S_e = \dfrac{1}{T}$. (9) then follows directly from (1) and from the jump condition (3) $\left[\dfrac{T_x}{v} \right] = 0$. Time-independent L^2

bounds for $v - \tilde{v}$ and $e - \tilde{e}$ are then obtained from (9) by expanding S about its value at $(\tilde{v}, \tilde{e})$ and controlling the first order terms via the hypothesis $v_0 - \tilde{v}, e_0 - \tilde{e} \in L^1$. These estimates, together with the smallness conditions, then enable us to bound various higher derivatives, so as to obtain time-independent pointwise bounds for v and e.

We remark that the condition $W_0(-\infty) = W_0(+\infty)$ in Theorem 2b is an essential one. Indeed, any global analysis of solutions of (1)–(2) is likely to include information about the asymptotic behavior of the solution, and this behavior can be quite complicated when $W_0(-\infty) \neq W_0(+\infty)$. One result in this direction is that of Hoff and Liu [3], in which we obtain both the asymptotic behavior $(t \to \infty)$ as well as the strong inviscid limit $(\epsilon \to 0)$ of solutions of the isentropic/isothermal version of (1) with Riemann shock data.

The proof of Theorem 2 is given in [2], which also includes the following result concerning continuous dependence on initial data:

THEOREM 3. *In addition to the hypotheses of Theorem 2, assume that $e_0 \in BV$ and that p and T satisfy the conditions of an ideal gas, $pv = const.$ T and $T = T(e)$. Then the solutions constructed in Theorem 2 depend continuously on their initial values in the sense that, given a time t_0, there is a constant C such that, if $W_i = \begin{bmatrix} v_i \\ u_i \\ e_i \end{bmatrix}$, $i = 1, 2$, are solutions of (1) as described in Theorem 2, then for $t \in [0, t_0]$,*

(10)
$$\|W_2(\cdot, t) - W_1(\cdot, t)\|_{L^2} + \sup_{b-a=1} Var[v_2(\cdot, t) - v_1(\cdot, t)]|_{(a,b)}$$
$$\leq C\big(\|W_2(\cdot, 0) - W_1(\cdot, 0)\|_{L^2} + \sup_{b-a=1} Var[v_2(\cdot, 0) - v_1(\cdot, 0)]|_{(a,b)}\big).$$

C depends on t_0, K, and on upper bounds for the norms in Theorem 2b of the solutions W_1 and W_2.

We remark that the local variation of $v_2 - v_1$ is included in the norm in (10) in order to deal with terms arising from the differencing of $\left(\frac{u_x}{v}\right)_x$ and $\left(\frac{T_x}{v}\right)_x$. On the other hand, given that v_1 and v_2 are discontinuous variables, it would no doubt be useful to prove continuous dependence in the L^2 norm alone.

Finally, we point out that the existence, regularity, and continuous dependence results of Theorems 1–3 can be effectively employed in the design and rigorous analysis of algorithms for the numerical computation of solutions of (1)–(2). Indeed, Roger Zarnowski [5] has applied the present analysis to prove convergence of certain finite difference approximations to discontinuous solutions of the isentropic/isothermal version of (1). His scheme can be implemented under mesh conditions essentially equivalent to the usual CFL conditions for the corresponding hyperbolic equations $(\epsilon = 0$ in (1)); and he proves that, for piecewise smooth initial data, the error is bounded by $\Delta x^{1/6}$ in the norm of (10). Observe that, while the convergence rate is somewhat low, the topology is quite strong, dominating the sup norm of the discontinuous variable v.

REFERENCES

[1] DAVID HOFF, *Discontinuous solutions of the Navier-Stokes equations for compressible flow*, (to appear in Arch. Rational Mech. Ana).

[2] DAVID HOFF, *Global existence and stability of viscous, nonisentropic flows*, (to appear).

[3] DAVID HOFF AND TAI-PING LIU, *The inviscid limit for the Navier-Stokes equations of compressible, isentropic flow with shock data*, (to appear in Indiana Univ. Math. J).

[4] DAVID HOFF AND JOEL SMOLLER, *Solutions in the large for certain nonlinear parabolic systems*, Ann. Inst. Henri Poincaré, Analyse Non linéare 2 (1985), 213–235.

[5] ROGER ZARNOWSKI AND DAVID HOFF, *A finite difference scheme for the Navier-Stokes equations of one-dimensional, isentropic, compressible flow*, (to appear).

NONLINEAR GEOMETRICAL OPTICS

JOHN K. HUNTER*

Abstract. Using asymptotic methods, one can reduce complicated systems of equations to simpler model equations. The model equation for a single, genuinely nonlinear, hyperbolic wave is Burgers equation. Reducing the gas dynamics equations to a Burgers equation, leads to a theory of nonlinear geometrical acoustics. When diffractive effects are included, the model equation is the ZK or unsteady transonic small disturbance equation. We describe some properties of this equation, and use it to formulate asymptotic equations that describe the transition from regular to Mach reflection for weak shocks. Interacting hyperbolic waves are described by a system of Burgers or ZK equations coupled by integral terms. We use these equations to study the transverse stability of interacting sound waves in gas dynamics.

0. Introduction. Geometrical Optics is the name of an asymptotic theory for wave motions. It is based on the assumption that the wavelength of the wave is much smaller than any other characteristic lengthscales in the problem. These lengthscales include: the radius of curvature of nonplanar wavefronts; the lengthscale of variations in the wave medium; and the propagation distances over which dissipation, dispersion, diffraction, or nonlinearity have a significant effect on the wave. When this assumption is satisfied, we say that the wave is a short, or high frequency, wave. For short waves, the wave energy propagates along a set of curves in space-time called rays. This is one reason why geometrical optics is such a powerful method: it reduces a problem in several space dimensions to a one dimensional problem. For a single weakly nonlinear hyperbolic wave, this one dimensional problem is the inviscid Burgers equation (1.6), as we explain in section 1.

When diffraction effects are important in some part of the wave field, one must modify the straightforward theory of geometrical optics. For linear waves, this modified theory is called the geometrical theory of diffraction. In section 2, we analyze the diffraction of weakly nonlinear waves. One obtains the ZK equation (2.2), which is a two dimensional Burgers equation. Unfortunately, little is known about the ZK equation, and this makes it difficult to develop a nonlinear geometrical theory of diffraction. As an example, we use the ZK equation to formulate asymptotic equations which describe the transition from regular to Mach reflection for weak shocks.

Unlike linear waves, nonlinear waves interact and produce new waves. For multiple waves, nonlinear geometrical optics leads to a coupled system of Burgers equations (3.3). In section 3, we formulate asymptotic equations (3.6) which describe the diffraction of interacting waves. We use these equations to study the transverse stability of interacting sound waves in gas dynamics.

Keller [18] reviews linear geometrical optics. Other reviews of geometrical optics for weakly nonlinear hyperbolic waves are given by Nayfeh [27], Majda [24], and Hunter [13].

---

*Department of Mathematics, Colorado State University, Fort Collins, CO 80523.
Present Address: Department of Mathematics, University of California, Davis, CA 95616.

1. Single Waves.

1.1 The eikonal and transport equations. We consider a hyperbolic system of conservation laws in $N + 1$ space-time dimensions.

$$(1.1) \qquad \sum_{i=0}^{N} f^i(x, u)_{x_i} = 0.$$

Short wave solutions of (1.1) are solutions which vary rapidly normal to a set of wavefronts $\phi(x) = $ constant . We call ϕ the phase of the short wave. We look for small amplitude, short wave solutions of (1.1), with an asymptotic approximation of the form.

$$(1.2) \qquad u(x; \epsilon) = \epsilon U[\epsilon^{-1}\phi(x), x] + 0(\epsilon^2).$$

The amplitude in (1.2) is of the order of the wavelength. We choose this particular scaling because it allows a balance between weakly nonlinear and nonplanar effects.

Multiple scale methods [14] show that the phase in (1.2) satisfies the eikonal equation associated with the linearized version of (1.1), namely

$$(1.3) \qquad \det\left[\sum_{i=0}^{N} \phi_{x_i} A^i(x)\right] = 0.$$

In (1.3), $A^i(x) = \nabla_u f^i(x, 0)$. We denote left and right null-vectors of the matrix in (1.3) by $\ell(x, \nabla\phi)$ and $r(x, \nabla\phi)$ respectively.

Associated with the phase is an N-parameter family of rays or bicharacteristics. The rays are curves in space-time with equation $x = X(s; \beta)$ where

$$\frac{dX_i}{ds} = \ell \cdot A^i r.$$

Here, $s \in \mathbf{R}$ is an arclength parameter along a ray, while $\beta \in \mathbf{R}^N$ is constant on a ray. We assume that the transformation between space-time coordinates x and ray coordinates (s, β) is smooth and invertible. This assumption is not true at caustics, and then the simple ansatz in (1.2) does not provide the correct asymptotic solution. Instead, diffractive effects must be included (see section 2 and [23], [15], [13]).

The explicit form of the asymptotic solution (1.2) is

$$(1.4) \qquad u = \epsilon a[\epsilon^{-1}\phi(x), x]r(x, \nabla\phi) + 0(\epsilon^2),$$

where the scalar function $a(\theta, x)$ is called the wave amplitude. The dependence of a on θ describes the wave-form. For oscillatory wavetrains, a is a periodic or an almost periodic function of θ; for pulses, a is compactly supported in θ; for wavefronts, the derivative of a with respect to θ jumps across $\theta = 0$, etc. The dependence of a on x describes modulation effects such as the increase in amplitude caused by focusing and the nonlinear steepening of the wave-form.

Multiple scale methods also imply that the wave amplitude satisfies a nonlinear transport equation,

(1.5)
$$a_s + Maa_\theta + Qa = 0.$$

In (1.5), $\partial/\partial s$ is a derivative along a ray,

$$\frac{\partial}{\partial s} = \sum_{i=0}^{N} \ell \cdot A^i r \frac{\partial}{\partial x_i}.$$

The coefficient M measures the strength of the waves quadratically nonlinear self-interaction, and is given by

$$M(s,\beta) = \sum_{i=0}^{N} \phi_{x_i} \ell \cdot \nabla_u^2 f^i(x,0) \cdot (r,r).$$

M is nonzero for genuinely nonlinear waves and M is zero for linearly degenerate waves. The coefficient Q describes the growth or decay of the amplitude due to focusing of the wave and nonuniformities in the medium. It is given by

$$Q(s,\beta) = \sum_{i=0}^{N} \ell \cdot \frac{\partial}{\partial x_i}(A^i r).$$

Since r depends on $\nabla\phi$, Q involves second derivatives of ϕ. It is therefore unbounded near caustics, where the curvature of the wavefronts is infinite.

There is one Burgers equation (1.5) for each ray. Solving them, together with appropriate initial data obtained from initial, boundary, or matching conditions, gives $a(\theta, s, \beta)$. Finally, evaluating θ at $\epsilon^{-1}\phi(x)$ in the result gives the asymptotic solution (1.4).

The transport equation (1.5) can be reduced to a standard form by the change of variables

$$\bar{u}(\bar{x}, \bar{t}; \beta) = E^{-1}(s,\beta)a(s,\beta,\theta)$$
$$\bar{x} = \theta,$$
$$\bar{t} = \int_0^s M(s',\beta)E(s',\beta)ds',$$

where

$$E(s,\beta) = \exp\left[-\int_0^s Q(s',\beta)ds'\right].$$

We assume that that $M \neq 0$. The result is that $\bar{u}(\bar{x}, \bar{t}; \beta)$ satisfies

(1.6)
$$\bar{u}_{\bar{t}} + \bar{u}\bar{u}_{\bar{x}} = 0.$$

Thus, (1.6) is the canonical asymptotic equation for a genuinely nonlinear, hyperbolic wave.

We remark that if weak viscous effects are included, then, instead of (1.6), one obtains a generalized Burgers equation,

(1.7)
$$\bar{u}_{\bar{t}} + \bar{u}\bar{u}_{\bar{x}} = \nu(\bar{t})\bar{u}_{\bar{x}\bar{x}}.$$

The viscosity ν is constant only for plane waves in a uniform medium. In that case, (1.7) can be solved explicitly by the Cole-Hopf transformation [32]. If ν is not constant, then (1.7) cannot be solved explicitly, and numerical or perturbation [28] methods are required.

1.2 Nonlinear geometrical acoustics. Sound waves in a compressible fluid are a fundamental physical application of the above ideas. The resulting theory is called nonlinear geometrical acoustics (NGA). For reviews of NGA, see [7], [8], [9].

The equations of motion of an inviscid, compressible fluid are

(1.8)
$$\rho_t + \text{div} \, (\rho \mathbf{u}) = 0,$$
$$(\rho \mathbf{u})_t + \text{div} \, (\rho \mathbf{u} \otimes \mathbf{u} - pI) = \rho \mathbf{f},$$
$$\left[\rho \left(\frac{1}{2} \mathbf{u} \cdot \mathbf{u} + e \right) \right]_t + \text{div} \, \left[\rho \mathbf{u} \left(\frac{1}{2} \mathbf{u} \cdot \mathbf{u} + e \right) - p \mathbf{u} \right] = 0.$$

Here, ρ is the fluid density, p is the pressure, e is the specific internal energy, and $\mathbf{u}$ is the fluid velocity. We include a given body force $\mathbf{f}(\mathbf{x}, t)$ and we neglect any heat sources. For simplicity, we consider a polytropic gas for which

$$e = \frac{1}{\gamma - 1} \frac{p}{\rho}.$$

Here, the constant $\gamma > 1$ is the ratio of specific heats. Similar results are obtained for general equations of state.

Suppose that

$$\rho = \rho_0(\mathbf{x}, t), \quad p = p_0(\mathbf{x}, t), \quad \mathbf{u} = \mathbf{u}_0(\mathbf{x}, t)$$

is a given smooth solution of (1.8). We denote the corresponding sound speed by $c = c_0(\mathbf{x}, t)$. The NGA solution for a sound wave propagating through this medium is

(1.9)
$$\begin{bmatrix} \rho \\ \mathbf{u} \\ p \end{bmatrix} = \begin{bmatrix} \rho_0 \\ \mathbf{u}_0 \\ p_0 \end{bmatrix} + \epsilon a [\epsilon^{-1} \phi(\mathbf{x}, t), \mathbf{x}, t] \begin{bmatrix} \rho_0 \\ c_0 \Omega^{-1} \mathbf{k} \\ \rho_0 c_0{}^2 \end{bmatrix} + o(\epsilon).$$

Here, we define the local frequency ω, the wavenumber $\mathbf{k}$, and the Doppler shifted frequency Ω by

$$\omega = -\phi_t, \quad \mathbf{k} = \nabla \phi, \quad \Omega = \omega - \mathbf{u}_0 \cdot \mathbf{k}.$$

The eikonal equation for the phase is

(1.10)
$$\left(\phi_t + \mathbf{u}_0 \cdot \nabla \phi \right)^2 = c_0{}^2 | \nabla \phi |^2.$$

Equation (1.10) states that ω and $\mathbf{k}$ satisfy the local, linearized dispersion relation $\omega = W(\mathbf{k}; \mathbf{x}, t)$, where

$$W(\mathbf{k}; \mathbf{x}, t) = \mathbf{u}_0(\mathbf{x}, t) \cdot \mathbf{k} \pm c_0(\mathbf{x}, t) | \mathbf{k} |.$$

The transport equation for the wave amplitude $a(\theta, \mathbf{x}, t)$ is

(1.11) $\quad a_t + \mathbf{C} \cdot \nabla a + \dfrac{\gamma + 1}{2} \Omega a a_\theta + \dfrac{\Omega}{2 \rho_0 c_0{}^2} \left\{ \left(\dfrac{\rho_0 c_0{}^2}{\Omega} \right)_t + \nabla \cdot \left(\dfrac{\rho_0 c_0{}^2}{\Omega} \mathbf{C} \right) \right\} a = 0.$

In (1.11), $\mathbf{C}$ is the group velocity,

$$\mathbf{C} = \nabla_{\mathbf{k}}W = u_0 + c_0{}^2\Omega^{-1}\mathbf{k}.$$

The "entropy" inequality for (1.11) is

$$\mathcal{A}_t + \nabla \cdot [\mathcal{A}\mathbf{C}] + \partial_\theta \left[\frac{1}{3}(\gamma + 1)\Omega a \mathcal{A}\right] \le 0,$$

with equality for smooth solutions. Here, $\mathcal{A}$ is the wave action density for the linearized equations,

$$\mathcal{A} = \text{acoustic energy density/Doppler shifted frequency} = \rho_0 c_0{}^2 a^2/\Omega.$$

Conservation of wave action is a fundamental law for dispersive waves that are governed by variational principles. For sound waves, shocks cause the wave action to decrease.

We consider two examples. The first example is outgoing spherical waves. Suppose that ρ_0 and c_0 are constant and $u_0 = 0$. A solution of the eikonal equation (1.10) is then

$$\phi = r - c_0 t,$$

where $r = [x_1^2 + \cdots + x_n{}^2]^{1/2}$ and n is the number of space dimensions. The transport equation for $a(\theta, r, t)$ is

(1.12) $$a_t + c_0 a_r + \frac{\gamma + 1}{2}c_0 a a_\theta + c_0 \frac{n - 1}{2r}a = 0.$$

The last term in (1.12) describes the geometrical attenuation of the wave.

To put (1.12) in the standard form (1.7), we define:

$$\bar{u}(\bar{x}, \bar{t} : \beta) = \frac{\gamma + 1}{2}c_0 r^{(\frac{n-1}{2})}a(\theta, r, t);$$

$$\bar{t} = \frac{2}{3 - n}r^{(\frac{3-n}{2})} \quad \text{if} \quad n \ne 3; \quad \bar{t} = \log r, \quad \text{if} \quad n = 3;$$

$$\bar{x} = \theta; \quad \beta = r - c_0 t.$$

For $n \le 3$, $\bar{t} \to +\infty$ as $r \to +\infty$, but for $n \ge 4$, $\bar{t} \to 0$ as $r \to +\infty$. It follows that geometrical attenuation does not prevent shock formation when $n \le 3$ (although in three space dimensions the propagation distance for shock formation is exponentially long in the initial slope of the wave). For $n \ge 4$, geometrical attenuation does prevent shock formation for initial data with sufficiently small slopes.

The second example illustrates the effect of nonuniformities on a sound wave. We suppose that the unperturbed fluid is exponentially stratified, meaning that $\rho_0 = \rho_* \exp(-x/H), c_0 = c_*$, and $u_0 = 0$. Here, ρ_*, c_*, and H are constants. We consider a wave propagating in the positive x-direction, when the appropriate phase is $\phi = x - c_* t$. The transport equation (1.11) is

$$a_t + c_* a_x + \frac{\gamma + 1}{2}c_* a a_\theta = \frac{c_*}{2H}a.$$

In the absence of nonlinear effects, the nonuniformity causes an exponential growth in the amplitude as the sound wave propagates into regions of lower density.

2. Diffraction.

2.1 Weak diffraction. The straightforward geometrical optics expansion, described in section 1, is based upon a locally one-dimensional approximation of the wave. The effects of diffraction are neglected. In this section, we describe a generalization which includes weak diffraction.

We look for asymptotic solutions of (1.1) which depend on three different length-scales,

$$(2.1) \qquad u(x;\epsilon) = \epsilon U\left[\epsilon^{-1}\phi(x), \epsilon^{-1/2}\psi(x), x\right] + 0\left(\epsilon^{3/2}\right).$$

The scaling chosen in (2.1) allows a balance between nonlinear, nonplanar, and diffractive effects. In order for (2.1) to solve (1.1) asymptotically, the phase ϕ must satisfy the eikonal equation (1.10), and ψ must be constant along the rays associated with ϕ [12]. Thus, the dependence of U on $\epsilon^{-1/2}\psi$ describes weak variations in the wave transverse to the group lines. One can obtain a system of equations that generalizes the transport equation (1.4). After a rescaling, similar to the one leading to (1.6), this system is

$$(2.2) \qquad \begin{aligned} \bar{u}_{\bar{t}} + \bar{u}\bar{u}_{\bar{x}} + \delta(\bar{t})\bar{v}_{\bar{y}} &= 0, \\ \bar{u}_{\bar{y}} - \bar{v}_{\bar{x}} &= 0. \end{aligned}$$

Here, $\bar{t}$ is a ray variable, $\bar{x}$ is the phase variable, evaluated at $\epsilon^{-1}\phi$, and $\bar{y}$ is the intermediate, transverse variable, evaluated at $\epsilon^{-1/2}\psi$. Eliminating $\bar{v}$ from (2.2), and dropping the "—" on all variables, gives

$$(2.3) \qquad \partial_x\left[u_t + uu_x\right] + \delta(t)u_{yy} = 0.$$

Equation (2.3) is a generalized Zabolotskaya-Khokhlov (GZK) equation. For plane waves in a uniform medium, δ is constant, and it can be normalized to one without loss of generality. This gives the ZK equation [33],

$$(2.4) \qquad \partial_x\left[u_t + uu_x\right] + u_{yy} = 0.$$

Equation (2.4) is also called the unsteady transonic small disturbance equation [5]. Including weak dissipative or dispersive effects in (2.4) gives

$$(2.5) \qquad \partial_x\left[u_t + uu_x - u_{xx}\right] + u_{yy} = 0,$$

$$(2.6) \qquad \partial_x\left[u_t + uu_x \pm u_{xxx}\right] + u_{yy} = 0.$$

Equation (2.5) was derived by Kuznetsov [21]. Equation (2.6) is the Kadomtsev-Petviashvili (KP) equation [17].

Equation (2.4) arises in many different situations. It was first derived by Timman in the context of transonic flows [31]. In nonlinear acoustics, it was derived by Zabolotskaya and Khokhlov [33], and is used to describe the diffraction of nonlinear acoustic beams [9]. Cramer and Seebass [6] used (2.4) to study caustics in nearly planar sound waves. They were motivated by the experiments of Sturtevant and Kulkarny [30] on the focusing of shocks. The same equation arises as a weakly nonlinear equation for cusped caustics [13]. Hunter [12] has also shown that (2.3) describes high-frequency waves near singular rays. Below, we use (2.4) to formulate asymptotic equations for the transition from regular to Mach reflection for weak shocks.

2.2 The parabolic approximation. The ZK equation (2.4) is a weakly non-linear version of the "parabolic approximation" (see [2] for some recent references). To give some insight into (2.4), we shall give a brief heuristic description of the parabolic approximation applied to the wave equation,

$$(2.7) \qquad u_{tt} = u_{XX} + u_{YY}.$$

We rewrite (2.7) as

$$(2.8) \qquad (\partial_t - \partial_X)(\partial_t + \partial_X)u - u_{YY} = 0.$$

For waves propagating in a direction close to the X-axis, $\partial_t \simeq -\partial_X$.
Using this in (2.8) gives

$$(2.9) \qquad 2\partial_X[u_t + u_X] + u_{YY} = 0.$$

After a Galilean transformation and a rescaling, $x = X - t, \quad y = 2^{1/2}Y$, equation (2.9) reduces to the linearization of (2.4), namely

$$(2.10) \qquad u_{xt} + u_{yy} = 0.$$

Equation (2.10) is the parabolic approximation of (2.7).

The name "parabolic approximation" is misleading, since (2.10) is hyperbolic, with x and t as characteristic directions. The reason for the name is that periodic solutions of (2.10), $u(x,y,t) = U(y,t)e^{ix}$, satisfy the Schrödinger equation,

$$(2.11) \qquad iU_t + U_{yy} = 0.$$

The advantage of this approximation is that it filters out all left-moving waves. As a result, (2.11) can be solved by "marching" in t. The disadvantage is that time becomes a characteristic direction, and this introduces some undesirable features. For example, the wave equation (2.7) has characteristic surfaces,

$$[X^2 + Y^2]^{1/2} = \pm t.$$

The corresponding characteristic surfaces of (2.10) are

$$(2.12) \qquad x = \pm\frac{y^2}{4t}.$$

Thus, the parabolic approximation turns a focusing circle into a collapsing parabola. The interior of the parabolas (2.12) are the domains of influence ($\pm = +, t > 0$) and dependence ($\pm = -, t < 0$) of $(x,y,t) = (0,0,0)$. As a consequence of the fact that t is a characteristic direction, these domains are unbounded.

2.3 The ZK equation. Despite its rather simple appearance, very little is known about the ZK equation. It has many of the essential difficulties of multi-dimensional hyperbolic systems of conservation laws. If, as seems likely, it correctly describes shock-focusing phenomena, like those observed by Sturtevant and Kulkarny [30], then it must possess solutions with complicated geometrical behavior.

One of the few known analytical solutions of (2.4) describes focusing or defocusing waves. To derive it, we observe that the change of variables

$$\bar{t} = \int_0^t |\rho|^{-1/2} dt, \quad \bar{x} = x + y^2/4\rho, \quad \bar{y} = y/\rho, \quad \bar{u} = |\rho|^{1/2} u,$$

where $\delta(t) = \rho'(t)$, transforms the GZK equation (2.3) into another GZK equation,

$$\partial_{\bar{x}}\left[\bar{u}_{\bar{t}} + \bar{u}\bar{u}_{\bar{x}}\right] + \bar{\delta}(\bar{t})\bar{u}_{\bar{y}\bar{y}} = 0.$$

The transformed coefficient is

$$\bar{\delta} = |\rho|^{-3/2}\delta.$$

In particular, if $\delta = 1$, then the transformation is

$$\bar{t} = 2|t|^{1/2}, \quad \bar{x} = x + y^2/4t, \quad \bar{y} = y/t, \quad \bar{u} = |t|^{1/2}u.$$

The transformed equation is the cylindrical ZK equation,

(2.13) $$\partial_{\bar{x}}\left[\bar{u}_{\bar{t}} + \bar{u}\bar{u}_{\bar{x}}\right] + 8\bar{t}^{-3}\bar{u}_{\bar{y}\bar{y}} = 0.$$

A particular solution of (2.13) is $\bar{u} = U(\bar{x},\bar{t})$, where U solves

$$U_{\bar{t}} + UU_{\bar{x}} = 0.$$

Thus, a solution of (2.4) is

(2.14a) $$u(x,y,t) = |t|^{-1/2}U\left(x + \frac{y^2}{4t}, 2|t|^{1/2}\right).$$

The corresponding solution for v in (2.2) is, dropping the " $-$ ",

(2.14b) $$v(x,y,t) = \frac{1}{2}|t|^{-3/2}yU\left(x + \frac{y^2}{4t}, 2|t|^{1/2}\right).$$

This solution has also been derived by Cates [3] using a similarity analysis of the Kuznetsov equation (2.5). For $t < 0$, (2.14) represents a focusing wave whose amplitude blows up as $t \to 0-$. The solution is then inconsistent with the approximations leading to the ZK equation. For $t > 0$, the solution represents an outgoing, defocusing wave.

A similar transformation applies to generalized KP or Kuznetsov equations (2.5), (2.6). Johnson [35] uses it to transform the KP equation (2.6) into the cylindrical KdV equation.

In other analytical work, Kodama and Gibbons [19], [20] have found a generalization of the hodograph transformation which applies to the ZK equation, as well as other equations. They use this transformation to construct a family of implicit solutions.

Since time is a characteristic direction, the formulation of appropriate auxiliary conditions for the ZK equation is not altogether straightforward. The simplest reasonable initial-boundary value problem is the following:

$$(2.15) \quad u_t + \left(\frac{1}{2}u^2\right)_x + v_y = 0, \quad u_y - v_x = 0; \qquad \text{PDE}$$

$$(2.16) \quad u(x,y,0) = u_0(x,y); \qquad \text{Initial Condition}$$

$$(2.17) \quad v(x,y,t) \to 0 \text{ as } x \to +\infty; \qquad \text{Boundary Condition}$$

$$(2.18) \quad (u^2)_t + \left(\frac{2}{3}u^3 - v^2\right)_x + (2uv)_y \leq 0. \qquad \text{Entropy Inequality}$$

The jump conditions for a shock located at $x = s(y,t)$ can be written as

$$(2.19) \qquad \begin{aligned} [v] + s_y[u] &= 0, \\ s_t &= <u> + s_y. \end{aligned}$$

In (2.19), $[u]$ and $[v]$ denote the jumps in u and v across the shock, and $<u>$ denotes the average of the values of u on either side of the shock. To obtain solutions of (2.15) - (2.18) with compact support, the initial data must satisfy

$$u_0(x,y) \text{ compactly supported,}$$
$$\int_{-\infty}^{+\infty} u_0(x,y)dx = 0.$$

In (2.17), it is important to impose the boundary condition on v in the limit $x \to +\infty$, rather than the limit $x \to -\infty$. This can be seen from a consideration of the domain of dependence of (2.15). Also, this condition eliminates the focusing solutions in (2.14).

The global existence and regularity of weak solutions of (2.15) – (2.18) is unknown. Because of focusing, solutions are unlikely to have bounded variation. Some kind of global existence and regularity result would support the use of the ZK equation as a description of weak shock focusing. This equation may also serve as a model problem for multi-dimensional hyperbolic conservation laws, analogous to Burgers equation for conservation laws in one space dimension.

2.4 Transition to Mach reflection for weak shocks. When a shock wave hits a wedge, a number of different reflection patterns are observed. The particular pattern which occurs depends on the shock strength and the angle of incidence of the shock on the wedge. The simplest patterns are regular reflection and single Mach reflection. Hornung [11] gives a recent review of this subject.

Lighthill [22] analyzed the reflection of a strong shock off a wedge with small angle. In that case, Mach reflection occurs. The transition from regular to Mach reflection takes place when the wedge angle is below a critical value which depends on the shock strength. This critical angle tends to zero as the shock strength tends to zero. Thus, to study the transition to Mach reflection by perturbation methods, we need to consider a limit in which the shock strength and the wedge angle both tend to zero. Specifically, the critical angle is of the order $\epsilon^{1/2}$ for dimensionless shock strengths of the order $\epsilon \ll 1$. In this limit, the incident and reflected shocks are nearly parallel. The shock pattern propagates primarily along the wedge, with slower variations normal to the wedge. The transition from regular to Mach reflection for weak shocks should therefore be described by the ZK equation. We shall formulate the appropriate asymptotic equations here.

We suppose that a shock moving from left to right is incidence on a wedge at $y = 0$. In the parabolic approximation, the corner of the wedge is at $x = -\infty$, so that the problem is posed on a half space, $y > 0$. The diffracted shock approaches the line $-\infty < x < 0$, $y = 0$ as $t \to 0+$. The initial data for (2.15) in the reflection problem is

$$(2.20) \qquad \begin{aligned} u, \ v &\to 0 \text{ as } t \to 0+ \text{ with } y > 0, \ x > \alpha y; \\ u &\to 1, \ v \to -\alpha \text{ as } t \to 0+ \text{ with } y > 0, \ x < \alpha y. \end{aligned}$$

In (2.20), $\alpha > 0$ parametrizes the angle of the incident shock to the wedge normal. We have normalized u ahead of the shock to zero and behind the shock to one without loss of generality. The value of v behind the incident shock follows from the jump condition (2.19b). The y-component of the fluid velocity is proportional to v, so the boundary condition on the wedge is

$$(2.21) \qquad v = 0 \text{ on } y = 0, \ t > 0.$$

Like the full gas dynamics problem, equations (2.15), (2.20) - (2.21) are self-similar (or "pseudo-stationary"). We therefore look for solutions which depend on the similarity variables

$$\xi = \frac{x}{t}, \quad \zeta = \frac{y}{t}.$$

Equation (2.15) implies that

$$(2.22) \qquad \begin{aligned} \xi u_\xi + \zeta u_\zeta - \left(\frac{1}{2}u^2\right)_\xi - v_\zeta &= 0, \\ v_\xi - u_\zeta &= 0, \end{aligned}$$

in $-\infty < \xi < +\infty$ and $\zeta > 0$. The initial and boundary conditions are

$$(2.23) \qquad \begin{aligned} u, \ v &\to 0 \text{ as } \zeta \to +\infty \text{ with } \xi/\zeta > \alpha; \\ u &\to 1, \ v \to -\alpha \text{ as } \zeta \to +\infty \text{ with } \xi/\zeta < \alpha; \\ v &= 0 \text{ on } \zeta = 0, \end{aligned}$$

The jump conditions for a shock located at $\xi = \sigma(\zeta)$ are

$$(2.24) \qquad \begin{aligned} [v] + \sigma_\zeta [u] &= 0, \\ \sigma &= \zeta \sigma_\zeta + \sigma_\zeta{}^2 + <u>. \end{aligned}$$

The shock is admissible provided that

(2.25)
$$\lim_{\xi \to \sigma-} u(\xi, \zeta) > \lim_{\xi \to \sigma+} u(\xi, \zeta).$$

These equations are considerably simpler than those for full gas dynamics, as can be seen by comparing them with the gas dynamics problem formulated in Chang and Hsiao [4]. The asymptotic problem has the same fundamental difficulties as the gas dynamics problem: it is a nonlinear, mixed type, free boundary value problem. It is easy to check that

$$\xi + \frac{\zeta^2}{4} > u \Rightarrow (2.22) \text{ hyperbolic,}$$

$$\xi + \frac{\zeta^2}{4} < u \Rightarrow (2.22) \text{ elliptic.}$$

However, (2.22) – (2.25) does not model all features of the gas dynamics problem. Complex Mach reflection and double Mach reflection are not observed for weak shocks, so (2.22) – (2.25) is unlikely to describe those phenomena.

A simple local analysis shows that regular reflection is impossible for $0 < \alpha < 2^{1/2}$. We can approximate a regularly reflected solution near the point where the incident and reflected shocks meet the wedge by a piecewise constant solution,

$$u = v = 0, \quad x > \alpha y + Vt, \quad y > 0;$$
$$u = 1, v = -\alpha, \quad -\beta y + Vt < x < \alpha y + Vt, \quad y > 0;$$
$$u = u_L, v = 0, \quad x < -\beta y + Vt, \quad y > 0.$$

The jump conditions (2.19) imply that

$$V = \frac{1}{2} + \alpha^2, \quad u_L = 1 + \frac{\alpha}{\beta},$$

where β is a solution of

(2.26)
$$\beta^3 - \left(\alpha^2 - \frac{1}{2}\right)\beta + \frac{1}{2}\alpha = (\alpha + \beta)\left(\beta^2 - \alpha\beta + \frac{1}{2}\right) = 0.$$

The reflected shock is admissible if $\beta > 0$. Equation (2.26) has two positive roots for β when $\alpha > 2^{1/2}$. The equation has no positive roots when $0 < \alpha < 2^{1/2}$.

One interesting explicit solution of (2.22) can be obtained from (2.14) with

$$U(\bar{x}, 0) = \pm|\bar{x}|^{1/2}, \quad \bar{x} < 0,$$
$$U(\bar{x}, 0) = 0, \quad \bar{x} \geq 0.$$

Taking the minus sign, the corresponding solution for u and v is

(2.27)
$$u = 1 - (1 - \rho)^{1/2}, \quad v = \frac{1}{2}\zeta\left[1 - (1 - \rho)^{1/2}\right], \quad \rho < 0,$$
$$u = v = 0, \quad \rho > 0.$$

Taking the plus sign, the solution is

$$u = 1 + (1 - \rho)^{1/2}, \quad v = \frac{1}{2}\zeta\left[1 + (1 - \rho)^{1/2}\right], \quad \rho < \frac{3}{4},$$
(2.28)
$$u = v = 0, \quad \rho > \frac{3}{4}.$$

Here, $\rho = \xi + \zeta^2/4$. Equation (2.27) describes an outgoing cylindrical expansion wave; (2.28) describes an outgoing cylindrical shock.

Equation (2.22), with different boundary conditions, also arises as a description of weak shocks at a singular ray [10], [12], [34]. This equation may serve as a model equation for two dimensional Riemann problems in general.

3. Diffraction of Interacting Waves.

3.1 Diffraction of interacting waves. The ZK equation is a generalization of Burgers equation that includes diffraction effects. Interacting hyperbolic waves are described asymptotically by a system of Burgers equations coupled by integral terms. In this section, we generalize these equations to include diffraction. The result is a coupled system of ZK equations.

An asymptotic theory for weakly nonlinear interacting hyperbolic waves is developed in [16], [25]. We shall briefly describe that theory in the simplest case. We consider a hyperbolic system of conservation laws in one space dimension,

$$(3.1) \qquad u_t + f(u)_x = 0.$$

Suppose that there are three interacting periodic waves which satisfy the resonance condition

$$\omega_1 + \omega_2 + \omega_3 = 0,$$
$$k_1 + k_2 + k_3 = 0,$$
$$\omega_j = \lambda_j k_j, \qquad j = 1, 2, 3.$$

Here, ω_j and k_j are the frequency and wavenumber of the jth wave, and λ_j is the linearized wave velocity.

The asymptotic solution for the interacting waves is then

$$(3.2) \qquad u = \epsilon \sum_{j=1}^{3} a_j\left[\epsilon^{-1}(k_j x - \omega_j t), t\right] r_j + 0(\epsilon^2),$$

as $\epsilon \to 0$ with $x, t = 0(1)$. In (3.2), r_j is a right eigenvector of $\nabla_u f(0)$ associated with the eigenvalue λ_j, and the wave amplitudes $a_j(\theta, t)$ are 2π-periodic functions of the phase variable θ. The amplitudes solve the following system of integro-differential equations,

$$(3.3) \quad a_{jt}(\theta, t) + M_j a_j(\theta, t) a_{j\theta}(\theta, t) + \Gamma_j \frac{1}{2\pi} \int_0^{2\pi} a_p(-\theta - \xi, t) a_{p\xi}(\xi, t) d\xi = 0,$$

where (j, p, q) runs through cyclic permutations of $(1, 2, 3)$. The coefficients are

$$M_j = \nabla_u \lambda_j(0) \cdot r = \ell_j \cdot \nabla_u{}^2 f(0) \cdot (r_j, r_j),$$
$$\Gamma_j = \ell_j \cdot \nabla_u{}^2 f(0) \cdot (r_p, r_q).$$

Here, ℓ_j is a left eigenvector of $\nabla_u f(0)$ associated with λ_j. It is normalized so that $\ell_j \cdot r_j = 1$.

To analyze the effects of wave diffraction, we consider a two dimensional version of (3.1), namely

$$(3.4) \qquad u_t + f(u)_x + g(u)_y = 0.$$

For simplicity, we assume that (3.4) is isotropic, meaning that is is invariant under $(x, y) \to O\ (x, y)$ for all orthogonal transformations O. The rays associated with the phase $\phi_j = k_j x - \omega_j t$ are then $\phi_j = $ constant, $y = $ constant. Thus, the transverse variable $\psi = y$ is constant on the rays associated with each phase ϕ_j. Complications arise when the transverse variable is not constant on all sets of rays. This case may occur for anisotropic waves, and we shall not consider it further here.

The generalization of (3.2) that includes weak diffraction in the y-direction is then

$$(3.5) \qquad u = \epsilon \sum_{j=1}^{3} a_j \left[\epsilon^{-1}(k_j x - \omega_j t), \epsilon^{-1/2} y, t \right] r_j + 0(\epsilon^{3/2}).$$

The amplitudes $a_j(\theta, \eta, t)$ satisfy

$$(3.6) \qquad \partial_\theta \bigg\{ a_{jt}(\theta, \eta, t) + M_j a_j(\theta, \eta, t) a_{j\theta}(\theta, \eta, t)$$
$$+ \Gamma_j \frac{1}{2\pi} \int_0^{2\pi} a_p(-\theta - \xi, \eta, t) a_{q\xi}(\xi, \eta, t) d\xi \bigg\} + \frac{\lambda_j}{2k_j} a_{j\eta\eta}(\theta, \eta, t) = 0.$$

For solutions which are independent of η, (3.6) reduces to (3.2), after an integration with respect to θ. For a single wave ($a_2 = a_3 = 0$), it reduces to the ZK equation for a_1.

3.2 Transverse stability of interacting waves in gas dynamics. There are three wave-fields in one dimensional gas dynamics. They are the left- and right-moving sound waves, and a stationary entropy wave. According to the asymptotic theory described in section 3.1, the entropy wave decouples from the sound waves. Consequently, the system of three equations in (3.3) reduces to a pair of equations for the sound wave amplitudes. These equations describe the resonant reflection of sound waves off a periodic entropy perturbation. After rescaling to remove inessential coefficients, the equations are

$$(3.7) \qquad \begin{aligned} u_t + u u_x + \frac{1}{2\pi} \int_0^{2\pi} K(x - \xi) v(\xi, t) d\xi &= 0, \\ v_t + v v_x - \frac{1}{2\pi} \int_0^{2\pi} K(-x + \xi) u(\xi, t) d\xi &= 0. \end{aligned}$$

In (3.7), K is a known kernel, which is proportional to the derivative of the entropy wave amplitude. The dependent variables $u(x,t)$ and $v(x,t)$ are proportional to the amplitudes of the right-moving and the left-moving sound waves. The sound-wave amplitudes and the kernel are 2π-periodic, zero-mean functions of the phase variable, x. These equations are derived in [25], and they are further analyzed in [26].

Pego [29] found an explicit smooth travelling wave solution of (3.7), in the special case of a sinusoidal kernel, $K(x) = \sin x$. His solution is

(3.8)
$$u = u_0(x - ct) = \sigma\left[c + bf(x - ct; \alpha)\right],$$
$$v = v_0(x - ct) = \sigma\left[c + bf(x - ct : \sigma\alpha)\right].$$

In (3.8), $\sigma \in \{-1, +1\}, \alpha \in [0, 1]$,

(3.9)
$$f(\theta; \alpha) = [1 + \alpha\cos\theta]^{1/2},$$

and

(3.10)
$$b(\alpha) = \frac{1}{\pi\alpha}\int_0^{2\pi} \cos\xi(1 + \alpha\cos\xi)^{1/2}d\xi,$$
$$c(\alpha) = -\frac{b}{2\pi}\int_0^{2\pi}(1 + \alpha\cos\xi)^{1/2}d\xi.$$

There are two families of travelling waves (3.8), depending on the choice of σ. They exist only up to a finite amplitude. The wave of maximum amplitude, corresponding to $\alpha = 1$, has a corner in its crest or trough. We shall show that the waves with $\sigma = +1$ are unstable to transverse perturbations when α is small and when α is close to one. We remark that the stability of these waves to one dimensional perturbations has not been studied. Our stability analysis is essentially the same as the use of the KP equation to study the transverse stability of KdV solitons [1], [17].

The generalization of (3.7), with $K(x) = \sin x$, that includes weak diffraction is

(3.11)
$$\partial_x\left\{u_t + uu_x + \mathcal{K}v\right\} + u_{yy} = 0,$$
$$\partial_x\left\{v_t + vv_x + \mathcal{K}u\right\} + v_{yy} = 0,$$

where

(3.12)
$$\mathcal{K}u(x, y, t) = \frac{1}{2\pi}\int_0^{2\pi}\sin(x - \xi)u(\xi, y, t)d\xi.$$

The choice $K(x) = \sin x$ simplifies some of the subsequent algebra. However, transverse perturbations of interacting waves for general kernels K can be analyzed in a similar way.

Let T denote translation by π in x. Then $T\mathcal{K} = \mathcal{K}T = -\mathcal{K}$. The change of variables $u \to -u$, $v \to -Tv$, $x \to -x$, $y \to y$, $t \to t$ maps the solution in (3.8) with $\sigma = -1$ onto the solution with $\sigma = +1$, and it transforms (3.11) to

(3.13)
$$\partial_x\left\{u_t + uu_x + \mathcal{K}v\right\} - u_{yy} = 0,$$
$$\partial_x\left\{v_t + vv_x + \mathcal{K}u\right\} - v_{yy} = 0.$$

We shall consider solutions of (3.11) or (3.13) with $u = v$. (This assumption does not alter the final result.) It therefore suffices to consider transverse perturbations of $u = u_0(x - ct)$, where

$$(3.14) \qquad \partial_x \{u_t + uu_x + \mathcal{K}u\} + \sigma u_{yy} = 0.$$

We seek an expansion for long-wavelength transverse perturbations of the travelling wave solution (3.8) in the form

$$(3.15) \qquad u = u_0(\theta) + \epsilon u_1(\theta, Y, T) + \epsilon^2 u_2(\theta, Y, T) + 0(\epsilon^3), \quad \epsilon \to 0.$$

In (3.15), the multiple scale variables are evaluated at

$$\theta = x - ct - \phi(Y, T),$$
$$Y = \epsilon y, T = \epsilon t.$$

The function ϕ describes transverse variations in the wavefronts of the travelling wave. We shall show that ϕ satisfies a wave or Laplace equation – see (3.20) below.

We use (3.15) in (3.14), expand in powers of ϵ, equate coefficients of ϵ to zero, and integrate the result with respect to θ. At leading order in ϵ, we obtain the equation

$$\mathcal{K}u_0 + (u_0 - c)u_{0\theta} = 0.$$

This is satisfied by u_0 given in (3.8) with $\sigma = +1$, $u_0(\theta) = c + bf(\theta; \alpha)$.

At order ϵ, we obtain

$$(3.16) \qquad \mathcal{K}u_1 + [(u_0 - c)u_1]_\theta = \phi_T u_{0\theta}.$$

At order ϵ^2, we obtain

$$(3.17) \qquad \mathcal{K}u_2 + [(u_0 - c)u_2]_\theta = \phi_T u_{1\theta} - u_{1T} - u_1 u_{1\theta} + \sigma \phi_{YY} u_0 - \sigma \phi_Y^2 u_{0\theta}.$$

We use $< \cdot, \cdot >$ and $\|\cdot\|$ to stand for the L^2-inner product and norm with respect to θ,

$$< g, h > = \frac{1}{2\pi} \int_0^{2\pi} g(\theta)h(\theta)d\theta, \quad \|g\|^2 = < g, g > .$$

Suppose that g is a smooth 2π-periodic function of θ with zero mean, such that the system

$$(3.18) \qquad \mathcal{K}u + [(u_0 - c)u]_\theta = g,$$

has a 2π-periodic, zero-mean solution for u. Taking the inner product of (3.18) with u_0, integrating by parts, and using the fact that $\mathcal{K}$ is skew-adjoint, implies the solvability condition

$$(3.19) \qquad < u_0, g > = 0.$$

The solvability condition (3.19) is automatically satisfied for (3.16). We write a solution of (3.16) as

$$(3.20) \qquad u_1 = \phi_T U,$$

where U is a solution of

$$(3.21) \qquad \mathcal{K}U + [(u_0 - c)U]_\theta = u_{0\theta}.$$

Since u_0 is even, U is also an even functions of θ, as we show explicitly below. We do not include a homogeneous solution in u_1, proportional to $u_{0\theta}$, since it does not affect the final equation for θ.

We use (3.20) in (3.17) and impose the solvability condition (3.19). Some of the inner products are zero, because they are integrals of odd functions. After some algebra, we obtain the following equation for ϕ,

$$(3.22) \qquad \kappa \phi_{TT} = \phi_{YY}.$$

The coefficient κ in (3.22) is

$$(3.23) \qquad \kappa = \sigma \|u_0\|^{-2} < u_0, U > .$$

If $\kappa > 0$, then (3.22) is a wave equation, and the interacting waves are linearly stable to long transverse perturbations. If $\kappa < 0$, then (3.22) is a Laplace equation, and the interacting waves are unstable. The ill-posedness of (3.22) at short wavelengths is not physically meaningful, since we used a long wave approximation to derive the equation.

To obtain an explicit expression for κ, we solve (3.21). Assuming that U is even, (3.12) implies that

$$(3.24) \qquad \mathcal{K}U = C \sin \theta,$$

where C is the first Fourier cosine coefficient of U,

$$(3.25) \qquad C = \frac{1}{2\pi} \int_0^{2\pi} U(\xi) \cos \xi \, d\xi.$$

Using (3.24) and (3.8) in (3.21), integrating, and rearranging the result gives

$$(3.26) \qquad U = \frac{C \cos \theta}{bf} + \frac{P}{bf} + 1,$$

where P is a constant of integration. Using (3.26) in (3.25), and requiring that U has zero mean, gives a pair of equations for C and P,

$$(3.27) \qquad \begin{aligned} IC + HP + b &= 0, \\ (J - b)C + IP &= 0. \end{aligned}$$

In (3.27), it is convenient to define

$$(3.28) \quad M = <1, f>, \quad H = <1, \frac{1}{f}>, \quad I = <1, \frac{\cos\theta}{f}>, \quad J = <1, \frac{\cos^2\theta}{f}>,$$

where f is given in (3.9). these integrals are functions of the amplitude parameter α, and they are related by

$$(3.29) \qquad I = \frac{1}{\alpha}[M - H], \quad J = \frac{1}{3\alpha^2}\left[(\alpha^2 + 2)H - 2M\right].$$

In addition, from (3.10),

$$(3.30) \qquad b = \frac{2}{3\alpha^2}\left[(\alpha^2 - 1)H + M\right], \quad c = -bM.$$

All these functions can be expressed in terms of complete elliptic integrals of the first and second kinds,

$$K(m) = \int_0^{\pi/2} [1 - m\sin^2\theta]^{-1/2} d\theta,$$

$$E(m) = \int_0^{\pi/2} [1 - m\sin^2\theta]^{1/2} d\theta.$$

In particular,

$$H = \frac{2}{\pi}(1 + \alpha)^{-1/2} K\left(\frac{2\alpha}{1+\alpha}\right), \quad M = \frac{2}{\pi}(1 + \alpha)^{1/2} E\left(\frac{2\alpha}{1+\alpha}\right).$$

The solution of (3.27) is

$$(3.31) \qquad \begin{aligned} C &= \frac{-bI}{I^2 + bH - HJ}, \\ P &= \frac{b(J - b)}{I^2 + bH - HJ}. \end{aligned}$$

Equations (3.26) and (3.31) are the explicit solution of (3.21). Using (3.26), (3.28), and (3.30) in (3.23) implies that

$$(3.32) \qquad \kappa = \frac{\sigma b^2(1 - M^2)}{P(1 - HM) - CIM},$$

For general values of α, (3.32) must be evaluated numerically. Here, we shall calculate κ for small amplitude waves ($\alpha \to 0$) and waves close to the limiting wave ($\alpha \to 1$).

For small amplitudes, (3.28) and (3.29) imply that

$$(3.33) \qquad H = 1 + \frac{3}{16}\alpha^2 + 0(\alpha^4), \quad M = 1 - \frac{1}{16}\alpha^2 + 0(\alpha^4).$$

Then, using (3.29) – (3.33), we find that

$$(3.34) \qquad\qquad \kappa = -32\sigma\alpha^{-2} + 0(1).$$

It follows that small amplitude travelling waves (3.8) with $\sigma = +1$ are unstable, while those with $\sigma = +1$ are linearly stable to long transverse perturbations.

For large amplitudes, standard asymptotic expansions of complete elliptic integrals imply that

$$(3.35) \qquad H = \frac{1}{2^{1/2}\pi} \log\left(\frac{32}{1-\alpha}\right) + o(1), \quad M = \frac{2^{1/2}2}{\pi} + o(1),$$

as $\alpha \to 1-$. After some algebra, (3.29) – (3.32) and (3.35) show that

$$(3.36) \qquad\qquad \kappa = -\frac{3\sigma\pi^2}{32}\left(\frac{3\pi^2 - 16}{\pi^2 - 8}\right) + o(1).$$

Thus, the wave with $\sigma = +1$ is also unstable at large amplitudes, while the wave with $\sigma = -1$ is stable.

Acknowledgements. This work was supported in part by the Institute for Mathematics and its Applications with funds provided by the NSF, and by the NSF under Grant Number DMS-8810782.

REFERENCES

[1] ABLOWITZ, M.J., AND SEGUR, H., *Solitons and the Inverse Scattering Transform*, SIAM, Philadelphia (1981).

[2] BAMBERGER, A., ENQUIST, B., HALPERN, L., AND JOLY, P., *Parabolic wave equations and approximations in heterogeneous media*, SIAM J. Appl. Math., 48 (1988), pp. 99-128.

[3] CATES, A., *Nonlinear diffractive acoustics*, fellowship dissertation, Trinity College, Cambridge, unpublished, (1988).

[4] CHANG, T., AND HSIAO, L., *the Riemann Problem and Interaction of Waves in Gas Dynamics*, Longman, Avon (1989).

[5] COLE, J.D., AND COOK, L.P., *Transonic aerodynamics*, Elsevier, Amsterdam (1986).

[6] CRAMER, M.S., AND SEEBASS, A.R., *Focusing of a weak shock at an arête*, J. Fluid Mech, 88 (1978), pp. 209-222.

[7] CRIGHTON, D.G., *Model equations for nonlinear acoustics*, Ann. Rev. Fluid Mech., 11 (1979), pp. 11-13.

[8] CRIGHTON, D.G., *Basic theoretical nonlinear acoustics*, in Frontiers in Physical Acoustics, Proc. Int. School of Physics "Enrico Fermi", Course 93 (1986), North-Holland, Amsterdam.

[9] HAMILTON, M.F., *Fundamentals and applications of nonlinear acoustics*, in Nonlinear Wave Propagation in Mechanics, ed. T.W. Wright, AMD-77 (1986), pp. 1-28.

[10] HARABETIAN, E., *Diffraction of a weak shock by a wedge*, Comm. Pure Appl. Math., 40 (1987), pp. 849-863.

[11] HORNUNG, H., *Regular and Mach reflection of shock waves*, Ann. Rev. Fluid Mech., 18 (1986), pp. 33-58.

[12] HUNTER, J.K., *Transverse diffraction and singular rays*, SIAM J. Appl. Math., 75 (1986), pp. 187-226.

[13] HUNTER, J.K., *Hyperbolic waves and nonlinear geometrical acoustics*, in Transactions of the Sixth Army Conference on Applied Mathematics and Computing, Boulder CO (1989), pp. 527-569.

[14] HUNTER, J.K., AND KELLER, J.B., *Weakly nonlinear high frequency waves*, Comm. Pure Appl. Math., 36 (1983), pp. 547-569.

[15] HUNTER, J.K., AND KELLER, J.B., *Caustics of nonlinear waves*, Wave motion, 9 (1987), pp. 429-443.

[16] HUNTER, J.K., MAJDA, A., AND ROSALES R.R., *Resonantly interacting weakly nonlinear hyperbolic waves, II: several space variables*, Stud. Appl. Math., 75 (1986), pp. 187-226.

[17] KADOMTSEV, B.B., AND PETVIASHVILI, V.I., *On the stability of a solitary wave in a weakly dispersing media*, Sov. Phys. Doklady, 15 (1970), pp. 539-541.

[18] KELLER, J.B., *Rays, waves and asymptotics*, Bull. Am. Math. Soc., 84 (1978), pp. 727-750.

[19] KODAMA, Y., *Exact solutions of hydrodynamic type equations having infinitely many conserved densities*, IMA Preprint # 478 (1989).

[20] KODAMA, Y., AND GIBBONS, J., *A method for solving the dispersionless KP hierarchy and its exact solutions II*, IMA Preprint # 477 (1989).

[21] KUZNETSOV, V.P., *Equations of nonlinear acoustics*, Sov. Phys. Acoustics 16 (1971), pp. 467-470.

[22] LIGHTHILL, M.J., *On the diffraction of a blast I*, Proc. R. Soc. London Ser. A 198 (1949), pp. 454-470.

[23] LUDWIG, D., *Uniform asymptotic expansions at a caustic*, Comm. Pure Appl. Math. 19 (1966), pp. 215-250.

[24] MAJDA, A., *Nonlinear geometrical optics for hyperbolic systems of conservation laws*, in Oscillation Theory, Computation, and Methods of Compensated Compactness, Springer-Verlag, New York, IMA Volume 2 (1986), pp. 115-165.

[25] MAJDA, A. AND ROSALES, R.R., *Resonantly interacting hyperbolic waves, I: a single space variable*, Stud. Appl. Math. 71 (1984), pp. 149-179.

[26] MAJDA, A., ROSALES, R.R., AND SCHONBEK, M., *A canonical system of integro-differential equations in resonant nonlinear acoustics*, Stud. Appl. Math., 79 (1988), pp. 205-262.

[27] NAYFEH, A., *A comparison of perturbation methods for nonlinear hyperbolic waves*, in Singular Perturbations and Asymptotics, eds. R. Meyer and S. Parter, Academic Press, New York (1980), pp. 223-276.

[28] NIMMO, J.J.C., AND CRIGHTON, D.C., *Nonlinear and diffusive effects in nonlinear acoustic propagation over long ranges*, Phil. Trans. Roy. Soc. London Ser. A 384 (1986), pp. 1-35.

[29] PEGO, R., *Some explicit resonating waves in weakly nonlinear gas dynamics*, Stud. Appl. Math., 79 (1988), pp. 263-270.

[30] STURTEVANT, B., AND KULKARNY, V.A., *The focusing of weak shock waves*, J. Fluid Mech., 73 (1976), pp. 1086-1118.

[31] TIMMAN, R., in Symposium Transsonicum, ed. K. Oswatitsch, Springer-Verlag, Berlin, 394 (1964).

[32] WHITHAM, G.B., *Linear and Nonlinear Waves*, Wiley, New York (1974).

[33] ZABOLOTSKAYA, E.A., AND KHOKHLOV, R.V., *Quasi-plane waves in the nonlinear acoustics of confined beams*, Sov. Phys.-Acoustics, 15 (1969), pp. 35-40.

[34] ZAHALAK, G.I., AND MYERS, M.K., *Conical flow near singular rays*, J. Fluid. Mech., 63 (1974), pp. 537-561.

[35] JOHNSON, R.S., *Water Waves and Korteweg-deVries Equations*, J. Fluid. Mech., 97 (1980), pp. 701-719.

GEOMETRIC THEORY OF SHOCK WAVES*

TAI-PING LIU†

Abstract. Substantial progresses have been made in recent years on shock wave theory. The present article surveys the exact mathematical theory on the behavior of nonlinear hyperbolic waves and raises open problems.

Key words. Conservation laws, nonlinear wave interactions, dissipation and relaxation.

AMS(MOS) subject classifications. 76N15, 35L65

1. Introduction. A large class of nonlinear waves in mechanics, gas dynamics, fluid mechanics and the kinetic theory of gases are nonlinear hyperbolic waves in that the behaviors of these waves are governed in a basic way by certain a-priori determined characteristic values. Of these the most important ones are the shock waves. The strong nonlinear nature of shock waves makes the theory interesting and rich. Because most physical models carrying hyperbolic waves are not scalar but systems, waves of different characteristic families interact. Understanding this nonlinear coupling of waves is the essence of the theory for hyperbolic conservation laws, which is described in the next section. With the inclusion of dissipative mechanism, as in the compressible Navier-Stokes equations, we have viscous conservation laws. The inclusion is due to the importance of dissipative mechanisms in the study of shock layer, initial and boundary layers, and wave interactions. It is also to check the validity of hyperbolic conservation laws through the zero dissipation limits. These issues are considered in Section 3. The phenomenon of relaxation occurs in many physical situations: gas dynamics with thermo-non-equilibrium elasticity with fading memory, kinetic theory of gases, etc. Conservation laws with relaxation are in some sense more singular a perturbation of hyperbolic conservation laws than that of viscous conservation laws. This and the dual nature of hyperbolicity and parabolicity for a relaxation system are explained in Section 4. Conservation laws with reaction and diffusion may be highly unstable. In Section 5 a class of such systems originated from the study of nozzle flow is described. Nonstrictly hyperbolic conservation laws are important in the study of MHD, elasticity and multiphase flows. Behavior of waves for such a system with or without the effect of damping or dissipation is discussed in Section 6. Finally several concluding remarks are made in the last section.

2. Hyperbolic Conservation Laws.

Hyperbolic conservation laws

$$(2.1) \qquad \frac{\partial u}{\partial t} + \frac{\partial f(u)}{\partial x} = 0, \qquad u \in \mathbf{R}^n \ ,$$

*This paper was written while the author was visiting the Institute for Mathematics and Its Applications, University of Minnesota, Minneapolis, MN 55455

†Courant Institute, New York University, 251 Mercer Street New York, NY 10012

are the basic model for shock wave theory. In this section we assume that it is strictly hyperbolic, i.e. $f'(u)$ has real and distinct eigenvalues $\lambda_1(u) < \lambda_2(u) < \cdots < \lambda_n(u)$. The compressible Euler equations have this property. Each characteristic value $\lambda_i(u)$ carries a family of waves. The interactions of these waves are the essence of the theory for (2.1). For system of two conservation laws the interactions of waves of different characteristic families are weaker because of the existence of the Riemann invariants for two equations. Mainly because of this, behavior of solutions whose initial data oscillating around a constant has been studied only for two conservation laws, Glimm-Lax [4], which are genuinely nonlinear in the sense of Lax [8]. For general systems for which each characteristic field is either genuinely nonlinear or linear degenerate, the large-time behavior of solutions with small total variation has been studied satisfactorily, Lin [11]. The regularity and large-time behavior for general systems not necessarily genuinely nonlinear are studied in Liu [12], though with no rate of convergence to asymptotic states. The above results are obtained through a principle of nonlinear superposition, Glimm [3], Glimm-Lax [4] and Liu [10].

There is no satisfactory uniqueness theory for any physical system. The best result so far is that of DiPerna [2], which shows that for genuinely nonlinear two conservation laws a piecewise smooth solution with no compression wave is unique within the class of solutions of bounded total variation. The problem is important because of the need of the entropy condition for the hyperbolic shock wave theory.

Existence theory for large data has been obtained for certain two conservation laws using the theory of compensated compactness, see articles on the theory in this volume. The problem makes sense in general only for certain physical systems, since it is easy to construct systems for which Riemann problem with large data is not solvable. It would be interesting to study the problem for the compressible Euler equations. Study of interaction of rarefaction waves, Lin [9], has led us to conjecture that when the data does not yield vacuum immediately then the solution does not contain vacuum and is of bounded local variation for any positive time.

3. Viscous Conservation Laws.
Consider the viscous conservation laws

$$(3.1) \qquad \frac{\partial u}{\partial t} + \frac{\partial f(u)}{\partial x} = \frac{\partial}{\partial x}\left(B(u)\frac{\partial u}{\partial x}\right) \quad , \ y \in \mathbf{R}^n \ .$$

For physical systems, the viscosity matrix is in general not positive. The system then becomes hyperbolic-parabolic and not uniformly parabolic. This has the effect that discontinuity in the initial data may not be smoothed out. A more important hyperbolic character of the system comes from the nonlinearity of the flux function $f(u)$, it is there even if $B(u)$ is positive. The behavior of a nonlinear wave can be detected through the characteristic values. Viscous shock waves are compressive and therefore is stable in a different sense from the expansion waves, Liu [13], Liu-Xin [15], and references therein. Diffusion waves are at worst weakly expansive and quite stable, Chern-Liu [1]. It is therefore natural to study these waves through a hyperbolic-parabolic technique, Liu [13]. The technique needs to be refined to study the stability of a general wave pattern consisting of both compression and expansion waves.

The central question for (3.1) is to understand the behavior of a general flow as the strength of the viscosity matrix $B(u)$ varies, in particular when it tends to zero. The interesting case, of course, is when the corresponding inviscid solution for (2.1) is not smooth. Hopf-Liu [6] solves the problem for a single shock wave. The interaction of initial and shock layers are studied there. For viscosity matrix not positive, discontinuities in the data propagate into the solution [6] and reference therein. Recently Goodman-Xin [5] uses the technique of matched asymptotic expansion and characteristic-energy method to show that for a given piecewise smooth inviscid solution there exists a sequence of viscous solutions converging to the given inviscid solution. While the inviscid solution in [5] is more general than that of [6], [5] does not address the formation and interaction of shock waves, and the initial data for viscous solutions are not fixed. The difference of results in these works provided interesting problems. One possible way to make further progress in this area is to refine and generalize the techniques in [13].

4. Relaxation. In many physical situations effects of nonequilibrium, delay, memory and relaxation are important. The mathematical models usually take the form of either hyperbolic conservation laws with integral or lower-order terms. Such an effect has a partial smoothing property much like the viscous conservation laws except that it is weaker and does not smooth out strong shock wave. Although equations in various physical situations, e.g. gas dynamics with thermo-non-equilibrium, elasticity with fading memory, kinetic theory of gases, take different forms, there are common features, see the mathematical analysis for a simple model in [14] and physical models in the references therein. Mathematical models in phase transition and multi-phases often are ill-posed. It is hoped that the inclusion of the nonequilibrium would make the equations well-posed. When succeeded, the question is then to study the stability and instability of nonlinear waves. These, however, remain challenges for future researches.

As mentioned above, hyperbolic conservation laws with relaxation have some dissipative behavior not dissimilar to viscous conservation laws. However, the former is less parabolic than the latter. There is a hierarchy of hyperbolicity and parabolicity. Mathematical analysis suitable to study general nonlinear behavior of these various degree of hyperbolicity and parabolicity of these physical models remains far from being complete. Here we mention a specific problem. Consider a simple model as in [14].

$$(4.1) \qquad \begin{aligned} \frac{\partial v}{\partial t} + f(v, w) &= 0 \\ \frac{\partial w}{\partial t} + \frac{\partial g}{\partial x}(v, w) &= h(v, w) \end{aligned}$$

where v represents the conserved quantity and w the relaxing quantity and $h(v, w)$ may take the form

$$h(v, w) \sim \frac{w^*(v) - w}{\tau(v)} ,$$

$\tau(v) > 0$ the relaxation time and $w^*(v)$ the equilibrium value for v. One may view the system as the perturbation of the equilibrium conservation law

$$(4.2) \qquad \frac{\partial v}{\partial t} + \frac{\partial}{\partial x} f(v, w^*(v)) = 0$$

The question is to show that solutions of 14.1) tend to solutions of (4.2) in the limit of the relaxation time $\tau(v) \to 0_+$. Because the perturbation (4.1) of (4.2) is more singular than that of (3.1) of (2.1), theory of compensated compactness has not been shown to work.

5. Convective Reaction-Diffusion. When reaction is present in viscous conservation laws, the system becomes convective reaction-diffusion equations and often takes the form

$$\frac{\partial u}{\partial t} + \frac{\partial f(u)}{\partial t} = \frac{\partial}{\partial x} \left(B(u) \frac{\partial u}{\partial x} \right) + g(x, t, u) \ .$$

Such a model occurs in important physical situations such as combustions. Nonlinear waves for the system can be highly unstable, see articles on combustions in this volume. In [7] a model arised from the theory for gas flow through a nozzle is studied. It turns out that the inviscid theory offers guide for the stability and instability of waves for such viscous model. The theory for nozzle flow, see (7) and references therein, provides the mathematical basis for the occurance of unstable waves for gas flows. For nozzle flow instability results from the geometric effect of the nozzle. In combustion the reactions are chemical and have highly unstable effects. Mathematical study of such behavior remains largely a challenge.

6. Nonstrictly Hyperbolicity. Studies of conservation laws which are nonstrictly hyperbolic are centered mostly on the important Riemann problem, see articles on this volume. The effects of viscosity on the behavior of overcompressive waves have been studied, see the article of Liu and Xin in the proceeding of the last IMA workshop on equations of mixed types. Recently overcompressive shock waves have shown to occur in MHD and nonlinear elasticity, see also articles in the aforementioned volume. It would be interesting to study the effects of viscosity for these systems as well as for other systems with crossing shocks.

7. Concluding Remarks. We have seen several types of equations which carry shock waves; and there are more. The classification of these equations into hyperbolic, parabolic or mixed types tells part of the story. Coupling of different models of waves and dissipation induced by nonlinearity, relaxation, viscosity etc. are also important elements in the shock wave theory. It is important to recognize the elementary modes of a general flow, whenever possible. Even though the progresses so far have been very substantial, many more fundamental questions remain to be answered. One hopeful sign is that several different approaches are available now. The present article emphasizes the geometric approach of shock waves. Undoubtedly new progress will be made based on the basic understanding of the available techniques illustrated in the articles in this volume.

REFERENCES

[1] CHERN, I.-L., AND LIU, T.-P., *Convergence to diffusion waves of solutions for viscous conservation laws*, Comm. Math. Phys. 110 (1987), 503–517.

[2] DI PERNA, R., *Uniqueness of solutions to hyperbolic conservation laws*, Indiana U. Math. J. 28 (1979), 244–257.

[3] GLIMM, J., *Solutions in the large for nonlinear systems of equations*, Comm. Pure Appl. Math., 18 (1965), 95–105.

[4] GLIMM, J. AND LAX, P.D., *Decay of solutions of nonlinear hyperbolic conservation laws*, Amer. Math. Soc. Memoir, No. 101 (1970).

[5] GOODMAN, J. AND XIN, Z., (preprint).

[6] HOPF, D. AND LIU, T.-P., *The inviscid limit for the Navier-Stokes equations of compressible isentropic flow with shock data*, Indiana U. J. (1989).

[7] HSU, S.-B. AND LIU, T.-P., *Nonlinear singular Sturm-Liouville problem and application to transonic flow through a nozzle*, Comm. Pure Appl. Math. (1989).

[8] LAX, P.D., *Hyperbolic systems of conservation laws, II*, Comm. Pure Appl. Math., 10 (1957), 537–566.

[9] LIN, L., *On the vacuum state for the equation of isentropic gas dynamics*, J. Math. Anal. Appl. 120 (1987), 406–425.

[10] LIU, T.-P., *Deterministic version of Glimm scheme*, Comm. Math. Phys., 57 (1977), 135–148.

[11] LIU, T.-P, *Linear and nonlinear large time behavior of general systems of hyperbolic conservation laws*, Comm. Pure Appl. Math., 30 (1977), 767–798.

[12] LIU, T.-P., *Admissible solutions of hyperbolic conservation laws*, Amer. Math. Soc. Memoir, No. 240 (1981).

[13] LIU. T.-P., *Nonlinear stability of shock waves for viscous conservation laws*, Memoirs, Amer. Math. Soc. No. 328 (1975).

[14] LIU, T.-P., *Hyperbolic conservation laws with relaxation*, Math. Phys. 108 (1987), 153–175.

[15] LIU, T.-P, AND XIN, Z., *Nonlinear stability of rarefaction waves for compressible Navier-Stokes equations*, 118 (1988), 415–466.

AN INTRODUCTION TO FRONT TRACKING

CHRISTIAN KLINGENBERG* AND BRADLEY PLOHR†

Abstract. In fluid flows one can often identify surfaces that correspond to special features of the flow. Examples are boundaries between different phases of a fluid or between two different fluids, slip surfaces, and shock waves in compressible gas dynamics. These prominent features of fluid dynamics present formidable challenges to numerical simulations of their mathematical models. The essentially nonlinear nature of these waves calls for nonlinear methods. Here we present one such method which attempts to explicitly follow (track) the dynamic evolution of these waves (fronts). Most of this exposition will concentrate on one particular implementation of such a front tracking algorithm for two space, where the fronts are one-dimensional curves. This is the code associated with J. Glimm and many co-workers.

Introduction. In fluid flows one can often identify surfaces of co-dimension one that correspond to prominent features in the flow. Examples are boundaries between different phases of a fluid or between two different fluids, slip surfaces, shock curves in compressible gasdynamics. All such surfaces are characterized by significant changes in the flow variables over length scales small compared to the flow scale. For example in oil reservoirs the oil banks have a size of 10 meters compared to an average length scale of 10 kilometers; or in compressible gas dynamics shock waves have a width of 10^{-5} cm compared to a length scale of 10 cm. The dynamics of such waves may be influenced by their internal structures. Whereas for shock waves the speed depends on the asymptotic states to the left and right, for two dimensional detonation waves the speed depends also on the chemistry and curvature, [B], [J]. There are situations where it is necessary to take these physical aspects of the flow into account when doing a numerical simulation.

A simple model for nonlinear wave propagation is Burger's equation

$$u_t + uu_x = \nu u_{xx} \ ,$$

where the state variable u is convected with characteristic speed u and diffused with viscosity ν. Because of the dependence of the characteristic speed on the state variable one obtains a focusing effect that leads to the formation of shock waves. Consider initially a wave of length L (see Fig. 1). The monotone decreasing part of the wave will steepen such that in a thin layer the solution rapidly decreases from a value u_l to u_r. The width w of this layer is about $\dfrac{\nu}{|u_l - u_r|}$, and this layer moves with speed $s = \dfrac{1}{2}(u_l - u_r)$. If $w \ll L$, we may approximate the layer by a jump from u_l to u_r and consider the inviscid limit by neglecting ν to obtain the inviscid Burger's equation

$$u_t + \left(\frac{u^2}{2}\right)x = 0 \ .$$

*Department of Applied Mathematics, University of Heidelberg, Im Neuenheimer Feld 294, D-6900 Heidelbert, Germany

†Department of Applied Mathematics and Statistics, SUNY at Stony Brook, Stony, Brook, NY 11794

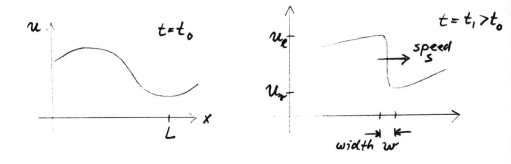

Fig. 1 The evolution of the initial data (left) under

$$u_t + uu_x = \nu u_{xx}$$

is given on the right.

Now data as in Fig. 1 leads to jumps in the solution, where the Rankine–Hugoniot conditions govern the relationship between the speed of the jump and its left and right asymptotic states.

When computing such a flow with very small viscosity ν, suppose we represent the state variables associated with points on a fixed underlying grid with spacing Δx. In this framework we would like to contrast two numerical methodologies: shock capturing and shock tracking. In the shock capturing methods ν is replaced by a numerical viscosity $\nu_{num} \gg \nu$. The width of a shock layer $w_{num} = \dfrac{\nu_{num}}{|u_l - u_r|} \sim 3\Delta x$, so that these waves are most accurate for weak waves. In a shock tracking method an additional moving grid point is introduced which serves as a marker for the shock position. The algorithm has to update its position and the asymptotic left and right states on the underlying fixed grid. For the moving of the shock point analytic information about it is necessary. Shock tracking corresponds to replacing ν by zero, so it is best for strong waves and gives a high resolution on relatively coarse grids.

The front tracking principle, which is not limited to conservation laws or to shocks, is that a lower dimensional grid gets fit to and follows the significant features in the flow. This is coupled with a high quality interior scheme to capture the waves that are not tracked.

In the following we talk only about front tracking in two space dimensions. First we describe tracking of a single wave and mathematical issues arising from this. Next we discuss tracking wave interactions and its mathematical issues. Then follows a section describing the data structure of a front tracking code. After a few numerical examples we give a conclusion.

Front tracking applied to a singe wave. Suppose we consider an expanding cylindrical shock wave for a certain time interval. Say this is modeled by the two dimensional Euler equations for polytropic gas dynamics where the outstanding feature of the flow is a shockwave with smooth flow in front and behind it. If the numerical simulation requires a high level of resolution on a moderate size grid, front tracking lends itself to this problem. To this end a one dimensional grid is fitted to the shock wave and follows its dynamic evolution. The smooth flow is captured using an underlying two dimensional grid, where in each time step an initial-boundary value problem is solved in each smooth component of the flow field.

The front is represented by a finite number of points along the curve, which carry with them physical data, in this case the left and right states and the fact that it is a hydrodynamic shock wave. Say the underlying grid is cartesian, which carries the associated state variables at each grid point. Each timestep consists of a front propagation and an interior update.

THE CURVE PROPAGATION is achieved by locally at each curve point rewriting the equation in a rotated coordinate system, normal and tangential to the front:

$$u_t + \hat{n}\big((\hat{n} \cdot \nabla)f(u)\big) + \hat{s} \cdot \big((\hat{s} \cdot \nabla)f(u)\big) = 0.$$

This then gets solved through dimensional splitting. The normal step reduces to a one dimensional Riemann problem, if one approximates the data to the left and right of the shock by constants.

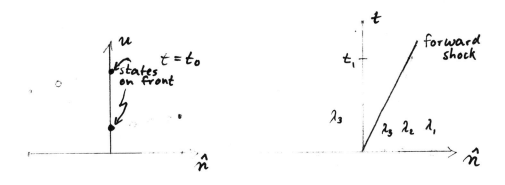

Fig. 2 A second order scheme for the normal propagation of a hydrodynamic shock wave, [CG].

This normal step can be made into a second order scheme in the following way [CG], see Fig. 2:

    – first solve Riemann problem to obtain speed and approximate states at $t = t_1$,

    – follow the characteristics from the left and right states at $t = t_1$ back to $t = t_0$ and use the data at the foot of them to obtain updated left and right states at $t = t_1$

– finally solve a Riemann problem at $t = t_1$ to improve states and speed there.

After the normal step has been implemented at all points representing the shock curve, the tangential step, which propagates surface waves, is done by a one dimensional finite difference scheme on each side of the front.

Because points on the front may move too far apart (or too close together) during propagation, a routine which redistributes the points along the curve is sometimes useful. One has to be cautious though, because this routine stabilizes the curve which may tend to become unstable due to physical or numerical effects.

THE INTERIOR SCHEME. The underlying principle is to solve an initial-boundary value problem on both sides of the front (the front is a moving boundary), and to never use states on the opposite side of the front. Away from the front this is readily achieved by using any finite difference scheme compatible with the resolution one needs in the interior. Near the front an algorithm which is consistent with the underlying partial differential equation has yet to be worked out. The following recipe has been implemented successfully (see Fig. 3): suppose the stencil gets cut off by the front. Use the states at the nearest crossing point (obtained through linear interpolation from the front states) and place them at the missing stencil points.

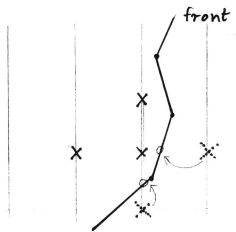

Fig. 3 A five point centered stencil near the front, where the states on the front are assigned to the two grid points on the opposite side of the front.

So far two papers have addressed the front-interior coupling problem in two space dimensions: [CC] suggest and implement a coupling which is conservative for gas dynamics. [KZ] have formulated a class of front tracking schemes for which they show stability.

Mathematical issues related to this. In the previous section we saw that this approach leads to the study of one dimensional Riemann problems. This is a special Cauchy problem of the type

$$u_t + f(u)_x = 0$$
$$u(0,) = \begin{cases} u_L, x < 0 \\ u_R, x > 0 \end{cases}$$

Since the equation and initial data are scale invariant

$$(x,t) \longrightarrow (\alpha x, \alpha t) \; , \; \alpha > 0$$

we may expect scale invariant solutions. These are well understood e.g. for the scalar equation and for gas dynamics.

There is a considerable research effort trying to understand the Riemann solutions of more complicated models. One example are the 2 × 2 systems with quadratic flux functions studied by various authors, e.g. [IM], [IT]. New interesting mathematical phenomena arise:

– non-classical waves

– non-contractible discontinuous waves, i.e. it is not possible to decrease the wave strength to zero while following a connected brach of the wave curve

– open existence and uniqueness questions.

Another example are Riemann solvers for equations describing conservation of mass, momentum and energy in real materials. Their effects on the wave structure has been studied , [MP]. In another approach the equation of state is tabulated (SESAME code at Los Alamos). Scheuermann used this for a Riemann solver by preprocessing the data.

Finally we mention certain waves where the internal structure of the waves play a role. Whereas say for shock waves of isentropic gas dynamics the two jump equations plus the three pieces of information given by the impinging characteristics determine the four state variables on both sides of the shock with its speed, for transitional shock waves not enough information impinges through the characteristics and one needs information from the internal structure in order to determine speed and states. The structure depends sensitively on the viscosity used in the parabolic approximation. These waves thus present a danger for finite difference schemes, which introduce their own brand of viscosity which is different for different schemes. Here a tracking algorithm which mimicks the structure with a Riemann solver lends itself naturally to this problem.

The front tracking method described so far could also be applied to more complex flow patterns than the expanding spherical shock wave by simply tracking a single front and capturing all other phenomena using a high quality interior scheme. An example are the Euler equations coupled with complex chemistry used to model the flow around a hypersonic projectile [Zhu]. Here the hydrodynamic bowshock is tracked and the flow with most of the chemistry concentrated right behind this shock is captured. This is an example where a tracking of the bowshock is necessary.

Wave interaction. One can also track interacting waves. To illustrate this consider a planar shock wave impinging on a curved ramp (Fig. 4), giving rise first to a regular and then to a Mach reflection. This is an example on how new curves may arise. For hydrodynamic shock waves this bifurcation may arise through the intersection of shocks with each other or with other "curves", or through compressive waves ("blow up" of the smooth solution). If one wants to incorporate these phenomena into a front tracking algorithm it is necessary to understand them mathematically. For example in the case of the planar shock impinging onto the wedge, one needs a criterion which gives for given shock strength the ramp-angle when a bifurcation from regular to Mach reflection occurs. If one wants to track all the waves, the algorithm needs to have this criterion built in.

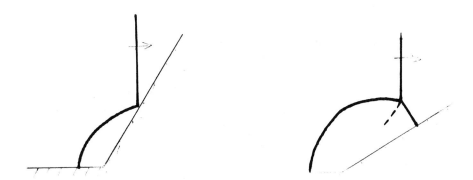

Fig. 4 A planar shock impinges onto a wedge, and, depending on the shock strength and wedge angle, give rise either to a regular reflection (left) or a Mach reflection (right). In the latter the reflected point has lifted off the wall to become a "triple point" from which a "Mach stem" connects to the wall.

This is an example of a two dimensional Riemann problem. In general, at the meeting point of more than two curves, if one approximates the curves by rays and the states nearby by constant states, these nodes are examples of two dimensional Riemann problems. As in one dimensional case, this is scale invariant Cauchy data $(x, y, t \longrightarrow \alpha, \alpha y, \alpha t \ , \ \alpha > 0)$ giving rise to a self similar solution $u = u\left(\dfrac{x}{t}, \dfrac{y}{t}\right)$. Thus front tracking may lead to two dimensional Riemann problems.

Mathematical issues related to this. There has been some progress on studying the qualitative behavior of two dimensional Riemann problems. For the equations of compressible inviscid, polytropic gas dynamics, in analogy to the one dimensional Riemann problem which is resolved by elementary waves, one expects that the two dimensional Riemann problem will evolve into a configuration containing several two dimensional elementary waves. This this end these elementary waves were completely classified [GK], some of them can already be found in [L].

For the scalar two dimensional conservation law the two dimensional Riemann problem could be solved much further. For

$$u_t + f(u)_x + g(u)_y = 0$$

with $f = g$ it was solved in [W] (f convex), [L1], [L2] (f one inflection point), [KO] (f any number of inflection points). For $f \neq g$ [W] (f close to g, f convex) and [KO], [CH] (f convex, g one inflection point) gave solutions.

Numerical implementation. This knowledge of two dimensional Riemann problems has been used in front tracking codes to some extent. The classification of elementary waves for gas dynamics gave a list of the generic node one can expect there, that is all generic meeting points of shock waves, contact discontinuities and centered rarefaction waves. The tracking of a node is the numerical solution of a subcase of the full Riemann problem, one has to determine the velocity and states associated with one specific elementary wave. for gas dynamics this has been done [GK], G1], [G2].

For the scalar two dimensional conservation law the resolution of the two dimensional Riemann problem caused by the crossing of two shocks has been implemented. Whereas in [K] the point is to solve the interaction of two scalar waves quite accurately, in [GG] the emphasis is on following scalar wave interaction within a complicated topology of curves in a robust fashion without an unacceptable proliferation of subcases. An approximate numerical solution to a general two dimensional Riemann problem was implemented by approximating the flux functions by piecewise linear functions [R].

Computer science issues related to front tracking. Here we briefly describe a package of subroutines which provides facilities for defining, building and modifying decompositions of the plane into disjoint components separated by curves. It is worth noting that ideas from conceptual mathematics, symbolic computation and computer science have been utilized, thereby going beyond the usual numerical analysis framework, see [GM].

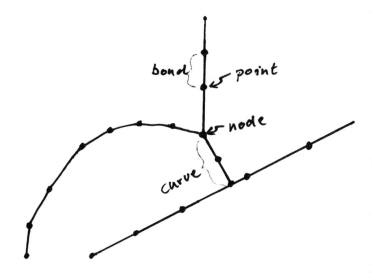

Fig. 5 The front tracking representation of a Mach reflection.

Taking the Mach reflection example (Fig. 4), we illustrate in Fig. 5 the representation of this particular flow. The front consists of piecewise linear curves at the endpoints of each linear piece we have associated quantities like states and wave types. Given this interface, the plane is decomposed into disjoint components. An integer component value is associated with each such component. Given any point x, y in the plane, the component value can be recovered. The underlying grid and possible interpolating grids near the front allow the definition of associated state variables in the interior.

There is a recursive data structure. It consists of

POINT:	which denotes the position of the grid points on the curve
BOND:	which denotes the piece of the curve between two adjacent points and previous bond
	by giving a start and an end point and having a pointer to the next
CURVE:	denoting usually a pice of the interface homotopic to an interval.
	A curve is a doubly-linked list of bonds given by a start and node (see below). It has a point to the first and last band.
NODE:	which is the position of a point on the interface where more than two curves meet. Its position is given with a list in and out curves.
INTERFACE:	is a list of nodes and curves

Then there are routines that operate on the interface structure. There are routines that allocate and delete the above structures, then those which add these to the interface, routines that split and join bonds and curves, all needed for example when there is a change in topology. Also one can traverse a list of the above structures.

The code has purposely been set up in such a way that this interface data structure can be dressed with the physics of a given problem containing curves. For gas dynamics one would associate with each point a left and right state, with each curve the wave type and at the node the state in each sector in order to have the set up for the Riemann problem.

This whole structure now needs routines which allows the interface to propagate from one timestep to the next. This is done by first moving the interface. This means moving bonds and nodes. Next the interior is updated. Then one has to handle possible interactions and bifurcations. These have to be detected, classified (they could be tangler of curves or two dimensional Riemann problems and then resolved. There is also a routine which redistribute points on the interface, in case they become to close together or too far apart.

Numerical examples. We shall give four examples out of many that have been calculated over the years with the code. Fig. 6 shows regular and Mach reflection, [GK]. Fig. 7 show an underwater explosion [G2]. Fig. 8 shows Rayleigh–Taylor instability [FG]. Fig. 9 shows an example from oil reservoir modelling [GG].

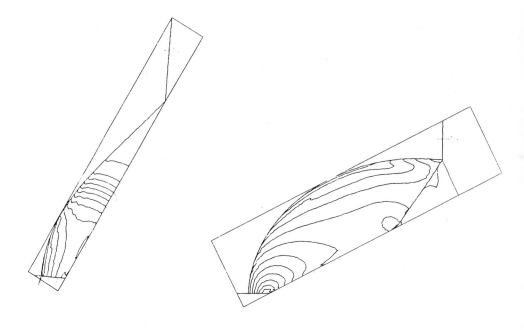

Fig. 6 On the left the numerical simulation of regular reflection, where the incident shock has Mach number 2.05 and the wedge angle is 63.4°. The calculation was performed on a 80 by 20 grid. The picture shows lines of constant density inside the bubble formed by the reflected shock.

On the right the numerical simulation of a Mach reflection, where the incident shock has Mach number 2.03 and the wedge is 27°. Inside the bubble formed by the reflected shock the calculated lines of constant density are shown. The calculations we performed on a 60 by 40 grid.

In both cases there is excellent agreement with experiments, [GK].

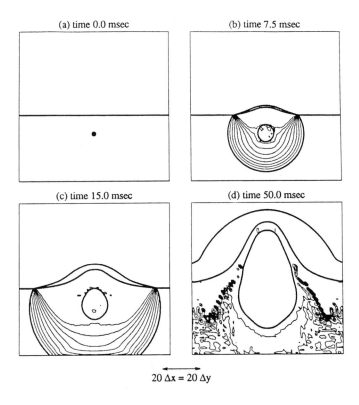

(a) time 0.0 msec   (b) time 7.5 msec

(c) time 15.0 msec   (d) time 50.0 msec

$20 \, \Delta x = 20 \, \Delta y$

Fig. 7 An underwater expanding shock wave diffracting through the water's surface. The internal pressure is 100 kbans and initial radius of 1 meter installed 10 meters below the water's surface. The tracked front in dark lines is super imposed over lines of constant pressure. The grid is 60 by 120.

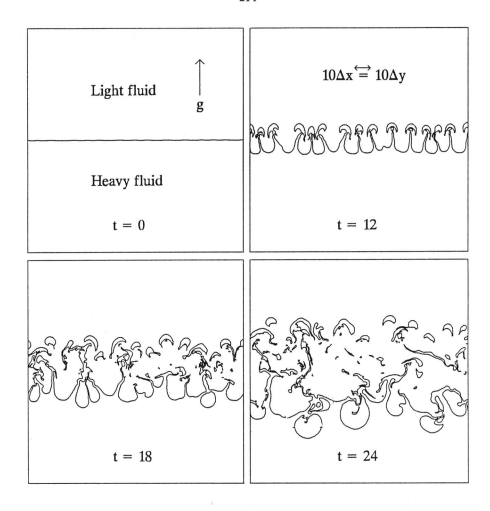

Fig. 8 Two compressible fluids of different densities, with gravitational forces (here positing upward) pushing lighter fluid into heavy one. The interface is initialized by 14 bubbles with different wave length and initial amplitude of 0.01. The density ratio is 10. The interface between these fluids is unstable and leads to a mixing layer, with bubbles of light fluid rising in the heavy fluid.

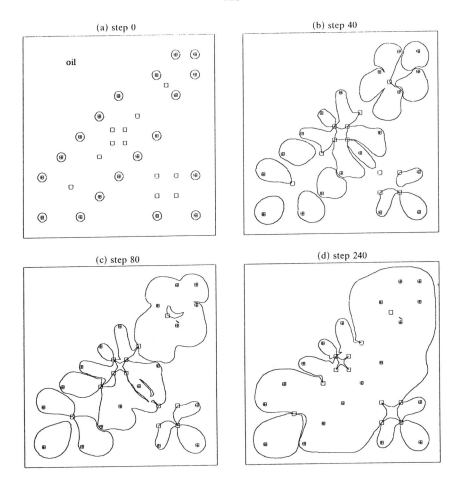

Fig. 9 A horizontal cross section of an oil reservoir modeled by the Buckley–Leverett equations. Water gets injected at 19 injection wells (cross squares), displacing the oil in the porous media, and oil get extracted at 12 producing wells (open squares). Plots of the fronts between water and oil are shown. The frontal mobility ratio for water displacing oil is 1.33.

Conclusion. It should have become clear that this numerical approach forces one to think hard about underlying physics and mathematics. If one is successful at penetrating the problem at hand, front tracking can give the correct simulation with very high resolution.

REFERENCES

[B] BUKIET, *The effect of curvature on detonation speed*, SIAM J. Appl. Math., 49 (1989).

[CH] CHANG, HSIAO, *The Riemann problem and interaction of waves in gas dynamics*, John Wiley, New York, 1989.

[CC] CHERN, COLELLA, *A conservative front tracking method for hyperbolic conservation laws*, Journal Comp. Physics, (1989).

[CG] CHERN, GLIMM, McBRYAN, PLOHR, YANIV, *Front tracking for gas dynamics*, J. Comp. Phys., 62 (1986).

[FG] FURTATO, GLIMM, GROVE, LI, LINDQUIST, MENIKOFF, SHARP, ZHANG, *Front tracking and the interaction of nonlinear hyperbolic waves*, NYU preprint (1988).

[GM] GLIMM, McBRYAN, *A computational model for interfaces*, Adv. Appl. Math., 6 (1985).

[GK] GLIMM, KLINGENBERG, McBRYANT, PLOHR, SHARP, YANIV, *Front tracking and two dimensional Riemann problems*, Adv. Appl. Math. 6 (1985).

[GG] GLIMM, GROVE, LINDQUIST, McBRYAN, TRYGGVASON, *The bifurcation of tracked scalar waves*, SIAM J. Sci. Stat. Comp. 9 (1988).

[G1] GROVE, *The interaction of shock waves with fluid interfaces*, Adv. Appl. Math (1990).

[G2] GROVE, *Anamolous reflection of a shock wave at fluid interfaces*, Los Alamos preprint LA UR (1989) 89–778.

[IM] ISAACSON, MARCHESIN, PLOHR, TEMPLE, *The classification of solutions of quadratic Riemann problems I*, MRC Report (1985).

[IT] ISAACSON, TEMPLE, *The classification of solutions of quadratic Riemann problems II, III*, to appear SIAM J. Appl. Math..

[J] JONES, *Asymptotic analysis of an expanding detonation*, NYU DOE report (1987).

[KO] KLINGENBERG, OSHER, *Nonconvex scalar conservation laws in one and two space dimensions*, Proc. 2nd Int. Conf. Nonlin. Hyp. Probl., ed. Ballmann, Jeltsch, Vieweg Verlag (1989).

[KZ] KLINGENBERG, ZHU, *Stability of difference approximations for initial boundary value problems applied to two dimensional front tracking*, Proc. 3rd Int. Conf. on Hyp. Problems, ed. Gustafsson (1990).

[L] LANDAU, LIFSHITZ, *Fluid Mechanics*, Addison Wesley (1959).

[L1] LINDQUIST, *The scalar Riemann problem in two space dimensions*, SIAM J. Anal. 17 (1986).

[L2] LINDQUIST, *Construction of solutions for two dimensional Riemann problems*, Adv. Hyp. PDE and Math. with Appl. 12A (1986).

[MP] MENIKOFF, PLOHR, *Riemann problem for fluid flow of real materials*, Los Alamos prepint LA UR–8849 (1988).

[R] RISEBRO, *The Riemann problem for a single conservation law in two space dimensions*, May 1988, Freiburg, Germany.

[W] WAGNER, *The Riemann problem in two space dimensions for a single conservation law*, SIAM J. Math. Anal. 14 (1983).

[Zhu] ZHU, CHEN, WARNATZ, *Same computed results of nonequilibrium gas flow with a complete model*, SFB123 Heidelberg University preprint 530 (July 1989).

ONE PERSPECTIVE ON OPEN PROBLEMS IN MULTI-DIMENSIONAL CONSERVATION LAWS

ANDREW J. MAJDA*

Introduction. It is evident from the lectures at this meeting that the subject of systems of hyperbolic conservation laws is flourishing as one of the prototypical examples of the modern mode of applied mathematics. Research in this area often involves strong and close interdisciplinary interactions among diverse areas of applied mathematics including

(1) Large (and small) scale computing

(2) Asymptotic modelling

(3) Qualitative modelling

(4) Rigorous proofs for suitable prototype problems

combined with careful attention to experimental data when possible. In fact, the subject is developing at such a rapid rate that new predictions of phenomena through a combination of theory and computations can be made in regimes which are not readily accessible to experimentalists. Pioneering examples of this type of interaction can be found in the papers of Grove, Glaz, and Colella in this volume as well as the recent work of Woodward, Artola, and the author ([1], [2], [3], [4], [5], [6]). In this last work, novel mechanisms of nonlinear instability in supersonic vortex sheets have been documented and explained very recently through a sophisticated combination of numerical experiments and mathematical theory.

Here I will discuss my own perspective on several open problems in the field of hyperbolic conservation laws which involve the interaction of ideas in modern applied mathematics. Since the audience at the meeting consisted largely of specialists in nonlinear P.D.E. and analysis, I will mostly emphasize open problems which represent rigorous proofs for prototype situations. I will concentrate on open problems in three different areas:

1) Self-similar patterns in shock diffraction;

2) Oscillations for conservation laws;

3) New phenomena in conservation laws with source terms.

In the first section, I will give the compressible Euler equations as the prototypical example of a system of conservation laws in several space variables and then describe several approximations such as isentropic, potential flows which yield other related hyperbolic conservation laws. I will also discuss the nature of these approximations in multi-dimensions. This material may not be well-known to the reader and provides background material for some of the open problems discussed in the remaining sections. Each of the remaining three sections is devoted to my own

*Department of Mathematics and Program in Applied and Computational Mathematics, Princeton University, Princeton, NJ 08544 partially supported by grants N.S.F. DMS 8702864, A.R.O. DAAL03-89-K-0013, O.N.R. N00014-89-J-1044

perspective on the open problems in the three areas mentioned earlier. It was clear from my lectures during the meeting that I regard the mathematical problems associated with turbulence and vorticity amplification and concentrations as extremely important for future research but they are not emphasized here. The interested reader can consult some of my other research/expository articles (see [7], [8]) for my perspective on these topics.

Section 1: The Compressible Euler equations and Related Conservation Laws.

A general $m \times m$ system of conservation laws in N-space variables is given by

$$(1.1) \qquad \frac{\partial \vec{u}}{\partial t} + \sum_{j=1}^{N} \frac{\partial}{\partial x_j}(F_j(\vec{u})) = \mathcal{S}(\vec{u}) \ .$$

The functions $F_j(\vec{u})$, $1 \leq j \leq N$, are the nonlinear fluxes and $\mathcal{S}(\vec{u})$, are the source terms. These are smooth nonlinear mappings from an open subset of $\mathbf{R}^M$ to $\mathbf{R}^M$. For convenience in notation, we have suppressed any explicit dependence on (x, t) of the coefficients in (1.1). The prototypical example of a system of homogeneous conservation laws is the system of $N+2$ conservation laws in N-space variables given by the *Compressible Euler Equations* expressing conservation of mass, momentum, and total energy and given by

$$(1.2) \qquad \begin{aligned} &\frac{\partial \rho}{\partial t} + \mathrm{div}\,(\vec{m}) = 0 \\ &\frac{\partial \vec{m}}{\partial t} + \mathrm{div}\,\left(\frac{\vec{m} \otimes \vec{m}}{\rho}\right) + \nabla p = 0 \\ &\frac{\partial E}{\partial t} + \mathrm{div}\,\left(\vec{m}\left(\frac{E}{\rho} + \frac{p}{\rho}\right)\right) = 0 \ . \end{aligned}$$

In (1.2), ρ is the density with $1/\rho = \tau$, the specific volume, $\vec{v} = {}^t(v_1, ..., v_N)$ is the fluid velocity with $\rho\vec{v} = \vec{m}$ the momentum vector, p is the scalar pressure, and $E = \frac{1}{2}(\vec{m} \cdot \vec{m})/\rho + \rho e(\tau, p)$ is the total energy with e the internal energy, a given function of (τ, p) defined through thermodynamics considerations. For an ideal gas, $e = (p\tau)/\gamma - 1$ with $\gamma > 1$, the specific heat ratio. The notation $\vec{a} \otimes \vec{b}$ denotes the tensor product of two vectors. It is well-known that for smooth solutions of (1.2), the entropy $S(\rho, E)$, is conserved along fluid particle trajectories, i.e.

$$(1.3) \qquad \frac{DS}{Dt} = 0 \text{ where } \frac{D}{Dt} = \frac{\partial}{\partial t} + \sum_{j=1}^{N} v_j \frac{\partial}{\partial x_j} \ .$$

The first simpler system of conservation laws which emerges as an approximation for solutions of the compressible Euler equations is probably well-known to the reader. If the entropy is initially a uniform constant and the solution remains smooth, then (1.3) implies that the energy equation can be eliminated. Furthermore, under standard assumptions on the equation of state, the pressure can be regarded as a function of density and entropy, $p(\rho, S)$. Thus, with constant initial entropy the smooth solution of (1.2) satisfies the system of $N + 1$ conservation laws in N-space variables given by the equations for *Isentropic Compressible Flow*

(1.4)
$$\frac{\partial \rho}{\partial t} + \text{div} \, (\vec{m}) = 0$$
$$\frac{\partial \vec{m}}{\partial t} + \text{div} \left(\frac{\vec{m} \otimes \vec{m}}{\rho} \right) + \nabla p(\rho, S_0) = 0 \, .$$

For an ideal gas law, $p(\rho) = A(S_0)\rho^\gamma$ with $\gamma > 1$. I remind the reader that solutions of the system in (1.4) are a genuine approximation to solutions of the system in (1.2) once shock waves form since the entropy increases along a shock to third order in wave strength for solutions of the compressible Euler equations while in (1.4), the entropy is constant.

Next, I present a conservation law which involves a further approximation to solutions of (1.2) beyond the isentropic approximation from (1.4); this approximation is well-known in transonic aerodynamics and is called the equation for time-dependent potential flow. First I consider smooth solutions of (1.4) that are irrotational, i.e.

(1.5)
$$\text{curl} \, v = 0 \, .$$

With $\vec{\omega} = \text{curl} \, v$ defining the vorticity, then the vorticity in a smooth solution of the 3-D Euler equations from (1.2) satisfies

(1.6)
$$\frac{D}{Dt} \left(\frac{\vec{\omega}}{\rho} \right) = \frac{\vec{\omega}}{\rho} \cdot \nabla \vec{v} + \frac{p_s(\rho, S)}{\rho^3} \nabla \rho \times \nabla S$$

where $p_s = \frac{\partial p}{\partial s}(\rho, S)$. The general formula in (1.6) is readily verified by taking the curl of the momentum equation and using vector identities. One immediate consequence of the equations in (1.6) and (1.3) is that a smooth solution of compressible 3-D Euler which is both isentropic and irrotational at time $t = 0$ remains isentropic and irrotational for all later times as long as this solution stays smooth; thus, the condition in (1.5) is reasonable for smooth solutions. Next, for smooth irrotational solutions of the equations for isentropic compressible flow, I will integrate the N-momentum equations in (1.4) through Bernoulli's law. With the condition, curl $v = 0$, the N-momentum equations in (1.4) assume the form

(1.7)A
$$\vec{v}_t + \frac{1}{2}\nabla(|\vec{v}|^2) + \nabla h = 0$$

where $h(\rho)$ satisfies

(1.7)B
$$h'(\rho) = \frac{dp}{d\rho}(\rho, S_0)/\rho > 0 \, .$$

On a simply connected space region, the condition, curl $\vec{v} = 0$, implies that

(1.8)
$$\vec{v} = \nabla \Phi$$

so that (1.7)A) determines the density from the potential Φ through the formula

$$(1.9) \qquad \rho(D\Phi) = h^{-1}(-(\Phi_t + \frac{1}{2}|\nabla\Phi|^2)) \ .$$

with $D\Phi = {}^t(\Phi_t, \nabla\Phi)$ and h^{-1}, the inverse function for h defined in (1.7)B). Thus, for smooth isentropic irrotational potential flow, the $N+2$ conservation laws for the compressible Euler equations reduce to a single conservation law, the conservation of mass, $\rho_t + \text{div}\ (\rho\vec{v}) = 0$; for these special flows the conservation of mass yields a second order conservation law for the potential Φ, the *Time Dependent Potential Flow Equation*:

$$(1.10) \qquad (\rho(D\Phi))_t + \sum_{j=1}^{N}(\Phi_{x_j}\rho(D\Phi))_{x_j} = 0$$

with $\rho(D\Phi)$ determined in (1.9). The steady state version of the equation in (1.10) involves looking at time-dependent solutions of (1.10) i.e. setting $\rho(D\Phi)_t \equiv 0$ and reduces to the celebrated steady potential flow equation of aerodynamics, (see [9], [10]). In applications in aerodynamics the second order conservation law in (1.10) is used for discontinuous solutions and the empirical evidence is that weak solutions of (1.10) are fairly good approximations to weak solutions for 3-D Euler provided

(1.11)
1) the shock strengths are small
2) the curvature of shock fronts is not too large
3) there is a small amount of vorticity in the region of interest

   I will comment briefly on the sources of error in the approximation of weak solutions of the 3-D compressible Euler equation by suitable weak solutions of the isentropic equations in (1.4) and the isentropic potential flow equations in (1.10). For planar shock waves, the Rankine-Hugoniot relations for either the equation in (1.4) or (1.10) agree to third order in shock strength with those for the full compressible Euler equations in (1.2). I have already given the reason for this agreement for the equation in (1.4). For plane waves (see Majda [11]) the potential flow equation in (1.10) is a 2×2 system with the same smooth solutions as the 2×2 system for plane waves derived from (1.4); thus, these different 2×2 systems have the same Riemann invariants so their respective shock solutions given a fixed pre-state agree to third order in wave strength. Two other effects are important in multi-dimensions in assessing the errors between solutions of (1.2) and (1.4) or (1.10) respectively. From (1.6), it follows that entropy gradients are an additional source of vorticity in smooth solutions of (1.2) which is completely ignored for both the equations in (1.4) and (1.10); of course, if the entropy gradients are small, this additional vorticity which is generated will also remain small at least in 2 space dimensions where vorticity does not amplify through self-stretching, i.e. $\vec{\omega} \cdot \nabla v = 0$. More subtle effects in the errors arise along a curved multi-dimensional shock front for solutions of the compressible Euler equations. Hayes ([12]) has given a beautiful simple derivation of a general formula which shows that additional vorticity is created across a shock in response to both curvature and differential changes in shock speed along the front. For simplicity, I consider the special case of these formulas when the preshock state is a uniform constant vector. If the shock front motion is specified in terms of the normal wave front velocity, $\vec{n}q_s$, then the vorticity created at the shock front is given by

$$(1.12) \qquad \omega|_S = -\vec{n} \times \nabla_t q_s \frac{(1-\delta)^2}{\delta}$$

with $\delta = (\rho_0/\rho_1)$ the compression ratio and ∇_t, the tangential gradient. If the preshock state is not constant, a more general formula applied which also includes the curvature of the front. The formula in (1.12) is independent of the equation of state and applies just as well to curved shock fronts for the isentropic equations in (1.4); in this case, the only errors at shock fronts in the vorticity that arise occur from errors in differential changes in wave speed of solutions of (1.4) as compared with those from the corresponding solution of (1.2) and these are multiplied by terms which are second order in shock strength – these are extremely small errors. The errors in using the time- dependent potential flow equations in (1.10) are larger even for initial data that is initially irrotational. The *second order wave equation in (1.10) always generates irrotational solutions* even when wave fronts are highly curved and the differential wave speed is changing *so the vorticity changes* along a wave front for solutions of compressible 3-D Euler as illustrated in (1.12) *are completely ignored* by solutions of (1.10); nevertheless, these errors are second order in wave strength and can remain quite small for suitable weak solutions of the full 3-D Euler equations in (1.2). I hope that the discussion in the preceding paragraph provides the reader with some quantitative justification for the empirical evidence regarding the use of weak solutions of (1.10) as approximations for suitable weak solutions of the compressible 3-D Euler equations in (1.2). In section 2, we use the equation in (1.10) to provide some similar but simpler problems in self-similar shock diffraction than a direct attack on the equations in (1.2); these solutions should be simpler mathematically but still physically relevant for (1.2) in the weak shock regime. In Section 3 we pose several open problems regarding the rigorous justification of the use of weak solutions of (1.10) as approximations for weak solutions of (1.2). The interested reader can consult the author's book ([11]) and the references there for the theory of multi-dimensional shock fronts for solutions of the equations in (1.2) and (1.4). A rigorous theory of shock fronts for the second order wave equation in (1.10) is developed in more recent work of Thomann and the author (see [13] and Thomann's lecture in this I.M.A. volume).

The Simplest Prototype Conservation Law in Several Space Variables

During the meeting, several speakers including Peter Lax and John Hunter have proposed their own versions of conservation laws in several space variables with simple nonlinear structure but having some of the features of more general systems. I advocate the second order wave equation from (1.10),

$$(1.13) \qquad \rho(D\Phi)_t + \sum_{j=1}^{N}(\Phi_{x_j}\rho(D\Phi))_{x_j} = 0$$

as the simplest nonlinear conservation law with prototypical systems behavior, i.e. $m \geq 2$. Here is a list of my reasons:

(1.14)

(1) Unidirectional plane wave solutions of (1.13) reduce to solutions of a 2×2 system of conservation laws with the structure of a wave equation.

(2) The linear structure of (1.13) is strictly hyperbolic with characteristics defined by a single light cone in several space variables. Neither Hunter's nor Lax's proposed systems have this property.

(3) Under reasonable thermodynamic assumptions such as an ideal gas law $p = A\rho^\gamma$, the system in (1.13) is genuinely nonlinear in all wave directions simultaneously and the corresponding multi-D shock fronts are uniformly stable ([11]). Lax's proposed system does not have this property.

(4) This system has the vorticity waves removed unlike the two other systems for fluid flow from (1.2) and (1.4) discussed earlier. Such vorticity waves are linearly degenerate wave fields but represent an enormous source of instability in multi-D through Kelvin-Helmholtz instability, etc.

Section 2: Self-Similar Shock Diffraction. An important prototype problem both for practical applications and for the theory of complex wave patterns in multi-D involves the problem of diffraction of a shock wave which is incident along an inclined ramp. The large scale computations, physical experiments, and new phenomena that arise are all discussed in the excellent article of Harland Glaz in this volume and I refer the reader to that article for detailed references. Here I will emphasize some of the unsolved problems of modern applied mathematics which are suggested by the detailed numerical simulations.

When a shock wave is incident along an inclined ramp, intuition based on linear theory leads one to guess a solution consisting of the incident wave pattern and a reflected curved wave much like that depicted in Figure 1(a). However, the large scale numerical simulations by Glaz and Colella ([14]) reveal remarkably complex unexpected wave patterns in the diffraction of shocks by ramps. Besides the expected wave pattern in Figure 1(a), which is called regular reflection (RR), more complex reflected wave patterns such as those in Figure 1(b)–1(d) occur for various angles and incident wave strengths. In Figure 1(b), the reflected shock is replaced by three shocks and a vortex sheet – this pattern is called simple Mach reflection (SMR), In Figure 1(d) two different Mach stem structures of this type occur in the reflected wave and this pattern is called double Mach reflection (DMR). In Figure 1(c) the diffracted pattern has a concave shock front so this case is called complex Mach reflection (CMR) because it is intermediate between SMR and DMR. Colella and Glaz have found even more of these complex wave patterns embedded in the reflected wave under different flow conditions and for different equations of state. These complex patterns have practical significance because there are higher pressures associated with the triple shock points. Such kinds of complex reflection have been documented experimentally. The interested reader can consult Glaz's article

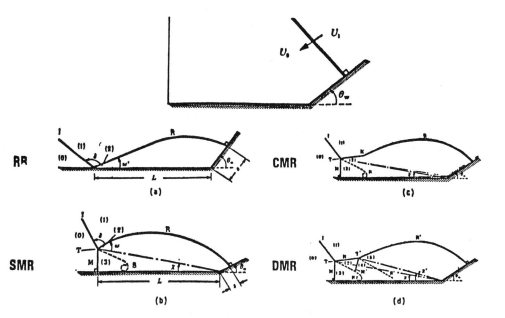

Figure 1: Oblique shock-wave reflection

in this volume or plates 235-238 of Van Dyke's book ([15]) which is a sequence of shadow graphs of experiments corresponding to the four cases depicted in Figure 1. Next, I turn to several open problems motivated by these phenomena.

Problem #1: Is there an Organizing Center for the Wave Bifurcations Associated with RR, SMR, CMR, and DMR?

This first problem is this theoretician's dream rather than being based on more substantial reasoning and represents a challenge for those readers interested in large scale computing. An experimentalist's extrapolation of the regions where the four wave patterns occur for various Mach numbers and ramp angles is presented in Figure 2. Most of the experimental

data are collected for the region very far away from the vicinity where the three bounding curves intersect and the remainder of the curves are plotted through crude extrapolation. My theoretician's dream is that all three curves might intersect at the same point,(M_S^*, θ_ω^*) and this would have great theoretical significance. If this conjecture is true, then all four of the different diffracted wave patterns would be present as *small perturbations* of the wave pattern given at (M_S^*, θ_w^*), which would be an organizing center for the wave bifurcations. Thus all these wave patterns should be amenable to an applied mathematics analysis which utilizes suitable *small amplitude perturbation theory* combined with numerical modelling. My challenge to the experts in large scale computation is to perform the delicate detailed numerical study in this parameter region to check the validity of this conjecture.

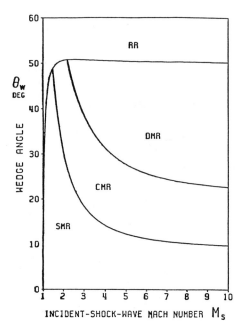

60

RR

50

θ_w
DEG

40

WEDGE ANGLE

30

DMR

CMR

20

SMR

10

0
1   2   3   4   5   6   7   8   9   10

INCIDENT-SHOCK-WAVE MACH NUMBER M_s

Figure 2: Approximate regions of various types of shock reflection for flow in air

Problem #2: Explain the appearance of the second Mach stem in the transition between CMR and DMR.

This is a problem for applied mathematicians who are willing to utilize all of the tools of modern applied mathematics which I listed earlier in the introduction – rigorous proofs on this problem seem far beyond our present techniques. Here is a sketch of one possible attack. In the vicinity of the point K in Figure 1c), the fluid flow undergoes a transition from subsonic to supersonic in an appropriate self-similar reference frame, i.e. the local Mach number changes from $M < 1$ to $M > 1$. I concentrate on the flow field in the vicinity of K and ignore the other parts of the wave structure needed to set this up. The steady flow in the vicinity of K can be described by an equation of mixed type with a free surface defined by the reflected shock front. The local flow structure has much in common with the local flow structure in steady transonic flow (see [10]) except that the air wing is replaced by the shock boundary here. I suspect that the associated linearized equation for the local flow field near K becomes ill-posed at a critical value of wave strength in a similar fashion as described by Morawetz in her pioneering work on perturbed shockless airfoils (see [16]) – this in itself is a challenging problem from linear P.D.E. motivated by this specific application. If this problem is ill posed at a critical wave strength, then the appearance of the second Mach stem in Figure 1d) might be explained by incorporating small amplitude nonlinear effects through appropriate nonlinear geometric optics (see Section 3 below and R. Rosales or J. Hunter's papers in this volume). The small amplitude nonlinear saturation of the linear instability predicted by the simplified asymptotic equation might yield the entire Mach stem structure; earlier work of Rosales and the author ([45]) achieves the prediction of an entire Mach stem structure through geometric optics for a much simpler problem.

Rigorous Prototype Problems for Self-Similar Shock Diffraction

The inviscid diffraction of shock waves from a wedge does not have a characteristic length scale and the wave patterns which emerge are self-similar, i.e. the solutions that emerge are functions of the variables

$$(2.1) \qquad \xi = x/t \,, \quad \eta = y/t$$

if the wedge corner is given at $(0,0)$. It is natural to introduce self-similar flow variables

$$(2.2) \qquad \tilde{u} = u - \xi \,, \quad \tilde{v} = v - \xi$$

where u and v are the x and y components of the velocity field. If one seeks solutions of the compressible 2-D Euler equations in (1.2) with the form, $\rho(\xi,\eta), \tilde{u}(\xi,\eta), \tilde{v}(\xi,\eta), E(\xi,\eta)$ the equations in (1.2) reduce to the

Compressible Euler Equations in Pseudo Stationary Coordinates

$$(2.3) \qquad \begin{aligned} (\rho\tilde{u})_\xi + (\rho\tilde{v})_\eta &= -2\rho \\ (\rho\tilde{u}^2 + p)_\xi + (\rho\tilde{u}\tilde{v})_\eta &= -3\rho\tilde{u} \\ (\rho\tilde{u}\tilde{v})_\xi + (\rho\tilde{v}^2 + p)_\eta &= -3\rho\tilde{v} \\ (\rho\tilde{u}\tilde{H})_\xi + (p\tilde{v}\tilde{H})_\eta &= -\rho(\tilde{u}^2 + \tilde{v}^2) - 2\rho\tilde{H} \end{aligned}$$

where $h = e + p/\rho$ is the enthalpy and $\tilde{H} = 1/2(\tilde{u}^2\tilde{v}^2) + h$. The wave patterns that emerge from inviscid numerical calculations involving RR, SMR, CMR, and DMR as depicted in Figure 1 all represent solutions of the system of equations in (2.3). Thus, the diffracted wave patterns reduce to complicated solutions of the nonlinear equations in (2.3). The solution outside the diffracted bubble is piecewise constant but the diffracted bubble itself is a free boundary defined by shock fronts in a region with corners defined by the diffracted wave boundaries and wedge corner. What is the type of the nonlinear system in (2.3)? This system is a novel physical example of equations of principal type: In subsonic regions where the Mach number M satisfies $M < 1$ there are two real and two complex characteristics while in supersonic regions with $M > 1$ there are four real characteristics. The equations change type along the sonic lines with $M = 1$ as occurs in transonic aerodynamics. Interesting work involving formal asymptotic solution of the equations in (2.3) for problems with shock diffraction has been given recently by Hunter and Keller ([17]) and Harabetian ([18]). With the complexity of the phenomena depicted in Figure 1, it is clear that the rigorous theory should first concentrate on the simplest regime of regular reflection, RR (see Figure 2). However, I would like to point out that even here, there are two types of regular reflection: in the first and simplest situation the flow remains subsonic throughout the entire diffracted bubble, i.e. $M < 1$;

in the second case there is a supersonic bubble with $M > 1$ located adjacent to the reflected wave pattern. Clearly, the first case is simpler mathematically because there is not a region of transition in the flow field which is sonic where the equations in (2.3) change type (the paper, [19], is a convenient reference for the elementary mathematics for this second type of behavior). Thus, the first natural rigorous prototype problem for self-similar shock diffraction is the following:

Problem #3: Find some subsonic regime for regular reflection where one can rigorously establish the existence of a subsonic diffraction pattern for the equations in (2.3) with the structure depicted in Figure 1a).

In this problem it is still necessary to deal with nonlinear equations of principal type involving transport of vorticity and entropy with source terms coupled with elliptic equations in a region with corners and a free boundary. I believe there is a simpler related problem to solve which would be extremely interesting, physically relevant, and removes the additional complexity associated with the transport of vorticity and entropy. I describe this problem next.

In the regime of subsonic regular reflection, the computations of Glaz and Colella ([14]) show that the overall density jumps are not very large and the vorticity generated in the flow field in the diffracted bubble is rather small as expected from the formula in (1.12). As I discussed at the end of Section 1, under these circumstances, it is reasonable to study self-similar shock diffraction from the same wedge for the time-dependent potential flow equations in (1.10). With the self-similar variables in (2.1), I introduce the potential $\Psi(x, y, t) = t\tilde{\Psi}\left(\frac{x}{t}, \frac{y}{t}\right)$ so that $u = \Psi_x = \tilde{\Psi}_\xi$ and $v = \Psi_y = \tilde{\Psi}_\eta$. The equation for *pseudo-stationary potential flow* is the second order equation for $\tilde{\Psi}$ given by

$$(2.4)A \qquad \frac{\partial}{\partial \xi}(\rho(\tilde{\Psi}_\xi - \xi)) + \frac{\partial}{\partial \eta}(\rho(\tilde{\Psi}_\eta - \eta)) = -2\rho$$

where $\rho(\tilde{\Psi}, \tilde{\Psi}_\xi, \tilde{\Psi}_\eta)$ is the nonlinear function determined by

$$(2.4)B \qquad \rho(\tilde{\Psi}, \tilde{\Psi}_\xi, \tilde{\Psi}_\eta) = \rho\left(-\left(\tilde{\Psi} - \xi\frac{\partial\tilde{\Psi}}{\partial\xi} - \eta\frac{\partial\tilde{\Psi}}{\partial\eta} + \frac{1}{2}\left(\frac{\partial\tilde{\Psi}}{\partial\xi}\right)^2 + \frac{1}{2}\left(\frac{\partial\tilde{\Psi}}{\partial\eta}\right)^2\right)\right)$$

and ρ on the right hand side of (2.4)B is the same nonlinear function defined in (1.7)B. The rigorous prototype problem for self-similar shock diffraction which I believe is accessible is the following

Problem #4: Establish the existence of the pseudo-stationary diffracted wave pattern for some regime of subsonic regular reflection for the potential flow equation in (2.4).

Next, I briefly list the reasons for my belief that this problem is accessible and much simpler than Problem #3.

(2.5)

1) In the regime of subsonic regular reflection, the equation in (2.4) defines a quasi-linear elliptic equation in the diffraction region.

2) The free surface boundary condition at the reflected shock boundary can be replaced by a fixed boundary condition for a related quasi-linear elliptic equation with a fixed boundary through the partial hodograph transformation introduced by Thomann and the author (see [13]).

3) An initial approximate solution for the nonlinear elliptic problem at weak wave strengths can be developed by utilizing the classical work of Blank and Keller ([20]) for linear diffraction from wedges and/or the nonlinear modifications of Hunter and Keller ([17]).

With all of the reductions in (2.5), Problem #4 becomes a boundary value problem for a quasi-linear elliptic equation on a fixed region with corners and the sophisticated approximate solution proposed in 3) of (2.5) suggests an attack utilizing the "method of continuity". What remains is the difficult task of obtaining suitable apriori estimates for this nonlinear elliptic boundary value problem to establish the existence of the solution.

Section 3: Oscillations for Conservation Laws. Here I discuss the propagation of oscillations for general systems of conservation laws. First, I discuss open problems regarding oscillations at small amplitudes which involve the method of nonlinear geometric optics then I discuss an important open problem for propagation of large amplitude oscillations.

Nonlinear Geometric Optics:

One of the most important asymptotic tools in understanding solutions of systems of conservation laws in several space variables such as

$$(3.1) \qquad u_t + \sum_{j=0}^{N} F_j(u)_{x_j} = 0$$

is the method of weakly nonlinear geometric optics. This method is discussed in detail in the articles of Rosales and Hunter in this symposium volume. There are other recent survey articles regarding these methods which I recommend ([21], [22], [23]); my own survey article in I.M.A. volume #2 ([24]) provides an introduction to both the methods and rigorous work on the subject. There are numerous recent sophisticated applications of these methods in explaining physical phenomena usually with other complexity involving free surface problems and strong nonlinearity in source terms. Examples are the author's work with Artola ([1], [2], [3]) regarding nonlinear instability in supersonic vortex sheets and the recent work of Almgren, Rosales, and the author ([25], [26]) on dynamic homogenization of reacting materials. Here I will discuss important but accessible open problems in the rigorous justification of these methods.

In the simplest case of a single propagating wave front, nonlinear geometric optics builds systematic asymptotic solutions of (3.1) with the form,

$$(3.2) \qquad u_0 + \epsilon \sigma \left(x, t, \frac{\phi}{\epsilon} \right) r(\nabla \phi) + 0(\epsilon^2)$$

for $\epsilon \ll 1$ where $u_0 \epsilon \mathbf{R}^M$ is a constant vector, $r \epsilon \mathbf{R}^M$ is an appropriate right eigenvector for the linearized problem at $u_0, \sigma(x, t, \theta)$ is an amplitude function, and ϕ is the phase function. The phase function ϕ solves the eikonal equation of linear geometric optics

$$(3.3) \qquad \phi_t + \lambda(\nabla \phi) = 0$$

where λ is one of the m wave speeds of the linearization at u_0. The amplitude σ solves a nonlinear transport equation

$$(3.4) \qquad D\sigma + b \left(\frac{1}{2} \sigma^2 \right)_\theta = 0$$

The operator D is a first order operator which is the linear transport operator of geometric optics given by differentiation along the bicharacteristic rays associated with ϕ and has the form,

$$(3.5) \qquad D = \frac{\partial}{\partial t} + \vec{a} \cdot \nabla + C$$

with $\vec{a}(x, t), b(x, t), C(x, t)$ determined from ϕ by explicit formulas. In the case of single propagating waves, provided $b \neq 0$ (which is always true for a genuinely nonlinear wave), there are elementary changes of variable which reduce (3.4) to the inviscid Burgers equation,

$$(3.6) \qquad \sigma_\tau + \left(\frac{1}{2} \sigma^2 \right)_\theta = 0 \ .$$

The advantage of utilizing geometric optics as an asymptotic tool in understanding phenomena in the complex general multi-D system in (3.1) is now evident. The solutions of (3.6) are known explicitly and provide general quantitative asymptotic approximations for (3.1) through the equations in (3.2)–(3.5). Obviously, it is an important theoretical problem to justify nonlinear geometric optics for discontinuous solutions of conservation laws. The only rigorous work thus far on this topic is that by Diperna and the author ([27]) for a class of systems in a single space variable which I will discuss briefly below. An outstanding and very important but accessible open problem is the following

Problem #1: Provide a rigorous justification of the single wave geometric optics expansion in (3.2) for discontinuous initial data.

I believe that this problem is ripe for solution for the following reasons: at this meeting, G. Metevier (see his paper in this volume) has announced the existence of shock front solutions of (3.1) for a uniform time T independent of ϵ as $\epsilon \downarrow 0$; the existence and structure of the discontinuous approximating solutions of (3.2)-(3.5) is completely understood in the genuinely nonlinear case because the discontinuous solutions of (3.6) are well known; the errors between the approximate and exact solution can probably be estimated through an appropriate multi-dimensional generalization of the estimates utilizing geometric measure theory developed by Diperna and the author ([27]) for systems in a single space variable.

Next, I turn to important open problems regarding the new phenomena which occur when one attempts to build multi-wave approximations for geometric optics with the form,

$$(3.7) \qquad u_0 + \epsilon \sum_{\rho=1}^{m} \sigma_\rho \left(x, t, \frac{\phi_\rho}{\epsilon} \right) r_\rho(\nabla \phi_\rho) + 0(\epsilon^2) \ .$$

In linear geometric optics the wave patterns in (3.7) superimpose and each amplitude solves the corresponding single wave transport equation of geometric optics. When does this happen in the nonlinear case? The formal asymptotic theory ([28], [29]) predicts that

(3.8)   the *single wave patterns of nonlinear geometric optics*
described in (3.2)-(3.5) *superimpose* and are *non-resonant provided*
that the amplitudes $\{\sigma_\rho(x,t,\theta)\}_{\rho=1}^{m}$ have
compact support in θ .

The *only* systematic *rigorous justification* for discontinuous solutions *for geometric optics has been developed* by Diperna and the author ([27]) in this *non-resonant case* in a *single space variable*. The main theorem from [27] requires the hypothesis that the initial amplitudes $\{\sigma_\rho(x,\theta)\}_{\rho=1}^{m}$ at time $t=0$ have compact support (thus, are non-resonant) and that all m wave fields are distinct and genuinely nonlinear. Under these assumptions Diperna and the author prove that

$$(3.9) \qquad \max_{0 \leq t \leq \infty} \|u(x,t) - (u_0 + \epsilon \sum \sigma_\rho \left(x, t, \frac{\phi_\rho}{\epsilon} \right) r_\rho)\|_{L^1} \leq C\epsilon^2$$

where $u(x,t)$ is the solution of the conservation laws with the same initial data constructed by Glimm's method. Here $\| \cdot \|_{L^1}$ is the L^1 norm. Thus, even for discontinuous initial data, geometric optics is valid uniformly for all time in this non-resonant situation – this is a surprising result !! Incidentally, one immediate corollary of this theorem is that for small amplitude initial data of compact support with size ϵ,

(3.10)   the discontinuous solutions of the isentropic flow equations
in (1.4) in one-space dimension and the discontinuous solutions
of the potential flow equations in (1.10) in one-space
dimension with the same initial data agree within $C\epsilon^2$
in the L^1 norm for all time

This result provides a rigorous justification for some of the approximations described in Section 1 in the special case of a single space variable. Since the isentropic flow equations in (1.4) and the potential flow equations in (1.10) have the same smooth solutions, it is an exercise to check that these two equations have the same single wave expansions for nonlinear geometric optics. With this fact, the corollary in (3.10) follows immediately from (3.9) and the triangle inequality. Some interesting and accessible open problems generalizing the results stated in (3.9) are described at the end of the author's survey article ([24]).

I return to the general multi-wave expansions of geometric optics and ask whether new phenomena occur when the non-resonance conditions from (3.8) are no longer satisfied? The answer is yes. Recent research of Hunter, Rosales, and the author ([29], [30]) employing a systematic development of nonlinear geometric optics reveals more complex effects beyond (3.8); general periodic or almost periodic wave trains do not superimpose but instead interact resonantly. The eikonal equations in (3.3) remain the same but the amplitudes, $\{\sigma_\rho(x,t,\theta)\}_{\rho=1}^m$ no longer solve simple decoupled transport equations like those in (3.4); in fact the different amplitudes resonantly exchange energy through nonlinear interaction and solve a coupled system of quasi-linear integro-differential equations provided that $m \geq 3$. Applications to the equations of compressible fluid flow from (1.2) and (1.4) are developed in detail in the above papers in both a single and several space dimensions. As regards the 3×3 system from (1.2) describing compressible fluid flow in one space variable, the resonant nonlinear interaction of small amplitude sound waves with small amplitude entropy waves produces additional sound waves which resonantly interact. After some elementary changes of variables, the two sound wave amplitudes $\sigma^\pm(\theta,t)$ satisfy the coupled system of resonant equations

$$(3.11) \qquad \sigma_t^+ + \left(\frac{1}{2}\sigma^+\right)_\theta^2 + \int_0^1 k(\theta - y)\sigma^-(y,t)dy = 0$$

$$\sigma_t^- + \left(\frac{1}{2}\sigma^-\right)_\theta^2 - \int_0^1 k(-\theta - y)\sigma^+(y,t)dy = 0$$

where I assume in (3.11) that σ^+, σ^-, and k are periodic with period one. The kernel k is a multiple of a rescaled derivative of the initial entropy perturbation; the asymptotics predicts that the entropy perturbation does not change to leading order in time. Recent papers ([31], [32]) which combine small scale numerical computation and several exact solutions reveal surprising new phenomena in the solutions of (3.11) through resonant wave interaction. Thus, the formal predictions from geometric optics for periodic wave trains at small amplitudes for 3×3 compressible fluid flow involve surprising new phenomena. The open problems which I suggest next are motivated by these new phenomena. Since conservation laws without source terms are scale invariant, I propose some open problems for the rigorous justification of nonlinear geometric optics for the resonant case by considering solutions of the $M \times M$ system of conservation laws in a single space variable

$$(3.12)A) \qquad\qquad u_t + F(u)_x = 0$$

with small amplitude periodic initial data

$$(3.12)B) \qquad\qquad u|_{t=0} = u_0 + \epsilon u_1^0(x) \ .$$

Here $u_0 \epsilon \mathbf{R}^M$ is a constant and $u_1^0(x)$ is a function with period one, i.e. $u_1^0(x+1) = u_1^0(x)$.

Problem #2: For a general system of conservation laws, let u^ϵ denote the weak solution with initial data in (3.12)B) and let u_w^ϵ denote the corresponding approximation from nonlinear geometric optics (involving resonant wave interaction for $m \geq 3$ in general). Show that there is a time, $T(\epsilon)$ with $\epsilon T(\epsilon) \to \infty$ as $\epsilon \to 0$ so that

$$(3.13) \qquad\qquad \max_{0 \leq t \leq T(\epsilon)} ||u^\epsilon - u_w^\epsilon||_{L^1} \leq o(\epsilon)$$

where L^1 denotes the L^1-norm of a one periodic function in x.

I make several remarks on this problem. For $m = 2$ where the resonant effects are absent and for a pair of genuinely nonlinear conservation laws the estimate in (3.13) has been proved in [27] with $T(\epsilon) = o(\epsilon^{-2})$; it would be interesting to know if this is sharp. Furthermore, there is an improved geometric optics formal approximation for large times due to Cehelsky and Rosales (see [33]) which accounts for accumulating phase shifts from wave interactions and this geometric optics approximation u_w^ϵ should be used in Problem #2. In fact, the result of Diperna and the author for periodic waves for pairs of conservation laws does not utilize this more refined geometric optics approximation with phase shift corrections for long times. An interesting and much more accessible technical problem than Problem #2 is to assess whether through the use of this refined geometric optics approximation for $m = 2$, the time of validity $T(\epsilon)$ becomes significantly larger than $T(\epsilon) = 0(\epsilon^{-2})$. One of the reasons that the work in [27] for the periodic case is restricted to $m = 2$ is that general existence theorems for small periodic initial data for conservation laws following Glimm's work are unknown for $m \geq 3$. A straightforward repeat of Glimm's proof shows that the solution u^ϵ of the system of conservation laws in (3.12)A) with general initial data exists for times of order $0(\epsilon^{-2})$. I conjecture that for a general system of conservation laws with genuinely nonlinear and linearly degenerate wave fields, this crude result is sharp; my conjecture is based on the unstable nature of solutions of the resonant asymptotic equations for a particular example system discussed in [30]. It would be very interesting to find out whether this conjecture is correct. One the other hand, I believe that there is global existence for the 3×3 system of compressible fluid flow, (1.2), for small amplitude periodic initial data as given in (3.12)B). This I list as

Problem #3: Show that for the specific 3×3 system of compressible fluid flow, Glimm's method yields the global existence of solutions for general small amplitude periodic initial data.

I believe that Problem #2 is too difficult to attack in full generality; the special and important case of 3×3 gas dynamics is already extremely interesting. For emphasis I state this as

Problem #4: For the 3×3 system of compressible fluid flow, let u_w^ϵ denote the resonant geometric optics approximation for the initial data in 3.12B) given through 3.11 (see [30]) but including the large time phase shift corrections of Cehelsky-Rosales ([33]). Let u be the weak solution of $(3.12)A$ with the same initial data that exists for times of order ϵ^{-2}. Find a time interval $T(\epsilon)$ with $\epsilon T(\epsilon) \nearrow \infty$ as $\epsilon \to 0$ so that u_w^ϵ differs from u^ϵ by $o(\epsilon)$ on that time interval.

I remind the reader that from my earlier comments the full solution of Problem #3 is not needed to study Problem #4 and any progress on Problem #4 would be very interesting.

Large Oscillations

This section involves the study of existence of solutions via the weak topology and the propagation of large amplitude oscillations for systems of conservation laws

$$(3.14) \qquad \begin{aligned} u_t + F(u)_x &= 0 \\ u(x,0) &= u_0 \end{aligned}$$

with large amplitude initial data, u_0. The use of the weak topology and the method of compensated compactness was introduced by Tartar ([34]) and applied to scalar conservation laws. Diperna ([35], [36]) carried out Tartar's program for pairs of conservation laws provided that both wave fields are genuinely nonlinear; in this case strong convergence was deduced from the apriori weak convergence so that no oscillations propagate. Rascle and P. Serre (see [37], [38], and Serre's paper in this volume) have studied pairs of conservation laws which are not genuinely nonlinear; for example, for a general nonlinear wave equation, they show that oscillations propagate but the nonlinear terms in the equations still converge and define a weak solution in the limit.

Given all of the phenomena deduced via geometric optics for propagation and interaction of oscillations at small amplitudes for $m \geq 3$ it is not surprising that the propagation of large amplitude oscillations and the use of the weak topology provide difficult questions for systems of conservation laws with $m \geq 3$. The most important and most accessible of these problems regards propagation of oscillations for 3×3 compressible fluid flow. In Lagrangian mass co-ordinates, these equations have the form

$$(3.15) \qquad \begin{aligned} \tau_t - v_x &= 0 \\ v_t + p_x &= 0 \\ \left(\frac{1}{2}v^2 + e(\tau,p)\right)_t + (pv)_x &= 0 \end{aligned}$$

where τ is the specific volume; the interval energy e is given by

$$(3.16) \qquad e = \frac{p\tau}{\gamma - 1}, \qquad \gamma > 1$$

for an ideal gas law. The first remark is that large amplitude oscillations do propagate in solutions of (3.15). Consider the rapidly oscillating exact solution sequence defined by contact discontinuities, i.e.

(3.17)
$$\begin{pmatrix} \tau^\epsilon \\ v^\epsilon \\ p^\epsilon \end{pmatrix} = \begin{pmatrix} \tau_0\left(\frac{x}{\epsilon}\right) \\ v_0 \\ p_0 \end{pmatrix}$$

where v_0, p_0 are fixed constants and τ_0 is a fixed positive 1-periodic function. Large amplitude oscillations propagate for this equation because the weak limit of $\tau_0\left(\frac{x}{\epsilon}\right)$ has a non-trivial Young measure but the velocity and presure converge strongly. Nevertheless, the weak limit is a solution of the equations in (3.15). The conjectured behavior is that these examples provide the worst possible situation. I present this as

Problem #5: Let $^t(\tau^\epsilon, v^\epsilon, p^\epsilon)$ be a sequence of weak solutions of the compressible fluid equations in (3.15). Assume the uniform bounds, $0 < \tau_- \le \tau^\epsilon \le \tau_+$, $0 < p_- \le p^\epsilon \le p_+$, $|v^\epsilon| \le V$ and as $\epsilon \to 0$, $^t(\tau^\epsilon, v^\epsilon, p^\epsilon)$ converges weakly to $^t(\tau, v, p)$. Is it true that (v^ϵ, p^ϵ) converges strongly to (v, p)?

Both C.S. Morawetz and D. Serre are currently working on this problem. In fact, Serre has remarked that if the conjecture in Problem #5 is true, then

(3.18)   for an ideal gas law, the limit is a weak solution of the equations for compressible flow.

With (3.16), the result in (3.18) is an easy exercise for the reader. Nevertheless, I have some doubts that this conjecture is true; some high quality numerical simulations could generate some important insight here.

Section 4: New Phenomena in Conservation Laws with Source Terms. In this section, I briefly discuss my perspective on conservation laws with source terms. I focus on solutions of systems in a single space variable with the form,

(4.1) $$u_t + F(u)_x = S(u) .$$

T.P. Liu has been the principal contributor to the study of conservation laws with a special class of source terms with x dependence which model physical problems such as the averaged duct equations for one dimensional fluid flow. He has discussed the stability and large time asymptotic behavior for a large class of problems with source terms. The interested reader can consult Liu's paper in this volume for a detailed list of references. Here I will discuss open problems for conservation laws with source terms which do not satisfy the hypotheses of Liu's work – the equations of reacting gas flow are a prototypical example. I will discuss some of the new phenomena that occur for these systems with source terms which have been discovered recently through numerical, asymptotic, and qualitative modelling and then I suggest some accessible open problems motivated by these new phenomena. I emphasize the phenomena for the compressible Euler equations of reacting gas flow as an example of (4.1) although I am confident that similar phenomena are likely to occur for the hyperbolic conservation laws with suitable source terms arising in multi-phase flow, retrograde materials, and other applications.

The compressible Euler equations for a reacting gas with simplified one-step irreversible kinetics are given by

(4.2)
$$\rho_t + (\rho v)_x = 0$$
$$(\rho v)_t + (\rho v^2 + p)_x = 0$$
$$(\rho E)_t + (\rho v E + pv)_x = q_0 K(T)\rho Z$$
$$(\rho Z)_t + (\rho v Z)_x = -K(T)\rho Z$$

where $E = \frac{v^2}{2} + e$ is the energy density, Z is the mass fraction of fuel, $e = \frac{p}{\rho(\gamma-1)}$, $q_0 > 0$ is the heat release, and $T = \gamma p/\rho$ is the temperature. In the discussion below we assume either the Arrhenius form for the rate function $K(T)$,

(4.3)A
$$K(T) = K\exp(-E^+/T)$$

or the ignition temperature law

(4.3)B
$$K(T) = \begin{cases} K & , T \geq T_i \\ 0 & , T \leq T_i \end{cases}$$

where T_i is a fixed reference ignition temperature. In (4.3) K is the rate constant while in (4.3)A, E^+ is the non-dimensional activation energy.

An important practical problem for both safety and enhanced combustion regarding the system in (4.2) is the initiation of detonation. Detonation waves are travelling wave solutions of (4.2) which have the structure of an ordinary fluid dynamic shock followed by chemical reaction; these exact solutions are readily determined by quadrature of a single O.D.E. (see Fickett-Davis [39], Majda [40]) and are called Z-N-D waves; the Z-N-D wave moving with the slowest velocity is called the C-J (Chapman-Jouget) wave. The problem of initiation involves an initial flow field with a small region of hot gas kept at constant volume and velocity. The main issue in initiation regards whether this perturbation will grow into a fully-developed Z-N-D wave in which case there is transition to detonation or whether this perturbation will die out as time evolves and the chemical reaction will be quenched so that there is failure? Both experimental data and detailed numerical computations display many complex features in examples illustrating both failure and initiation. The recent paper by V. Roytburd and the author, [41], contains a discussion and documentation of these phenomena as well as a large list of background references. While these phenomena in initiation are becoming understood through a combination of experiments and numerical computation, the rigorous theory of such phenomena for solutions of the equations in (4.2) seems beyond reach. In fact, a very interesting preliminary open problem is the following

Problem #1: Establish the existence of solutions for the reacting Euler equations in (4.2) for appropriate initial data by a modification of Glimm's method.

At the meeting, D. Wagner (personal communication) has announced some major progress toward solving Problem #1.

Next, I present an interesting qualitative-quantitative model for high Mach number combustion and then I indicate some beautiful prototype problems for the initiation of detonation which are accessible in this model. It is not surprising given the complexity of the phenomena described in the preceding paragraph that there is an interest in simpler models which qualitatively mimic some of the features in solutions of (4.2) in various regimes. One such model for high Mach number combustion was proposed and studied by the author in [42] and then derived by Rosales and the author [40], [43] in a slightly modified form from the equations of reacting gas flow in (4.2) through nonlinear geometric optics as a quantitative asymptotic limiting equation of (4.2). This qualitative-quantitative model for high Mach number combustion is the following 2×2 system

$$(4.4) \qquad \begin{aligned} u_t + (\frac{1}{2}u^2)_x &= q_0 K(u) Z \\ Z_x &= K(u) Z \end{aligned}$$

where the rate function $K(u)$ has either of the forms in (4.3). The function Z is the mass fraction of reactant and the function u appearing in (4.4) is the amplitude of an acoustic wave moving to the right; when the reaction terms vanish so that $Z K(u) \equiv 0$, we get Burgers equation as expected from the theory of geometric optics sketched in (3.2)-(3.6). The coordinate x appearing in (4.4) is not physical space but instead is a suitable space-time distance to the reaction zone. Thus, the natural data for (4.4) is a signalling problem: $u_0(x)$ and $Z_0(t)$ are prescribed with

$$(4.5) \qquad u(x,t)|_{t=0} = u_0(x) \text{ and } \lim_{t \to \infty} Z(x,t) = Z_0(t) .$$

For simplicity in exposition, we assume below that $Z_0(t) \equiv 1$. From my discussion above, it is evident that the model equations in (4.4) retain the nonlinear interactions between the right sound wave and combustion but ignore all other multi-wave interactions that are present in solutions of (4.2). The model equations in (4.4) have a transparent analogue of Z-N-D waves and C-J waves (see [40], [42]) and also an analogue of the initiation problem. To mimic the initiation problem, I take the ignition temperature form from (4.3)B) for the rate function $K(u)$ in (4.4) and consider a pulse in the initial data for u given by

$$(4.6) \qquad u_0(x) = \begin{cases} u_1^0 & , \ 0 < x < d \\ u_2^0 & , \ x \le 0 \text{ or } x > d \end{cases}$$

where $u_1^0 > u_i$ and $u_2^0 < u_i$ and u_i is the ignition value in the model. The initial data in (4.6) is the analogue in the model of the hot spot mentioned earlier in the initiation problem. The solution of (4.4), (4.5) with the initial data in (4.6) was studied by Roytburd and the author through numerical computations in the paper, [44], from I.M.A. volume 12. Also, numerical solutions of initiation with (4.4), (4.5), (4.6) were compared with simulations of the full reacting gas system in (4.2) and as expected the solutions of (4.4)-(4.6) have good qualitative agreement with those in (4.2) provided the initiation process in solutions of (4.2) does not involve complex multi-wave gas dynamic interactions. In [44], Roytburd and the author found that depending on the parameters for the initial data u_1^0, u_2^0, d, and the heat release q_0 and the rate constant K in (4.3)B), the solution of (4.2) either was quenched and tended rapidly to zero so that there was failure or the solution grew (sometimes in a highly non-monotone fashion) to a fully-developed $C - J$ wave so that strong initiation occurs. The equations in (4.4) have both the attenuating effects on u of the spreading of rarefaction waves and the amplifying effects of exothermic heat release which compete to produce either outcome. A discussion of these competing effects is given in [44]. The main rigorous prototype problem which I propose in this section is the following

Problem #2: For a fixed K, q_0, and u_i characterize those initial data u_0 given in (4.6) so that either 1) the asymptotic solution of (4.4) as $t \nearrow \infty$ is a C-J wave or 2) the solution tends rapidly to zero as $t \nearrow \infty$.

I remark that the global existence of solutions for (4.4), (4.5) has been established by V. Roytburd (unpublished) through a constructive proof utilizing finite difference schemes. I believe that Problem #2 demonstrates very interesting new phenomena and also is extremely accessible to a rigorous analysis. One natural strategy would be to implement a version of the random choice scheme for the equations in (4.4) together with an appropriate version of Liu's wave tracing ideas to assess the ultimate growth or failure of the wave pattern.

I end this section with some additional comments regarding the equations of reacting gas flow as a source of new phenomena in conservation laws with source terms. Those familiar with homogeneous hyperbolic conservation laws know that shock waves in these systems are asymptotically stable at large times as $t \nearrow \infty$. The analogues of shock wave solutions for (4.2) are the Z-N-D- travelling waves mentioned earlier. It is both an experimental and numerically documented fact that in appropriate regimes of heat release, overdrive, and reaction rate, the Z-N-D waves lose their stability to time-dependent wave patterns with either regular or sometimes even chaotic pulsations. These facts and a corresponding asymptotic theory together with numerical calculations are mentioned in the paper by V. Roytburd in this volume. I would like to mention here that such effects cannot be found in solutions of the equations in (4.4); the full multi-wave structure of the gas dynamic equations in (4.2) is needed to produce these pulsation instabilities. An asymptotic analysis by Roytburd and the author to appear in a forthcoming publication confirms this.

Concluding Remarks: I have presented several problems in the modern applied mathematics of hyperbolic conservation laws. I have emphasized phenomena for the equations of compressible flow in several space variables. However, I believe that many of the phenomena and problems which I discuss here also have analogues in other applications such as dynamic nonlinear elasticity, magneto fluid dynamics, and multi-phase flow. I would like to thank Harland Glaz for the use of two of his graphs and also for interesting conversations regarding Section 2 of this paper.

REFERENCES

[1] M. ARTOLA AND A. MAJDA, *Nonlinear development of instabilities in supersonic vortex sheets I: the basic kink modes*, Physica 28D, pp. 253-281, 1988.

[2] M. ARTOLA AND A. MAJDA, *Nonlinear development of instabilities in supersonic vortex sheets II: resonant interaction among kink modes*, (in press S.I.A.M. J. Appl. Math., to appear in 1989).

[3] M. ARTOLA AND A. MAJDA, *Nonlinear kink modes for supersonic vortex sheets*, in press, Physics of Fluids A, to appear in 1989.

[4] P. WOODWARD, in *Numerical Methods for the Euler Equations of Fluid Dynamics*, eds. Angrand, Dewieux, Desideri, and Glowinski, S.I.A.M. 1985.

[5] P. WOODWARD, in *Astrophysical Radiation Hydrodynamics*, eds. K.H. Winkler and M. Norman, Reidel, 1986.

[6] P. WOODWARD AND K.H. WINKLER, *Simulation and visualization of fluid flow in a numerical laboratory*, preprint October 1988.

[7] A. MAJDA, *Vorticity and the mathematical theory of incompressible fluid flow*, Comm. Pure Appl. Math 39, (1986), pp. S 187-220.

[8] A. MAJDA, *Mathematical Fluid Dynamics: The Interaction of Nonlinear Analysis and Modern Applied Math*, Centennial Celebration of A.M.S., Providence, RI, August 1988 (to be published by A.M.S. in 1990).

[9] R. COURANT AND K. FRIEDRICHS, *Supersonic Flow and Shock Waves*, Springer-Verlag, New York, 1949.

[10] C.S. MORAWETZ, *The mathematical approach to the sonic barrier*, Bull. Amer. Math. Soc., 6, #2 (1982), pp. 127-145.

[11] A. MAJDA, *Compressible Fluid Flow and Systems of Conservation Laws in Several Space Variables*, Appl. Math. Sciences 53, Springer-Verlag, New York 1984.

[12] W. HAYES, *The vorticity jump across a gas dynamic discontinuity*, J. Fluid Mech. 2 (1957), pp. 595-600.

[13] A. MAJDA AND E. THOMANN, *Multi-dimensional shock fronts for second order wave equations*, Comm. P.D.E., 12 (1987), pp. 777-828.

[14] H. GLAZ, P. COLELLA, I.I. GLASS, AND R. DESCHAMBAULT, *A detailed numerical, graphical, and experimental study of oblique shock wave reflections*, Lawrence Berkeley Report, April 1985.

[15] M. VAN DYKE, *An Album of Fluid Motion*, Parabolic Press, Stanford, 1982.

[16] C.S. MORAWETZ, *On the non-existence of continuous transonic flows past profiles, I, II, III*, Comm. Pure Appl. Math. 9,, pp. 45-68, 1956; 10, pp. 107-132, 1957; 11, pp. 129-144, 1958.

[17] J. HUNTER AND J.B. KELLER, *Weak shock diffraction*, Wave Motion 6 (1984), pp. 79-89.

[18] E. HARABETIAN, *Diffraction of a weak shock by a wedge*, Comm. Pure Appl. Math. 40 (1987), pp. 849-863.

[19] D. JONES, P. MARTIN, AND C. THORNHILL, *Proc. Roy. Soc. London A*, 209, 1951, pp. 238-247.

[20] J.B. KELLER AND A. A. BLANK, *Diffraction and reflection of pulses by wedges and corners*, Comm. Pure Appl. Math. 4 (1951), pp. 75-94.

[21] J. HUNTER, *Hyperbolic waves and nonlinear geometrical acoustics*, in *Proceedings of 6th Army Conference on Applied Mathematics and Computations*, Boulder, CO, May 1988 (to appear).

[22] D.G. CRIGHTON, *Basic theoretical nonlinear acoustics*, in *Frontiers in Physical Acoustics*, Proc. Int. School of Physics Enrico Fermi, Course 93, North-Holland, Amsterdam (1986).

[23] D.G. CRIGHTON, *Model equations for nonlinear acoustics*, Ann. Rev. Fluid. Mech. 11 (1979), pp. 11-33.

[24] A. MAJDA, *Nonlinear geometric optics for hyperbolic systems of conservation laws*, in *Oscillation Theory, Computation, and Methods of Compensated Compactness*, IMA Volume 2, 115-165, Springer-Verlag, New York, 1986.

[25] A. MAJDA AND R. ROSALES, *Nonlinear mean field-high frequency wave interactions in the induction zone*, S.I.A.M. J. Appl. Math., 47 (1987), pp. 1017-1039.

[26] R. ALMGREN, A. MAJDA AND R. ROSALES, *Rapid initiation through high frequency resonant nonlinear acoustics*, (submitted to Combustion Sci. and Tech., July 1989).

[27] R. DIPERNA AND A. MAJDA, *The validity of nonlinear geometric optics for weak solutions of conservation laws*, Commun. Math. Physics 98 (1985), pp. 313-347.

[28] J. K. HUNTER AND J.B. KELLER, *Weakly nonliner high frequency waves*, Comm. Pure Appl. Math. 36 (1983), pp. 543–569.

[29] J.K. HUNTER, A. MAJDA, AND R.R. ROSALES, *Resonantly interacting weakly nonlinear hyperbolic waves, II: several space variables*, Stud. Appl. Math. 75 (1986), pp. 187–226.

[30] A. MAJDA AND R.R. ROSALES, *Resonantly interacting weakly nonlinear hyperbolic waves, I: a single space variable*, Stud. Appl. Math. 71 (1984), pp. 149–179.

[31] A. MAJDA, R. ROSALES, M. SCHONBEK, *A canonical system of integro- differential equations arising in resonant nonlinear acoustics*, Studies Appl. Math. in 1989 (to appear).

[32] R. PEGO, *Some explicit resonanting waves in weakly nonlinear gas dynamics*, Stud. Appl. Math. in 1989 (to appear).

[33] P. CEHELSKY AND R. ROSALES, *Resonantly interacting weakly nonlinear hyperbolic waves in the presence of shocks: A single space variable in a homogeneous time independent medium*, Stud. Appl. Math. 74 (1986), pp. 117–138.

[34] L. TARTAR, *Compensated compactness and applications to partial differential equations*, in *Research Notes in Mathematics, Nonlinear Analysis and Mechanics: Heriot-Watt Symposium, Vol. 4*, R. Knops, ed., Pitman, London, 1979.

[35] R. DIPERNA, *Convergence of approximate solutions to conservation laws*, Arch Rat. Mech. Anal. 82 (1983), pp. 27–70.

[36] R. DIPERNA, *Convergence of the viscosity method for isentropic gas dynamics*, Comm. Math. Phys. 91 (1983), pp. 1-30.

[37] M. RASCLE AND D. SERRE, *Comparite par compensation et systemes hyperboliques de lois de conservation*, Applications C.R.A.S. 299 (1984), pp. 673–679.

[38] D. SERRE, *La compucite par compensation pour les systems hyperboliques nonlineaires de deux equations a une dimension d'espace*, J. Maths. Pures et Appl., 65 (1986), pp. 423–468.

[39] W. FICKET AND W. DAVIS, *Detonation*, Univ. California Press, Berkeley, 1979.

[40] A. MAJDA, *High Mach number combustion*, in *Reacting Flows: Combustion and Chemical Reactors*, AMS Lectures in Applied Mathematics, 24, 1986, pp. 109–184.

[41] A. MAJDA AND V. ROYTBURD, *Numerical study of the mechanisms for initiation of reacting shock waves*, submitted to S.I.A.M. J. of Sci. and Stat. Computing in May 1989.

[42] A. MAJDA, *A qualitative model for dynamic combustion*, S.I.A.M. J. Appl. Math., 41 (1981), pp. 70–93.

[43] R. ROSALES AND A. MAJDA, *Weakly nonlinear detonation waves*, S.I.A.M. J. Appl. Math., 43 (1983), pp. 1086–1118.

[44] A. MAJDA AND V. ROYTBURD, *Numerical modeling of the initiation of reacting shock waves*, in *Computational Fluid Mechanics and Reacting Gas Flows*, B. Engquist et al eds., I.M.A. Volumes in Mathematics and Applications, Vol. 12,, 1988, pp. 195–217.

[45] A. MAJDA AND R. ROSALES, *A theory for spontaneous Mach stem formation in reacting shock fronts, I: the basic perturbation analysis*, S.I.A.M. J. Appl. Math. 43, (1983), pp. 1310–1334.

STABILITY OF MULTI-DIMENSIONAL WEAK SHOCKS

GUY MÉTIVIER*

Abstract. In this paper we discuss the stability of weak shocks for a class of multi-dimensional systems of conservation laws, containing Euler's equations of gas dynamics; we study the well-posedness of the linearized problem, and study the behaviour of the L^2 estimates when the strength of the shock approaches zero.

AMS(MOS) subject classifications. 35L65 - 76L05 - 35L50.

1. Introduction. In this lecture, we are concerned with the linearized stability of multi-dimensional weak shocks. Let us first recall that A. Majda has defined the notion of "uniform stability" for shock front solutions to systems of free boundary mixed hyperbolic problems, this stability condition is the natural "uniform Lopatinski condition" for the linearized problem. However, the analysis in [Ma 1] relies on the fact that the front of the shock is non-characteristic while, for weak shocks, the front is "almost" characteristic, i.e. the boundary matrix has a small eigenvalue; in fact the estimates given in [Ma 1] blow up when the strength of the shock tends to zero. In this context, our main goal is to make a detailed study of the behaviour of the L^2 estimates that are valid for the linearized equations, when the strength of the shock tends to zero. Another interesting point we get as a by-product of our analysis, is that, in rather general circumstances, any weak shock that satisfies Lax' shock conditions, is uniformly stable (this was already noted in [Mét 1] for 2×2 systems). The details of the proofs are given in [Mét 3].

2. Equations of shocks. Let us consider a system of conservation laws;

$$\tag{2.1} \sum_{0 \le j \le n} \partial_j f_j(u) = 0$$

where the space-time variables are called $x = (x_0, \ldots, x_n)$, and the unknowns $u = (u_1, \ldots, u_N)$. The functions f_j are supposed to be C^∞ on the open set $\Omega \subset \mathbb{R}^N$, and denoting by A_j the jacobian matrix of f_j, the quasilinear form of (2.1) is:

$$\tag{2.2} \sum_{0 \le j \le n} A_j(u)\partial_j u = 0$$

The typical example we keep in mind all along this paper, is Euler's system of gas dynamics:

$$\tag{2.3} \begin{cases} \partial_t \rho + \text{div } (\rho v) = 0 \\ \partial_t(\rho v) + \text{div } (\rho v \otimes v) + \text{grad } p = 0 \\ \partial_t(\rho E) + \text{div } (\rho E v + pv) = 0 \end{cases}$$

*IRMAR URA 0305 CNRS, Université de Rennes I, Campus Beaulieu, 35042 Rennes Cedex, France.

with ρ the density, p the pressure, v the velocity and $E = \frac{1}{2}|v|^2 + e(p,\rho)$; the unknowns are $u = (\rho, v, s)$, s being the entropy; as usual, we assume that p together with the temperature T are given function of (ρ, s), which satisfy the second law of thermodynamics: $de = Tds + \rho^{-2}d\rho$.

Going back to general notations (2.2), we will always assume that the system is symmetric hyperbolic with respect to the time variable $t = x_0$, (for instance assuming that it admits a strictly convex entropy) that is:

ASSUMPTION 1. *There is a matrix $S(u)$, which depends smoothly on $u \in \Omega$, such that all the matrices SA_j are symmetric, with SA_0 definite positive.*

A shock front solution of (2.1) is, to begin with, a piecewise smooth weak solution u which is discontinuous across an hypersurface $\sum$, say of equation $\phi(x) = 0$; the restrictions $u^\pm$ of u to each side of $\sum$ are smooth solutions of (2.1), and asserting that u is a weak solution is equivalent to the Rankine-Hugoniot jump conditions:

$$(2.4) \qquad \sum_{0 \le j \le n} \partial_j \phi[f_j(u)] = 0$$

where $[f]$ denotes the jump of the function f across $\sum$.

Recall from [Lax] the following lemma, which allows the construction of planar shock fronts (solutions where u^+ and u^- are constant and $\sum$ is the hyperplane of equation $\sigma t = x.\zeta$):

LEMMA 1. *Let $\lambda(u, \zeta)$ be, for $u \in \mathbf{R}^N$ and $\zeta \in \mathbf{R}^n \backslash 0$, a simple eigenvalue of*

$$(2.5) \qquad A'(u, \zeta) = \sum_{1 \le j \le n} \zeta_j A_0^{-1}(u) A_j(u)$$

Then there is a (Rankine-Hugoniot) "curve" of solutions to the jump equations:

$$(2.6) \qquad \sigma\{f_0(u^+) - f_0(u^-)\} = \sum_{1 \le j \le n} \zeta_j\{f_j(u^+) - f_j(u^-)\}$$

$u^+ = U(\varepsilon, u^-\zeta), \sigma = S(\varepsilon, u^-\zeta)$, ($\varepsilon$ being the parameter on the curve, $|\varepsilon|$ remaining small) such that:

$$(2.7) \qquad \begin{cases} u^+ = u^- + \varepsilon r(u^-, \zeta) + O(\varepsilon^2) \\ \sigma = \lambda(u^-, \zeta) + \frac{1}{2}\varepsilon r.d_u\lambda(u^-, \zeta) + O(\varepsilon^2) \end{cases}$$

where $r(u, \zeta)$ denotes a right eigenvector associated to the eigenvalue λ.

3. Structure of the problem. The starting point is to consider equations (2.2) for u^+ together with the jump condition (2.4), as a free boundary value problem, and in this context, the boundary matrix (the coefficient of the normal derivative to $\sum$ in (2.1)) plays an important role:

$$(3.1) \qquad M(u, d\phi) = \sum_{0 \le j \le n} \partial_j \phi.A_j(u)$$

The first requirement is:

a) *the front* $\sum$ *is non-characteristic.* That means that both matrices $M(u^+, d\phi)$ and $M(u^-, d\phi)$ are invertible. When looking at the example (2.7), this is well known to be equivalent to a ***genuine non − linearity*** assumption, i.e. $d_u\lambda(u, \zeta).r(u, \zeta) \neq 0$, in which case we impose the standard normalization:

$$(3.2) \qquad\qquad d_u\lambda(u, \zeta).r(u, \zeta) = 1$$

Also recall that, when the eigenvalue is linearly degenerate $(d_u\lambda(u, \zeta).r(u, \zeta) = 0)$, one falls into the completely different category of contact discontinuities, which in the multi-D case are not yet understood.

The next thing to check is

b) *the number of boundary conditions.* For that purpose one has to look at the number of characteristics impinging the boundary (number of positive and negative eigenvalues of $M(u^\pm, d\phi)$); note that, here, we have N boundary conditions for $2N + 1$ unknowns u^+, u^- and ϕ). In fact, requiring that our problem possesses the right number of boundary conditions leads to the familiar **Lax' shock conditions** ([Lax]). In the example (2.7), with the normalization (3.2), that means we need restrict ourselves to the case:

$$(3.3) \qquad\qquad \varepsilon < 0$$

In the one-D case, Lax shock conditions are sufficient to ensure the well posedness of the problem (at least for weak enough shocks). In the multi-D case the situation is quite more complicated, and further conditions are needed to ensure the well posedness of the problem.

c) *Stability condition.* The strongest way for an hyperbolic boundary value problem to be well posed is to satisfy a so called "uniform Lopatinski" condition (see Kreiss [Kr] or for instance Chazarain-Piriou [Ch-Pi]), and introducing the analogue for the free boundary (2.2) (2.4) led A. Majda ([Ma1]) to his ***uniform stability condition***. Also recall that Majda gave a detailed study of this stability condition for Euler's equations.

4. The problem of weak shocks. As immediately seen from example (2.7), when the strength of the shock is small (ε small), the boundary matrix (3.1) has a small eigenvalue, and the first consequence is that the analysis of [Ma1] collapses when the strength of the shock tends to zero. Now the first thing to do, is to convince oneself that ***something does happen*** and this is easily seen if one looks at the smoothness of the front $\sum$. Indeed, it is part of the uniform stability condition, that for uniformly stable shocks, the Rankine-Hugoniot conditions form an *elliptic system* in ϕ, with coefficients depending on (the traces of) u; in these conditions, we see that the front $\sum$ is one smoother than (the traces of) u. On the other hand, the limit problem of weak shocks, is the problem of ***sound waves*** (see [Mét 2] for a general study of such waves) and then the only relation that exists between $\sum$ and u is an eikonal equation of the form $\partial_t\phi + \lambda(u, \partial_x\phi) = 0$. In that case, we get that the front $\sum$ is just as smooth as (the traces of) u. Clearly then, some smoothness of the front is lost when one passes from shocks to sound waves, and in fact the loss is even worse than what was suggested above because there is also a loss of smoothness of the traces, due to the fact that one passes from a non-characteristic problem to a characteristic one. In a rough way, the fact is: *the weaker is the shock, the less stable it is*.

Another interesting problem is to know whether Lax shock conditions are sufficient to imply the uniform stability of weak shock. Of course this is not always true, as immediately seen by looking at a multi-D scalar law, but as we shall explain in the sequel, this is true in a rather general context.

5. Assumptions. Let us go back to system (2.2). The matrix $A'(u, \zeta)$ was introduced in (2.5). Because of hyperbolicity, its eigenvalues are real. Let $\lambda(u, \zeta)$ be a *simple eigenvalue* of $A'(u, \zeta)$ (for $\zeta \neq 0$); we call $r(u, \zeta)$ an associated right eigenvector. Let $\mathcal{C}$ be an open cone in $\mathbb{R}^n \backslash 0$; then we assume:

ASSUMPTION 2. *(genuine nonlinearity) for all $\zeta \in \mathcal{C}$ and all $u \in \Omega$:*

$$d_u \lambda(u, \zeta).r(u, \zeta) \neq 0.$$

ASSUMPTION 3. *("convexity") For $\zeta \in \mathcal{C}$ and $u \in \Omega$, the Hessian matrix $\lambda''_{\zeta\zeta}(u, \zeta)$ is semi-definite (either positive or negative), and its kernel is exactly $\mathbb{R}\zeta$.*

(because of the homogeneity in ζ, ζ is always in the kernel of $\lambda''_{\zeta\zeta}(u, \zeta)$).

The role of genuine nonlinearity has been recalled in section 3. Assumption 3 plays two roles: first, it implies that, for weak shocks, the Rankine-Hugoniot conditions form an *elliptic* system in φ (as mentioned in the introduction). Next, assumption 3 leads to an important technical simplification: for the limit characteristic problem, it corresponds to the simplest case considered by A. MAJDA and S. OSHER (see assumption 1.6 of [Ma-Os]).

In particular, if λ is an extreme eigenvalue, assumption 3 is a condition of strict convexity for the cone of hyperbolic directions, which is already known to be convex.

At last, in order to avoid further technical complications, we shall restrict our analysis to two cases:

ASSUMPTION 4. *Either $\lambda(u, \zeta)$ is an extreme eigenvalue, i.e. the smallest or the largest eigenvalue of $A'(u, \zeta)$, or the system is strictly hyperbolic, i.e. all the eigenvalues of $A'(u, \zeta)$ are simple.*

Example. With notations as in (2.3), the eigenvalues of Euler's equations are ζ, v and $\zeta, v \pm c|\zeta|$ with c the sound speed, so that the assumptions above are satisfied if one consider one of the extreme eigenvalue $\zeta, v \pm c|\zeta|$ which is known to be simple, genuinely non-linear, and which obviously satisfies assumption 3.

6. The linearized problem. Following A. Majda ([Ma 1]), we consider the following "linearized" problem:

$$(6.1) \qquad \partial_t v + \sum_{j=1}^{n-1} A_0^{-1}(u) A_j(u) \partial_j v + \frac{1}{\kappa} G(u, \theta) \partial_n v = f$$

$$(6.2) \qquad \partial_t \varphi[f_0(u)] + \sum_{j=1}^{n-1} \partial_j \varphi[f_j(u)] = [A_0(u) G(u, \theta) v] + g$$

where:

$$(6.3) \qquad
\begin{aligned}
G(u, \theta) &= A_0^{-1}(u) \left\{ A_n(u) - \sum_{j=0}^{n-1} \theta_j A_j(u) \right\} \\
&= A'(u, -\theta_1, \ldots, -\theta_{n-1}, 1)) - \theta_0 Id
\end{aligned}$$

Recall that (6.1) (6.2) are obtained from (2.2) (2.4) after a change of variables of the form $x_n = \phi(x_0, ., x_{n-1}, \tilde{x}_n)$ which straightens out the front $\sum$ into $\{\tilde{x}_n = 0\}$, and after linearization around a couple of functions (u, ϕ); in that case $(\theta, \kappa) = (\theta_0, \theta_1, \ldots, \theta_{n-1}, \kappa) = (\partial_0 \theta, ., \partial_{n-1} \phi, \partial_n \phi)$, but this particular form of (θ, κ) plays no role here. We also refer the reader to S. Alinhac [Al] (see also [Mét 1]) for a justification of (6.1) as the linearized equation of (2.2) with respect to the couple of variables (u, ϕ).

It is to be understood that equation (6.1) holds on each side $x_n > 0$ and $x_n < 0$, while the jump condition (6.2) holds on $x_n = 0$. We will denote by v^+ and v^- the restrictions of a function v to the half spaces $x_n > 0$ and $x_n < 0$, and by $\Gamma v = (\Gamma v^+, \Gamma v^-)$ their traces on $\{x_n = 0\}$.

In the sequel we consider u and (θ, κ) as given functions, and we are looking for L^2 estimates for the linear hyperbolic problem (6.1) (6.2).

Before stating the precise conditions we shall impose on (u, θ, κ), we introduce a few more notations: for $\theta = (\theta_0, \theta_1, \ldots, \theta_{n-1}) = (\theta_0, \theta')$, we note $\theta^\# = (-\theta', 1) = (-\theta_1, \ldots, -\theta_{n-1}, 1)$, $\lambda^\#(u, \theta') = \lambda(u, \theta^\#)$ and $r^\#(u, \theta) = r(u, \theta^\#)$; $r^\#(u, \theta)$ is an eigenvector of $G(u, \theta)$ associated to the eigenvalue:

$$(6.4) \qquad h(u, \theta) = \lambda(u, \theta^\#) - \theta_0$$

In the sequel, we fix compact subsets $\mathcal{U}$ of Ω and $\mathcal{K}$ of $\mathbf{R}^{n-1}$. We assume the coordinates and $\mathcal{K}$ so chosen that

$$(6.5) \qquad \theta' \in \mathcal{K} \quad \Rightarrow \quad \theta^\# = (-\theta', 1) \in \mathcal{C}$$

At last we fix a constant K and we suppose that $u^\pm$ and $(\theta^\pm, \kappa^\pm)$ are C^2 functions on the half spaces $\overline{\mathbf{R}}^{n+1}_\pm = \{\pm x_n \geq 0\}$ such that $u^\pm$ take their values in $\mathcal{U}, \kappa^\pm > 0$ and:

$$(6.6) \qquad \|u^\pm\|_{W^{2,\infty}} + \|(\theta^\pm, \kappa^\pm)\|_{W^{2,\infty}} + \|(\kappa^\pm)^{-1}\|_{W^{2,\infty}} \leq K$$

where $||.||_{W^{2,\infty}}$ denotes the norm in the usual Sobolev space $W^{2,\infty}$ over $\mathbf{R}_+^{n+1}$ or $\mathbf{R}_-^{n+1}$. We also assume that $\Gamma\theta'$ takes its values in $\mathcal{K}$ and that the jumps and traces satisfy for some $\varepsilon > 0$:

(6.7) $$[\theta] = 0$$

(6.8) $$[u] = -\varepsilon\hat{u} \quad \text{with } \hat{u} = e^a r^{\#}(\Gamma u^+, \Gamma\theta) + \varepsilon a_1$$

(6.9) $$\Gamma h^{\pm} = \lambda(\Gamma u^{\pm}, \Gamma\theta^{\#}) - \Gamma\theta_0 = \mp\varepsilon e^{b^{\pm}}$$

with $\hat{u}, a, b^{\pm}$ of class C^1 and a_1 bounded on $\mathbf{R}_y^n = \{x_n = 0\}$, such that:

(6.10) $$||\hat{u}||_{W^{1,\infty}} + ||a||_{W^{1,\infty}} + ||b^{\pm}||_{W^{1,\infty}} + ||a_1||_{L^{\infty}} \leq K$$

the norms being now taken over $\mathbf{R}_y^n$. These conditions mean that at each point of the boundary, $\Gamma u^+, \Gamma u^-$ and $(-\Gamma\theta_0, \Gamma\theta^{\#})$ almost satisfy (2.7) with the sign of ε reversed, so that Lax' shock conditions now correspond to $\varepsilon > 0$.

7. The main result. For any real number $\lambda > 0$, we note $|.|_{0,\lambda}$ the weighted L^2 norm:

(7.1) $$|v|_{0,\lambda} = \left\{\int e^{-2\lambda t}|v(x)|^2 dx\right\}^{1/2}$$

We keep the same notation $|.|_{0,\lambda}$ for the similar norm of functions on the "boundary" $\{x_n = 0\}$. We shall also make use of the norms:

(7.2) $$|v|_{1,\lambda} = |\partial_y v|_{0,\lambda} + \lambda|v|_{0,\lambda}$$

where $\partial_y = (\partial_0, \ldots, \partial_{n-1})$.

THEOREM 1. *Consider a system (2.1) satisfying assumptions 1 to 4 above. Let $\mathcal{U}$ and $\mathcal{K}$ be a compact subsets of Ω and $\mathbf{R}^{n-1}$ satisfying (4.8) and let K be a given constant. Then there are $\varepsilon_1 > 0, \lambda_1 > 0$ and $C > 0$, such that:*

If (u, θ, κ) are functions which satisfy conditions (6.5) to (6.10) with $0 < \varepsilon \leq \varepsilon_1$, if $v^{\pm}$ and $f^{\pm}$ belong to $C_0^1(\overline{\mathbf{R}}_{\pm}^{n+1})$ and satisfy equation (6.1), if ϕ and g belong to $C_0^1(\mathbf{R}^n)$ and satisfy together with the traces $\Gamma v^{\pm}$ of $v^{\pm}$ on $\{x_n = 0\}$, the boundary condition (6.2), then for any $\gamma \geq \gamma_1$, one has:

(7.3) $$\gamma^{1/2}|v|_{0,\gamma} + \varepsilon^{1/2}|\Gamma v|_{0\gamma} + \gamma\varepsilon^{1/2}|\varphi|_{0,\gamma} + \varepsilon|\varphi|_{1,\gamma} \leq C\{\gamma^{-1/2}|f|_{0,\gamma} + \varepsilon^{-1/2}|g|_{0,\gamma}\}$$

Remarks.

1. Taking $u^{\pm}, \theta$ and $\kappa^{\pm}$ constant, the proof includes that for any $\varepsilon > 0$ small enough, the problem (6.1) (6.2) is uniformly stable as defined by A. Majda. Such a fact was already noticed for 2×2 systems in the appendix of [Mét 1].

2. Estimate (7.3) just makes precise the dependence on ε in the estimates given by A. Majda ([Ma 1]). In particular, existence of solutions (v, φ) for the problem (6.1) (6.2) with data (f, g) in L^2, follows from [Ma 1] as well as estimates and existence in domains $\{t < T\}$.

3. The reader might be worried by the term $\varepsilon^{-1/2}|g|_{0,\gamma}$ in the right hand side of (7.3). Indeed, in the Rankine-Hugoniot condition (2.4) each term is $0(\varepsilon)$, so in the forthcoming applications, linearizing (2.4) will yield a term g which will contain a factor ε in front of it.

8. Several reductions. In the three last sections we shall give a few indications concerning the proof of theorem 1, assuming for simplicity that *the eigenvalue λ under consideration is the smallest one.* First, one can perform several reductions:

a) localize estimate (7.3), making use of a partition of unity.

b) after a (local) change of variables, one can assume that condition (6.9), $\Gamma h^\pm = \lambda(\Gamma u^\pm, \Gamma\theta^\#) - \Gamma\theta_0 = \mp\varepsilon e^{b^\pm}$, holds not only on $x_n = 0$, but also on both side $\pm x_n \geq 0$.

c) next, one can diagonalize the boundary matrix, getting a problem of the following form:

$$(8.1) \qquad J^\pm \partial_n v^\pm + \sum_{j=0}^{n-1} B_j^\pm \partial_j v^\pm = f^\pm$$

$$(8.2) \qquad J^+ \Gamma v^+ = M J^- \Gamma v^- + \varepsilon X\varphi + G$$

where

$$J^\pm = \begin{bmatrix} \mp\varepsilon & 0 \\ 0 & Id_{N-1} \end{bmatrix}; X = \sum_{j=0}^{n-1} b_j \partial_j; M = Id_n + 0(\varepsilon)$$

Thanks to assumption 1, the matrices B_j can be assumed to be symmetric, and B_0 definite positive.

The next lemma is a consequence of (6.8) and of assumption 3, but it is crucial in the understanding of the structure of the problem:

LEMMA 2. $b_j = -e^{a+b+/2} \times$ first column of $B_j + 0(\varepsilon)$, and X is elliptic.

Remark. It is a good exercise to go to the limit in the boundary conditions (8.2) (assuming that $g = \varepsilon h$). Indeed, in the first row, one can factor out ε, and the limit is of the form:

$$(8.3) \qquad \Gamma v_1^+ = \Gamma v_1^- + X_1\varphi + h_1$$

and the limit of the $N-1$ other equations is simply:

$$(8.4) \qquad \Gamma\hat{v}^+ = \Gamma\hat{v}^-$$

(8.3) is nothing but the linearized equation of the eikonal equation corresponding to the limit problem of sound waves mentioned in section 4, while (8.4) are the natural transmission conditions for the linearized equations of sound waves. In these conditions, weak shocks appear as *singular perturbations* of sound waves, the perturbation being singular in two aspects: first, the boundary becomes non characteristic and second, the boundary conditions become elliptic with respect to φ.

d) Denoting by v_1 [resp $\hat{v}$] the first component of v [resp the vector of the $N-1$ last components], theorem 1 is a consequence of the more precise following estimates:

THEOREM 2. *Under the same circumstances as in theorem 1, one has:*

$$(8.5) \quad \begin{aligned} \gamma^{1/2}|v|_{0,\gamma} + \varepsilon^{1/2}|\Gamma v_1|_{0,\gamma} + |\Gamma\hat{v}|_{0,\gamma} + \gamma\varepsilon^{1/2}|\varphi|_{0,\gamma} + \varepsilon|\varphi|_{1,\gamma} \leq \\ C\{\gamma^{-1/2}|f|_{0,\gamma} + \varepsilon^{-1/2}|g_1|_{0,\gamma} + |\hat{g}|_{0\gamma}\} \end{aligned}$$

e) Because λ is the smallest eigenvalue, the problem lying on the side $x_n \leq 0$ is symmetric hyperbolic and well posed without any boundary condition, so that for γ large enough:

$$(8.6) \quad \gamma^{1/2}|v^-|_{0,\gamma} + \varepsilon^{1/2}|\Gamma v_1^-|_{0,\gamma} + |\Gamma\hat{v}^-|_{0,\gamma} \leq C\gamma^{-1/2}|f^-|_{0,\gamma}$$

f) A direct analysis of the boundary conditions shows that:

$$(8.7) \quad \begin{aligned} \gamma\varepsilon^{1/2}|\varphi|_{0,\gamma} + \varepsilon|\varphi|_{1,\gamma} \leq C\{\varepsilon^{1/2}|\Gamma v_1|_{0,\gamma} + |\Gamma\hat{v}|_{0,\gamma}\} \\ + C\{\varepsilon^{-1/2}|g_1|_{0,\gamma} + |\hat{g}|_{0,\gamma}\} \end{aligned}$$

g) Therefore, it suffices to provide an estimate for $v^\pm$, and in fact, because

$$(8.8) \quad \gamma^{1/2}|v^+|_{0,\gamma} + \varepsilon^{1/2}|\Gamma v_1^+|_{0,\gamma} + C\gamma^{-1/2}|f^+|_{0,\gamma}$$

it suffices to give an estimate of Γv^+. More precisely, forgetting the $+$'s in (8.1), we consider the following problem:

$$(8.9) \quad J\partial_n v + \sum_{j=0}^{n-1} B_j \partial_j v = f$$

$$(8.10) \quad J\Gamma v = \varepsilon X\varphi + g$$

and it remains to prove an estimate of the form:

$$(8.11) \quad \begin{aligned} \varepsilon^{1/2}|\Gamma w_1|_{0,\gamma} + |\Gamma\hat{w}|_{0,\gamma} \leq \\ \leq C\{\gamma^{-1/2}|f|_{0,\gamma} + |\hat{g}|_{0,\gamma} + \varepsilon^{-1/2}|g_1|_{0,\gamma} + \gamma^{-1/2}|w|_{0,\gamma}\} + \\ + C\{\varepsilon + \gamma^{-1}\}\{\varepsilon^{1/2}|\Gamma w_1|_{0,\gamma} + |\Gamma\hat{w}|_{0,\gamma} + \gamma\varepsilon^{1/2}|\varphi|_{0,\gamma} + \varepsilon|\varphi|_{1,\gamma}\} \end{aligned}$$

Indeed, with (8.6) (8.7) and (8.8) estimate (8.5) follows immediately if γ is large enough and $\varepsilon > 0$ is small.

9. Symmetrizors. As usual, theorem 2 is proved by using suitable symmetrizors and "integrations by part", but, as shown in [Ma 1], the nature of the boundary conditions (8.2) or (8.10) forces us to introduce pseudo-differential symmetrizors; however, there is a difficulty due to the lack of smoothness of the coefficients (u, θ, κ), and the classical calculus does not apply. To overcome this, there exists a convenient modification of the pseudo-differential calculus which was introduced by J.M. Bony ([Bo]), and which he called the "para-differential" calculus. In fact, we need a version "with parameter γ" of the calculus, similar to the one which was used in [Mét 1]. We will not enter into the details here, referring the reader to [Mét 3] for a precise description of the calculus and also for a complete proof of the theorems. Instead, we would like to explain a little what happens at the symbolic level, and for that purpose, say that $(u^-, u^+, \theta, \kappa)$ **are constant**; for instance the reader may think of (6.1) (6.2) or (8.1) (8.2) or (8.9) (8.10), as the linearized equations of (2.2) (2.4) around a **weak planar shock**.

In that case, a natural way to study (8.9) (8.10) is to perform a partial Fourier-Laplace transform with respect to the tangential variables $y = (t, y') = (t, x_1, ., x_{n-1})$. Let us call $\eta = (\tau, \eta')$ the dual variables; as usual in this kind of problem, τ is complex, with $Im\tau = -\gamma < 0$ and η' remains real. So, after this transformation we are led to the following system:

$$(9.1) \qquad Lv = JD_n v + Pv = f$$

$$(9.2) \qquad J\Gamma v = \varepsilon X \varphi + g$$

where $D_n = \frac{1}{i}\partial_n$ and P and X are matrices which depend linearly on η:

$$P = \begin{bmatrix} \mu & {}^t\sigma \\ \sigma & P' \end{bmatrix}; \qquad X = \begin{bmatrix} \mu \\ \sigma \end{bmatrix} + O(\varepsilon|\eta|)$$

P and X are real when $\gamma = 0, P'$ is of dimension $N - 1$, and:

$$(9.3) \qquad \partial_\tau P \text{ is definite positive, and in particular } \partial_\tau \mu > 0$$

The following fact is a consequence of assumption 3, and implies that X is elliptic as stated in lemma 2: there is $c > 0$ such that:

$$(9.4) \qquad |\mu| + |\sigma| \geq c|\eta|$$

Moreover, the $O(\varepsilon|\eta|)$ term in X can be neglected because it only yields error terms in the right hand side of (8.11); so in the sequel we just drop it.

The construction of the symmetrizor $S = S(\eta)$ relies on the following formula, which holds as soon as SJ is hermitian:

$$(9.5) \quad \langle SJv(0), v(0) \rangle + \int \langle Im(SP)v, v \rangle dx_n = \int Im\langle SLv, v \rangle dx_n \leq ||Lv||_0 ||v||_0$$

where $\langle, \rangle$ denotes the scalar product on $\mathbf{C}^N$ and $||.||_0$ the L^2-norm on $[0, +\infty]$. Classically, two facts are needed (see [Kr] or [Ch-Pi]):

$$(9.6) \qquad Im(SP) \geq c\gamma Id$$

for some constant $c > 0$, and:

$$(9.7) \qquad \langle SJw, w \rangle \geq c\{\varepsilon|w_1|^2 + |\widehat{w}|^2\} - C\{\varepsilon^{-1}|g_1|^2 + |\widehat{g}|^2\}$$

whenever w satisfies the boundary condition $Jw = \varepsilon X\varphi + g$.

Now, the choice of S depends on whether $|\mu| \geq |\sigma|$ or $|\mu| \leq |\sigma|$.

Case I: $|\mu| \geq C|\eta|$. In that case, it suffices to take the standard symmetrizor $S = -Id$: one has:

    ∗ because of (9.3) $Im(SP) \geq c(-Im\tau) = c\gamma$ for some constant $c > 0$

    ∗ the boundary term is:

$\langle SJw, w\rangle = \varepsilon|w_1|^2 - |\widehat{w}|^2 = \varepsilon|w_1|^2 + |\widehat{w}|^2 - 2|\widehat{g} + \varepsilon\sigma\varphi|^2$ and because $|\sigma| \le C|\mu|$, this term is bigger than

$$\varepsilon|w_1|^2 + |\widehat{w}|^2 - 4|\widehat{g}|^2 - C|\varepsilon\mu\varphi|^2 \ge$$
$$\varepsilon|w_1|^2 + |\widehat{w}|^2 - 4|\widehat{g}|^2 - C\varepsilon^2|w_1|^2 - C|g_1|^2$$

which certainly implies (9.7) if ε is small enough.

Case II: $|\mu| \le \delta|\eta|$, with δ small. The main ingredient is the following one:

LEMMA 3. *There are invertible matrices $W(\eta)$ and $V(\eta)$ such that:*

$$WJV = J; \quad \prod \overset{\text{def}}{=} WPV = \begin{bmatrix} \tilde{\mu} & \tilde{\sigma} & 0 \\ \tilde{\sigma} & \tilde{\rho} & 0 \\ 0 & 0 & \prod'' \end{bmatrix}$$

with $\prod''$ of dimension $N - 2$, $\prod$ real when τ is real, and $\tilde{\mu} = \mu + O(\varepsilon|\eta|)$.

Setting $v = Vw$ and $w = (w_1, w_2, w'') = (w', w'')$, we see that (9.1) decouples into:

(9.8)
$$D_n w'' + \prod{}'' w'' = f''$$

(9.9)
$$\begin{bmatrix} -\varepsilon & 0 \\ 0 & 1 \end{bmatrix} D_n w' + \begin{bmatrix} \tilde{\mu} & \tilde{\sigma} \\ \tilde{\sigma} & \tilde{\rho} \end{bmatrix} w' = f'$$

Furthermore, neglecting $O(\varepsilon|\eta|\varphi)$ terms, the boundary conditions also decouple:

(9.10)
$$\Gamma w'' = g''$$

(9.11)
$$\begin{cases} -\varepsilon\Gamma w_1 = \varepsilon\tilde{\mu}\varphi + g_1 \\ \Gamma w_2 = \varepsilon\tilde{\sigma}\varphi + g_2 \end{cases}$$

The study of (9.8) (9.10) is easy, and in fact we can skip it because, as said in section 8, it suffices to provide estimates for the traces and this is trivial from (9.10).

So it suffices to study the 2×2 system (9.9) (9.11); the first step is to solve (9.9) with the boundary condition $\Gamma w_2 = g_2$, and subtracting this solution to the solution of (9.9) (9.11), one reduces oneself to the case where $g_2 = 0$. Eliminating φ in (9.11) leads to a boundary condition of the form:

(9.12)
$$\varepsilon\tilde{\sigma}\Gamma w_1 + \tilde{\mu}\Gamma w_2 = g$$

Now, it remains to get estimates for the traces of solutions of systems like (9.9) (9.12), and this will be performed in the next and last section.

10. The 2×2 problem. Let us stop for a while, to give a typical and very simple example of system (9.9) (9.12), which may help the reader to understand the problem. It is a differential example in which the normal variable and the space-time tangential variables are respectively called x, and $(t, y) \in \mathbf{R}^2$; the equations in the half-space $x > 0$, are

$$(10.1) \qquad \begin{cases} -\varepsilon \partial_x w_1 + \partial_t w_1 + \partial_y w_2 = f_1 \\ \partial_x w_2 + \partial_t w_2 + \partial_y w_1 = f_2 \end{cases}$$

and the boundary condition on $x = 0$ is:

$$(10.2) \qquad \varepsilon \partial_y w_1 + \partial_t w_2 = g$$

In that case the matrix P of (9.1) is simply:

$$(10.3) \qquad P = \begin{bmatrix} \tau & \eta \\ \eta & \tau \end{bmatrix}$$

Let us go back now to a general system (9.9) (9.12); however, for simplicity, we drop the tildes from the notations, and we set $w = (w_1, w_2)$. The first idea is to use a new weight function, and more precisely we introduce:

$$(10.4) \qquad z = e^{-\alpha \gamma x_n / \varepsilon} w \quad \text{and} \quad f^\circ = e^{-\alpha \gamma x_n / \varepsilon} f$$

where $\alpha > 0$ is a small parameter to be determined. (9.9) is transformed into:

$$(10.5) \qquad L^\circ z \equiv J D_n z + P z - i \alpha \gamma \varepsilon^{-1} J z = f^\circ$$

while the boundary condition is unchanged:

$$(10.6) \qquad \varepsilon \sigma \Gamma z_1 + \mu \Gamma z_2 = g$$

It is important to note that $\|f^\circ\| \leq \|f\|$ (because $\alpha \geq 0$), but that it is equivalent to estimate the traces of z or those of w. In order to do so, we introduce the symmetrizor:

$$(10.7) \qquad S = \begin{bmatrix} 1 & -2q \\ 2\varepsilon^{-1}\bar{q} & -2(\varepsilon^{-1} + 1)|q|^2 + 1 \end{bmatrix}$$

with $q = \sigma^{-1}\mu$ (recall that we are working in the domain $|\mu| \leq \delta|\eta|$, and hence that $|\mu| \leq \delta'|\sigma|$ by (9.4)).

With this choice, it is clear that SJ is hermitian, and that:

$$(10.8) \qquad \langle SJz, z \rangle = \varepsilon|z_1|^2 + (1 - 2|q|^2)|z_2|^2 - \varepsilon^{-1}|\varepsilon z_1 + q z_2|^2$$

Therefore, if the parameter δ is small enough, $|q|$ is small, so the boundary condition (10.6) implies that:

$$(10.9) \qquad \langle SJz, z \rangle = \varepsilon|z_1|^2 + \frac{1}{4}|z_2|^2 - \varepsilon^{-1}C|g|^2$$

On the other hand:

$$Im(SP) = \begin{bmatrix} -\mu & \ell \\ \ell & m \end{bmatrix}$$

with $\ell = O(\gamma)$ and:

$$m = \frac{1}{\varepsilon} Im \left\{ 2\frac{\sigma}{\sigma}\bar{\mu} - 2\rho(1+\varepsilon)|q|^2 + \varepsilon\rho \right\}$$

We now remark that the condition $|\mu| \leq \delta|\eta|$ implies that $\gamma \leq \delta|\eta|$; because σ is real when $\gamma = 0$ and $|q|$ is small when δ is small, and because $Im\bar{\mu} = -Im\mu \geq c\gamma$, we see that if δ and ε are small enough, then $Imm \geq c\varepsilon^{-1}\gamma$ and therefore:

$$(10.10) \qquad \langle Im(SP)z, z \rangle \geq c\gamma\{|z_1|^2 + \varepsilon^{-1}|z_2|^2\}$$

With (10.8), we conclude that, *if α is small enough,* then:

$$(10.11) \qquad \langle Im\{S(P - i\alpha\gamma\varepsilon^{-1}J)z\}, z \rangle \geq c\gamma\{|z_1|^2 + \varepsilon^{-1}|z_2|^2 + \varepsilon^{-2}|qz_2|^2\}$$

At last, we note the trivial estimate:

$$(10.12) \qquad \begin{aligned} |\langle f^\circ, z \rangle| &\leq \{|f_1| + \varepsilon^{1/2}|f_2| + |qf_2|^2\} \times \\ &\qquad \{|z_1| + \varepsilon^{-1/2}|z_2| + \varepsilon^{-1}|z_2|\} \\ &\leq C\|f^\circ\|_0 \{|z_1|^2 + \varepsilon^{-1}|z_2|^2 + \varepsilon^{-1}|qz_2|^2\}^{1/2} \end{aligned}$$

With a formula similar to (9.5), we see that estimates (10.10), (10.11) and (10.12) are exactly what we need in order to conclude and get an estimate for z and its traces. In fact, because the symmetrizor (10.7) is singular as $\varepsilon \to 0$, the actual calculus with operators is slightly more complicated than the calculus on symbols we have sketched above (several terms have a ε^{-1} coefficient). In particular remainders deserve a great attention, but again, we refer the reader to [Mét 3] for complete proofs.

REFERENCES

[Al] S. ALINHAC, *Existence d'ondes de raréfaction pour des systèmes quasilinéaires multidimensionnels*, Comm. in Partial Diff. Equ., 14 (1989), pp. 173-230.

[Bo] J.M. BONY, *Calcul symbolique et propagation des singularités pour les équations aux dérivées partielles non linéaires*, Ann. Sc. E.N.S., 14 (1981), pp. 209-246.

[Ch-Pi] J. CHAZARAIN - A. PIRIOU, *Introduction à la théorie des équations aux dérivées partielles*, Bordas (Dunod), Paris, 1981, English translation Studies in Math. and its Applications, vol 14, North Holland (1982).

[Co-Me] R. COIFMAN - Y. MEYER, *Au delà des opérateurs pseudodifférentiels*, Astérisque 57 (1978).

[Kr] H.O. KREISS, *Initial boundary value problems for hyperbolic systems*, Comm. Pure and Appl. Math., 23 (1970), pp. 277-298.

[Lax] P. LAX, *Hyperbolic systems of conservation laws*, Comm. on Pure and Appl. Math., 10 (1957), pp. 537-566.

[Ma1] A. MAJDA, *The stability of multidimensional shock fronts*, Memoirs of the Amer. Math. Soc., n° 275 (1983).

[Ma2] A. MAJDA, *The existence of multidimensional shock fronts*, Memoirs of the Amer. Math. Soc., n° 281 (1983).

[Mét 1] G. MÉTIVIER, *Interaction de deux chocs pour un système de deux lois de conservation en dimension deux d'espace*, Trans. Amer. Math. Soc., 296 (1986), pp. 431-479.

[Mét 2] G. MÉTIVIER, *Ondes soniques*, Séminaire E.D.P. École Polytechnique, année 1987-88, exposé n° 17 & preprint à paraître, J. Math. Pures et Appl..

[Mét 3] G. MÉTIVIER, *Stability of weak shocks*, preprint.

NONLINEAR STABILITY IN NON-NEWTONIAN FLOWS*

J. A. NOHEL†‡, R. L. PEGO†# AND A. E. TZAVARAS†##

1. Introduction. In this paper, we discuss recent results on the nonlinear stability of discontinuous steady states of a model initial-boundary value problem in one space dimension for incompressible, isothermal shear flow of a non-Newtonian fluid between parallel plates located at $x = \pm 1$, and driven by a constant pressure gradient. The non-Newtonian contribution to the shear stress is assumed to satisfy a simple differential constitutive law. The key feature is a non-monotone relation between the total steady shear stress and steady shear strain rate that results in steady states having, in general, discontinuities in the strain rate. We explain why every solution tends to a steady state as $t \to \infty$, and we identify steady states that are stable; more details and proofs will be presented in [8].

We study the system

$$(1.1) \qquad v_t = S_x , \qquad S := T + f\,x, \qquad T := \sigma + v_x ,$$

$$(1.2) \qquad \sigma_t + \sigma = g(v_x) ,$$

on $[0,1] \times [0,\infty)$, with f a fixed positive constant. We impose the boundary conditions

$$(1.3) \qquad S(0,t) = 0 \quad \text{and} \quad v(1,t) = 0, \qquad t \geq 0 ,$$

and the initial conditions

$$(1.4) \qquad v(x,0) = v_0(x), \quad \sigma(x,0) = \sigma_0(x), \qquad 0 \leq x \leq 1;$$

accordingly, $S(x,0) = S_0(x) := \sigma_0(x) + v_{0x}(x) + fx$. The function $g : \mathbf{R} \to \mathbf{R}$ is assumed to be smooth, odd, and $\xi\,g(\xi) > 0$, $\xi \neq 0$. In the context of shear flow, v, the velocity of the fluid in the channel, and T, the shear stress, are connected through the balance of linear momentum (1.1). The shear stress T is decomposed into a non-Newtonian contribution σ, evolving in accordance with the simple differential constitutive law (1.2), and a viscous contribution v_x. The coefficients of density and Newtonian viscosity are taken as 1, without loss of generality. The flow

---

*Supported by the U. S. Army Research Office under Grant DAAL03-87-K-0036 and DAAL03-88-K-0185, the Air Force Office of Scientific Research under Grant AFOSR-87-0191; the National Science Foundation under Grants DMS-8712058, DMS-8620303, DMS-8716132, and a NSF Post Doctoral Fellowship (Pego).

†Center for the Mathematical Sciences, University of Wisconsin– Madison, Madison, WI 53705.
‡Also Department of Mathematics.
#Department of Mathematics, University of Michigan.
##Also Department of Mathematics.

is assumed to be symmetric about the centerline of the channel. Symmetry dictates the following compatibility restrictions on the initial data:

$$(1.5) \qquad v_0(1) = 0, \qquad S_0(0) = 0, \quad and \quad \sigma_0(0) = 0;$$

they imply that $\sigma(0,t) = v_x(0,t) = 0$, and symmetry is preserved for all time.

The system (1.1)–(1.4) admits steady state solutions $(\overline{v}(x), \overline{\sigma}(x))$ satisfying

$$(1.6) \qquad \overline{S} := g(\overline{v}_x) + \overline{v}_x + fx = 0, \qquad \overline{\sigma} = g(\overline{v}_x)$$

on the interval $[0,1]$. In case the function $w(\xi) := g(\xi) + \xi$ is not monotone, there may be multiple values of $\overline{v}_x(x)$ that satisfy (1.6) for some $x's$, thus leading to steady velocity profiles with jumps in the steady velocity gradient $\overline{v}_x$. Our objective is to study the stability of such steady velocity profiles; we also study well-posedness and the convergence of solutions of (1.1)–(1.4) to steady states as $t \to \infty$.

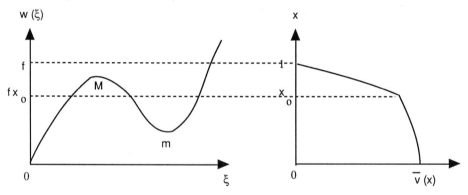

Fig. 1: w vs. ξ.

Fig. 2: Velocity profile with a kink; $w(-\overline{v}_x(x)) = fx$.

For simplicity, the function $w(\xi)$ is assumed to have a single loop. The graph of a representative $w(\xi)$ is shown in Fig. 1; in the figure m and M stand for the levels of the bottom and top of the loop, respectively. Our results and techniques can be easily generalized to cover the case when $w(\xi)$ has a finite number of loops. Steady state velocity profiles are constructed as follows: First solve $w(\overline{u}(x)) = fx$ for each $x \in [0,1]$, where $\overline{u} = -\overline{v}_x$. This equation admits a unique solution for $0 \le fx < m$ or $fx > M$, and three solutions for $m < fx < M$; let $\overline{u}(x)$, $0 \le x \le 1$, be a solution. Setting

$$(1.7) \qquad \overline{v}(x) = \int_x^1 \overline{u}(y)dy, \qquad \overline{\sigma}(x) = g(-\overline{u}(x)),$$

then $(\overline{v}(x), \overline{\sigma}(x))$ satisfy (1.6) and (1.3) for a.e. $x \in [0,1]$ and give rise to a steady state. Clearly, if $f < m$ there is a unique smooth steady state; if $m < f < M$, there is a unique smooth velocity profile and a multitude of profiles with kinks; finally, if $f > M$, all steady state velocity profiles have kinks. An example of a velocity profile with kinks is shown in Fig. 2.

Problem (1.1)–(1.4) captures certain key features of a class of viscoelastic models that have been proposed to explain the occurence of "spurt" phenomena in non-Newtonian flows. Specifically, for a particular choice of the function g in (1.2), the system under study has the same steady states as the more realistic systems studied in [6] and [7]; the latter, derived from a three-dimensional setting that is restricted to one-dimensional shearing motions, produce non-monotone steady shear stress vs. strain-rate relations of the type shown in Fig. 2. The phenomenon of spurt was apparently first observed by Vinogradov *et al.* [13] in the flow of highly elastic and very viscous non-Newtonian fluids through capillaries or slit-dies. It is associated with a sudden increase in the volumetric flow rate occuring at a critical stress that appears to be independent of the molecular weight. It has been proposed by Hunter and Slemrod [5], using techniques of conservation laws, and more recently by Malkus, Nohel, and Plohr [6] and [7], using numerical simulation and extensive analysis of suitable approximating dynamic problems (motivating the present work), that spurt phenomena may be explained by differential constitutive laws that lead to a non-monotone relation of the total steady shear stress versus the steady shear strain-rate. In this framework, the increase of the volumetric flow rate corresponds to jumps in the strain rate when the driving pressure gradient exceeds a critical value. We conjecture that our stability result discussed in Sec. 3 below can be extended to these more complex problems.

2. Preliminaries. In this section, we discuss preliminary results that are essential for presenting the stability result; further details and proofs can be found in [8].

A. Well-Posedness.

We use abstract techniques of Henry [4] to study *global existence of classical solutions for smooth initial data of arbitrary size, and also existence of almost classical, strong solutions with discontinuities in the initial velocity gradient and in the stress components.* The latter result allows one to prescribe discontinuous initial data of the same type as the discontinuous steady states studied in this paper. Existence results of this type are established in [8] for a general class of problems that serve as models for shearing flows of non-Newtonian fluids; the total stress is decomposed into a Newtonian contribution and a finite number of stress relaxation components, viewed as internal variables that evolve in accordance with differential constitutive laws frequently used by rheologists (for discussion, formulation and results, see [11], [7], also the Appendix in [8]). Existence of classical solutions may also be obtained by using an approach based on the Leray - Schauder fixed point theorem (cf. Tzavaras [12] for existence results for a related system). Other existence results were obtained by Guillopé and Saut [2], and for models in more than one space dimension in [3].

As a consequence of the general theory, one obtains two global existence results (see Theorems 3.1, 3.2, 3.5, and Corollary 3.4 in [8]):

(i.) the existence of a unique classical solution $(v(x,t), \sigma(x,t))$ of (1.1)– (1.5) on $[0,1] \times [0,\infty)$ for initial data $(v_0(x), \sigma_0(x))$, not restricted in size, that satisfy: $S_0(x) := v_{0x}(x) + \sigma_0(x) + fx \in H^s[0,1]$ for some $s > 3/2$, with $S_0(0) = 0$, $v_0(1) = S_{0x}(1) = 0$, and $\sigma_0 \in C^1[0,1]$, where H^s denotes the usual interpolation space.

(ii.) the existence and uniqueness of a strong, "semi-classical" solution of (1.1)–(1.5), obtained by a different choice of function spaces, for initial data $(v_0(x), \sigma_0(x))$ that satisfy: $S_0(x) \in H^1[0,1]$ with $S_0(0) = 0$, $v_0(1) = S_{0x}(1) = 0$, and $\sigma_0 \in L^\infty[0,1]$.

Result (ii.) yields solutions in which σ and v_x may be discontinuous in x, but S_x and v_t are continuous, and σ is C^1 as a function of t for every x. Thus all derivatives appearing in the system may be interpreted in a classical sense as long as the equation is kept in conservation form. A result of this type was obtained by Pego in [10] for a different problem by a similar argument.

B. A Priori Bounds and Invariant Sets.

To discuss global boundedness of solutions, let σ, v be a classical solution on an arbitrary time-interval, and note that system (1.1)–(1.5) is endowed with the differential energy identity

(2.1)
$$\frac{d}{dt}\{1/2 \int_0^1 v_t^2 \, dx + \int_0^1 [W(v_x) + x f v_x] dx\}$$
$$+ \int_0^1 [v_t^2 + v_{xt}^2] dx = 0 \ .$$

The function $W(\xi) := \int_0^\xi w(\zeta) d\zeta$ plays the role of a stored energy function; by the assumption on g, W is not convex. This fact is the main obstacle in the analysis of stability.

(i.) *Boundedness of S.* Since $\xi g(\xi) > 0$, it follows that $\int_0^\xi g(\zeta) d\zeta \geq 0$ for $\xi \in \mathbb{R}$, and $W(\xi)$ satisfies the lower bound

(2.2)
$$W(\xi) + fx\xi \geq 1/4\xi^2 - f^2 \ , \qquad \xi \in \mathbb{R}, \quad 0 \leq x \leq 1 \ .$$

Standard energy estimates based on (2.1) and (2.2) coupled with integration of (1.1) with respect to x yield a global a priori bound for S:

(2.3)
$$|S(x,t)| \leq C \qquad 0 \leq x \leq 1, \quad 0 \leq t < \infty \ ,$$

where C is a constant depending only on data but not t.

(ii.) *Invariant Sets for a Related ODE.* Control of S enables us to take advantage of the special structure of Eq. (1.2) and determine suitable invariant regions. For this purpose, it is convenient to introduce the quantity $s := \sigma + fx$. Then, Eqs. (1.2), (1.3) readily imply that s satisfies

(2.4)
$$s_t + s + g(s - S) = f x \ .$$

For a fixed x, it is convenient to view Eq. (2.4) as an ODE with forcing term $S(x, \bullet)$. Also, observe that at a steady state $(\bar\sigma, \bar v_x)$, one has $\bar S = 0$, and consequently,

$\overline{s} = -\overline{v}_x$ is an equilibrium solution of (2.4) (with $S = 0$). If $S \equiv 0$ in (2.4), the hypothesis concerning g implies that the ODE admits positively invariant intervals for each fixed x. We sketch how this property is preserved in the presence of a priori control of S as provided by (2.3); more delicate bounds are essential in the proof of stability in Sec .3.

To fix ideas, let $t_0 > 0$ be given, and assume that

$$(2.5) \qquad\qquad |S(x,t)| \leq \rho , \quad 0 \leq x \leq 1, 0 \leq t \leq t_0 ,$$

for some $\rho > 0$. For x fixed in $[0,1]$, we use the notation $S(t) := S(x,t)$ and conveniently rewrite (2.4) as

$$(2.6) \qquad\qquad s_t + w(s - S(t)) = f\,x - S(t) .$$

We state the following result on invariant intervals; its proof is obvious.

Proposition 2.1. *Let S satisfy the uniform bound (2.5) for $0 \leq t \leq t_0$. For x fixed, $0 \leq x \leq 1$, assume there exist s_-, s_+ such that $s_- < s_+$ and*

$$(2.7) \qquad\qquad w(s_- - \lambda) < f\,x - \lambda \quad , \quad |\lambda| \leq \rho$$

$$(2.8) \qquad\qquad w(s_+ - \lambda) > f\,x - \lambda \quad , \quad |\lambda| \leq \rho$$

Then the compact interval $[s_-, s_+]$ is positively invariant for the ODE (2.6) on the time interval $0 \leq t \leq t_0$.

Invariant intervals are generated by solution sets of the inequalities (2.7) and (2.8) as functions of ρ and x. In particular, since $\lim_{\xi \to \pm\infty} w(\xi) = \pm\infty$, given any x and ρ, one easily determines s_{0+} large, positive and s_{0-} large, negative such that if $s_- < s_{0-}$ and $s_+ > s_{0+}$, then s_- and s_+ satisfy (2.7) and (2.8), respectively, and the compact interval $[s_-, s_+]$ is positively invariant for the ODE (2.6).

More discriminating choices of invariant intervals occur if one restricts attention to small values of ρ; the analysis becomes more delicate. For a function $w(\xi)$ with a single loop, the most interesting case arises when $f\,x - \rho$, $f\,x$ and $f\,x + \rho$ each intersects the graph of $w(\xi)$ at three distinct points. Referring to Fig. 3, the abscissae of the points of intersection are denoted by $(\alpha_-, \beta_-, \gamma_-)$, $(\alpha_0, \beta_0, \gamma_0)$ and $(\alpha_+, \beta_+, \gamma_+)$, respectively. It turns out that for x fixed and ρ small enough, there are discriminating invariant intervals of the type shown in Fig. 3. However, in contrast to the large invariant intervals discussed in the previous paragraph, the more discriminating ones degenerate as we approach the top or bottom of the loop (when x varies).

For the stability of discontinuous steady states in Sec. 3., it is crucial to construct compact invariant intervals that are of uniform length (see Corollary 2.2 in [8]). The latter is accomplished by taking ρ sufficiently small and by avoiding the top and bottom of the loop in Fig. 3. Of specific interest is the situation in which $\overline{s}(x)$ is a piecewise smooth solution of

$$(2.9) \qquad\qquad w(\overline{s}(x)) = f\,x ,$$

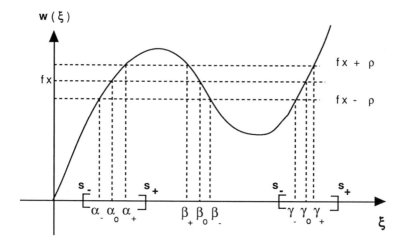

Fig. 3: Invariant Intervals.

defined on $[0, 1]$ and admitting jump discontinuities at a finite number of points $x_1, ..., x_n$ in $[0, 1]$. Recall that $\overline{s}(x)$ is a steady solution of the ODE (2.6) corresponding to the steady state $(\overline{\sigma}, \overline{v}_x)$. In addition, suppose that $\overline{s}(x)$ takes values in the monotone increasing parts of the curve $w(\xi)$ and that it avoids jumping at the top or bottom of the loop, i.e.,

$$(2.10) \qquad w'(\overline{s}(x)) \geq c_0 > 0, \qquad x \in [0, 1] \setminus \{x_1, ..., x_n\},$$

for some constant c_0. A delicate construction in [8] yields compact, positively invariant intervals of (2.6) of uniform length, centered around $\overline{s}(x)$ at each $x \in [0, 1] \setminus \{x_1, ..., x_n\}$.

(iii.) *Boundedness of σ and v_x.* As an easy application of Sec. 2 *(ii)*, choose a compact interval $[s_-, s_+]$ that is positively invariant for (2.6) and valid for all $x \in [0, 1]$. By virtue of the global bound (2.3) satisfied by $S(x, t)$, we conclude that

$$(2.11) \qquad |s(x, t)| \leq C, \qquad 0 \leq x \leq 1, t \geq 0$$

which, in turn, using (1.3) and (2.11), implies

$$(2.12) \qquad |v_x(x, t)| \leq C, \qquad 0 \leq x \leq 1, t \geq 0,$$

for some constant C depending only on the data. The definition of s also implies that σ is uniformly bounded.

(iv.) *Convergence to steady states.* Let $(v(x,t), \sigma(x,t))$ be a classical solution of (1.1) –(1.5) defined on $[0,1] \times [0,\infty]$. We discuss the behavior of this solution as $t \to \infty$.

The first result indicates that $S = \sigma + v_x + f\,x$ converges to its equilibrium value.

Proposition 2.2. *Under the assumptions of the existence results,*

$$(2.13) \qquad \lim_{t \to \infty} S(x,t) = 0,$$

uniformly for $x \in [0,1]$.

The proof is a consequence of Sobolev embedding applied to the following a priori estimates that are derived from the system (1.1)–(1.4) by standard techniques:

$$(2.14) \qquad \int_0^\infty \int_0^1 S_t^2 \, dx d\tau \leq C,$$

$$(2.15) \qquad \int_0^\infty \int_0^1 S^2 \, dx d\tau \leq C,$$

$$(2.16) \qquad \int_0^1 S_x^2(x,t) dx \leq C,\ 0 \leq t < \infty,$$

where C is a positive constant depending only on data.

Use of (2.13) enables us to identify the limiting behavior of solutions of (2.4) as $t \to \infty$. The following result is analogous to Lemma 5.5 in Pego [10]; its elementary proof is given in Lemma 4.2 of [8].

Proposition 2.3. *Let $s(x, \bullet) \in C^1[0, \infty)$ be the solution of (2.10), where $S(x, \bullet)$ is continuous and satisfies (2.13), $0 \leq x \leq 1$. Then $s(x, \bullet)$ converges to $s^\infty(x)$ as $t \to \infty$ and $s^\infty(x)$ satisfies*

$$(2.17) \qquad s^\infty(x) + g(s^\infty(x)) = f\,x \quad, \quad 0 \leq x \leq 1.$$

In view of the shape of $w(\xi) = \xi + g(\xi)$, equation (2.17) has one solution for $0 \leq f\,x < m$ or $f\,x > M$ and three solutions for $m < f\,x < M$.

Let $(v(x,t), \sigma(x,t))$ be a classical solution of (1.1)–(1.4) on $[0,1] \times [0,\infty)$. Recalling the definition of s, Proposition 2.3 implies

$$(2.18) \qquad \sigma^\infty(x) = \lim_{t \to \infty} \sigma(x,t) = s^\infty(x) - f\,x.$$

Also, combining (1.1), (2.13) and (2.18) yields

$$(2.19) \qquad v_x^\infty(x) := \lim_{t \to \infty} v_x(x,t) = \lim_{t \to \infty} (S(x,t) - s(x,t)) = -s^\infty(x),$$

and

$$(2.20) \qquad S^\infty(x) = v_x^\infty(x) + \sigma^\infty(x) + f x = 0 \,.$$

Finally, noting that

$$(2.21) \qquad v(x, t) = - \int_x^1 v_x(x, t)dx \,,$$

$v^\infty(x)$ is Lipschitz continuous and satisfies

$$(2.22) \qquad v^\infty(x) := \lim_{t \to \infty} v(x, t) = \int_x^1 s^\infty(\xi)d\xi \,.$$

We conclude that any solution of (1.1)–(1.4) converges to one of the steady states. If $0 \le f < m$, then there is a unique smooth steady state which is the asymptotic limit of any solution. However, if $m < f$, then there are multiple steady states and thus a multitude of possible asymptotic limits. In Sec. 3, we identify stable steady states. Also note from (2.20) that in a discontinuous steady state, the discontinuities in $\overline{\sigma}$ and $\overline{v}_x$ cancel.

Observe that in case $w(\xi)$ is monotone the above arguments yield that every solution converges to the unique steady state. Moreover, the above results can be routinely generalized to the case that the function $w(\xi)$ has multiple loops but the graph of w has no horizontal segments.

3. Stability of Steady States.

The purpose is to study the stability of velocity profiles with kinks. To fix ideas, let $(\overline{v}(x), \overline{\sigma}(x))$ be a steady state of (1.1)–(1.3) such that $\overline{v}(x)$ has a finite number of kinks located at the points $x_1, \ldots, x_n$ in $(0, 1)$; accordingly, $\overline{v}_x(x)$ and $\overline{\sigma}(x)$ have a finite number of jump discontinuities at the same points. Recall that, if we set $\overline{u}(x) = -\overline{v}_x(x)$,

$$(3.1) \qquad w(\overline{u}(x)) = f x, \quad x \in [0, 1], \, x \ne x_1, \ldots, x_n$$

and $\overline{\sigma}(x) = g(-\overline{u}(x))$.

Given smooth initial data $(v_0(x), \sigma_0(x))$, there is a unique smooth solution $(v(x, t)$, $\sigma(x, t))$ of (1.1)–(1.4). As $t \to \infty$, the solution converges to one of the steady states, not a-priori identifiable. We now restrict attention to initial data that are close to $(\overline{v}(x), \overline{\sigma}(x))$, except on the union $\mathcal{U}$ of small subintervals centered around the points $x_1, \ldots, x_n$. $\mathcal{U}$ can be thought of as the location of transition layers separating the smooth branches of the steady state. Roughly speaking, it turns out that the steady state is "asymptotically stable" under smooth perturbations that are close in energy, provided $(\overline{v}(x), \overline{\sigma}(x))$ takes values in the monotone increasing parts of $w(\xi)$; the stable solutions are local minimizers of an associated energy functional (see (3.8) below). The interesting problem of finding the domain of attraction of a stable steady solution appears to be a difficult task. Our main result is:

Theorem 3.1. *Let $(\overline{v}(x), \overline{\sigma}(x))$ be a steady state solution as described above and satisfying*

$$(3.2) \qquad w'(\overline{v}_x(x)) \geq c_0 > 0, \ x \in [0,1], \qquad x \neq x_1, \ldots, x_n$$

for some positive constant c_0. If the measure of $\mathcal{U}$ is sufficiently small, there is a positive constant δ_0 depending on $\mathcal{U}$ such that, if $\delta < \delta_0$, then for any initial data $(v_0(x), \sigma_0(x))$ satisfying

$$(3.3) \qquad \sup_{0 \leq x \leq 1} |S_0(x)| < \delta,$$

$$(3.4) \qquad \int_0^1 v_t^2(x, 0) dx < \frac{1}{2}\delta^2$$

and

$$(3.5) \qquad |v_{0x}(x) - \overline{v}_x(x)| < \delta \ , \quad x \in [0,1] \setminus \mathcal{U}$$

the corresponding solution $(v(x,t), \sigma(x,t))$ approaches the steady state $(\overline{v}(x), \overline{\sigma}(x))$
as
$t \to \infty$, in the sense,

$$(3.6) \qquad v_x(x, t) \to \overline{v}_x(x),$$

$$(3.7) \qquad \sigma(x, t) \to \overline{\sigma}(x),$$

for all $x \in [0,1] \setminus \mathcal{U}$.

The above result is similar, in spirit and approach, to the analysis of Andrews and Ball [1], and particularly to stability results established by Pego [10] for motions of one-dimensional viscoelastic materials of rate type with a non-monotonic stress-strain relation. Current work of Novick-Cohen and Pego [9] on spinodal decomposition involve a similar stability analysis.

Because the argument is lengthy and delicate, we can only indicate the main idea of the proof for the case that $\overline{u}(x) = -\overline{v}_x(x)$ has one single jump discontinuity located at x_0, $m < fx_0 < M$, and for $\mathcal{U} = (x_0 - \varepsilon, x_0 + \varepsilon)$ for some small ε. Minor modifications are needed to account for the general case. Technical details can be found in [8], Theorem 5.1. The proof is based on exploiting the energy identity (2.1), which, upon setting $u(x,t) = -v_x(x,t)$ and integrating with respect to t, yields the inequality

$$(3.8) \quad \begin{aligned} &\frac{1}{2} \int_0^1 v_t^2(x,t) dx + \int_0^1 [(W(u(x,t)) - xfu(x,t)) - \Phi(x)] dx \\ &\leq \frac{1}{2} \int_0^1 v_t^2(x,0) dx + \int_0^1 [(W(u_0(x)) - xfu_0(x) - \Phi(x)] dx; \end{aligned}$$

note that the integral of the function Φ has been subtracted from both sides of (3.8). The function $\Phi(x)$ is associated with the particular choice of the discontinuous steady state $\overline{u}(x)$, the stability of which is being tested. The function Φ identifies a basin of attraction of the state $\overline{u}(x)$. Roughly speaking, the goal is to find Φ so that the second integral on the left side of (3.8) is positive, and at the same time the right side can be made sufficiently small. The construction of Φ is delicate because W in (3.8) is not convex, and because the double-well potential $W(u) - xfu$ depends explicitly on x. Note that for each fixed x, the function $\overline{u}(x)$ is the horizontal coordinate of the bottom of one of the wells; the left well if $x < x_0$, the right well if $x > x_0$. To insure that the construction of Φ produces the property desired, this part of the analysis makes crucial use of invariant intervals of the ODE (2.6) that are of uniform length as discussed in Sec. 2(ii) above.

REFERENCES

1. G. ANDREWS AND J. BALL, " Asymptotic Stability and Changes of Phase in One-Dimensional Nonlinear Viscoelasticity:," *J. Diff. Eqns.* **44** (1982), pp. 306–341.

2. C. GUILLOPÉ AND J.-C. SAUT , "Global Existence and One-Dimensional Nonlinear Stability of Shearing Motions of Viscoelastic Fluids of Oldroyd Type ," *Math. Mod. Numer. Anal.*, 1990. To appear.

3. C. GUILLOPÉ AND J.-C. SAUT, "Existence Results for Flow of Viscoelastic Fluids with a Differential Constitutive Law," *Math. Mod. Numer. Anal.*, 1990. To appear.

4. D. HENRY, *Geometric Theory of Semilinear Parabolic Equations*, Lecture Notes in Mathematics, vol. 840 Springer-Verlag, New York, 1981.

5. J. HUNTERAND M. SLEMROD, "Viscoelastic Fluid Flow Exhibiting Hysteretic Phase Changes," *Phys. Fluids* **26** (1983), pp. 2345–2351.

6. D. MALKUS, J. NOHEL, AND B. PLOHR, "Dynamics of Shear Flow of a Non-Newtonian Fluid," *J. Comput. Phys.*, 1989. To appear.

7. D. MALKUS, J. NOHEL, AND B. PLOHR, "Analysis of New Phenomena In Shear Flow of Non-Newtonian Fluids," in preparation, 1989.

8. J. NOHEL, R. PEGO, AND A. TZAVARAS, "Stability of Discontinuous Steady States in Shearing Motions of Non-Newtonian Fluids," *Proc. Roy. Soc. Edinburgh, Series A*, 1989. submitted.

9. A. NOVICK-COHEN AND R. PEGO, "Stable Patterns in a Viscous Diffusion Equation," *preprint*, 1989. submitted.

10. R. PEGO, "Phase Transitions in One-Dimensional Nonlinear Viscoelasticity: Admissibility and Stability," *Arch. Rational Mech. and Anal.* **97** (1987), pp. 353–394.

11. M. RENARDY, W. HRUSA, AND J. NOHEL, *Mathematical Problems in Viscoelasticity*, Pitman Monographs and Surveys in Pure and Applied Mathematics, Vol. 35, Longman Scientific & Technical, Essex, England, 1987.

12. A. TZAVARAS, "Effect of Thermal Softening in Shearing of Strain-Rate Dependent Materials," *Arch. Rational Mech. and Anal.* **99** (1987), pp. 349–374.

13. G. VINOGRADOV, A. MALKIN, YU. YANOVSKII, E. BORISENKOVA, B. YARLYKOV, AND G. BEREZH-NAYA, "Viscoelastic Properties and Flow of Narrow Distribution Polybutadienes and Polyisoprenes," *J. Polymer Sci., Part A-2* **10** (1972), pp. 1061–1084.

A NUMERICAL STUDY OF SHOCK WAVE REFRACTION AT A CO_2 / CH_4 INTERFACE†

ELBRIDGE GERRY PUCKETT‡

Abstract. This paper describes the numerical computation of a shock wave refracting at a gas interface. We study a plane shock in carbon dioxide striking a plane gas interface between the carbon dioxide and methane at angle of incidence α_i. The primary focus here is the structure of the wave system as a function of the angle of incidence for a fixed (weak) incident shock strength. The computational results agree well with the shock polar theory for regular refraction including accurately predicting the transition between a reflected expansion and a reflected shock. They also yield a detailed picture of the transition from regular to irregular refraction and the development of a precursor wave system. In particular, the computations indicate that for the specific case studied the precursor shock weakens to become a band of compression waves as the angle of incidence increases in the irregular regime.

Key words. shock wave refraction, conservative finite difference methods, Godunov methods, compressible Euler equations

AMS(MOS) subject classifications. 35L65, 65M50, 76L05

1. The Problem. In this work we consider a plane shock wave striking a plane gas interface at angle of incidence $0° < \alpha_i < 90°$. This is a predominantly two dimensional, inviscid phenomenon which we model using the two dimensional, compressible Euler equations with the incident shock wave and gas interface initially represented by straight lines.

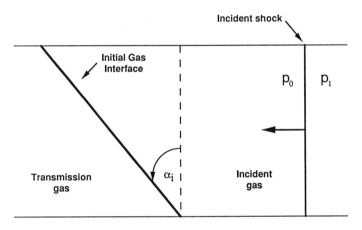

Figure 1 A diagram of the problem

†Work performed under the auspices of the U.S. Department of Energy at the Lawrence Livermore National Laboratory under contract number W-7405-ENG-48 and partially supported by the Applied Mathematical Sciences subprogram of the Office of Energy Research under contract number W-7405-Eng-48 and the Defense Nuclear Agency under IACRO 88-873.
‡Applied Mathematics Group, Lawrence Livermore National Laboratory, Livermore, CA 94550.

A diagram of the problem is shown in figure 1. The shock wave travels from right to left in the incident gas striking the interface from the right. This causes a shock wave to be transmitted into the transmission gas and a reflected wave to travel back into the incident gas. The reflected wave can either be a shock, an expansion, or a band of compression waves. Depending on the strength of the incident shock, the angle of incidence, and the densities and sound speeds of the two gases these three waves may appear in a variety of distinct configurations. In the simplest case the incident, transmitted, and reflected waves all meet at a single point on the interface and travel at the same speed along the interface. This is known as regular refraction. A diagram depicting regular refraction appears in figure 2.

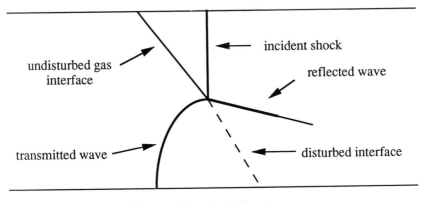

Figure 2 Regular Refraction

When the sound speed of the incident gas is less than that of the transmission gas the refraction is called *slow-fast*. In this case the transmitted wave can break away from the point of intersection with the incident and reflected waves and move ahead of them, forming what is known as a *precursor*. The incident shock can also form a stem between its intersection with the interface and its intersection with the reflected wave, similar to the well known phenomenon of Mach reflection. When the sound speed of the incident gas is greater than that of the transmission gas the refraction is called *fast-slow*. In this case the transmitted shock will lean back toward the interface. In this paper we restrict ourselves to the study of a specific sequence of slow-fast refractions. See Colella, Henderson, & Puckett [1] for a description of our work with fast-slow refraction.

For the purposes of modeling this phenomenon on a computer we assume the two gases are ideal and that each gas satisfies a γ-law equation of state,

$$p = A\rho^{\gamma}.$$

Here p is the pressure, ρ is the density, γ is the ratio of specific heats, and the coefficient A depends on the entropy but not on p and ρ. Note that γ is a constant for each gas but different gases will have different γ.

Given these assumptions the problem depends on the following four parameters: the angle of incidence α_i, the ratio of molecular weights for the two gases μ_i/μ_t, the ratio of the γ for the two gases γ_i/γ_t, and the inverse incident shock strength $\xi_i = p_0/p_1$ where p_0 (*respectively* p_1) is the pressure on the upstream (*respectively* downstream) side of the shock. In this paper we consider the case when the incident gas is CO_2, the transmission gas is CH_4, the inverse incident shock strength is $\xi_i = 0.78$ and only the angle of incidence α_i is allowed to vary. Thus $\gamma_i = 1.288$, $\gamma_t = 1.303$, $\mu_i = 44.01$, $\mu_t = 16.04$, and the incident shock Mach number is 1.1182.

For this choice of parameters we find three distinct wave systems depending on α_i. These are: i) regular refraction with a reflected expansion, ii) regular refraction with a reflected shock, and iii) irregular refraction with a transmitted precursor. These wave systems appear successively, in the order listed, as α_i increases monotonically from head on incidence at $\alpha_i = 0°$ to glancing incidence at $\alpha_i = 90°$. In this paper we examine this sequence of wave patterns computationally much as one would design a series of shock tube experiments. This particular case has been extensively studied both experimentally and theoretically by Abd-el-Fattah & Henderson [2]. This has enabled us to compare our results with their laboratory experiments thereby providing us with a validation of the numerical method. See Colella, Henderson, & Puckett [1, 3] for a detailed comparison of our numerical results with the experiments of Abd-el-Fattah & Henderson. Once we have validated the numerical method in this manner we can use it to study the wave patterns in a detail heretofore impossible due to the limitations of schlieren photography and other experimental flow visualization techniques.

Early work on the theory of regular refraction was done by Taub [4] and Polachek & Seeger [5]. Subsequently Henderson [6] extended this work to irregular refractions, although a complete theory of irregular refraction still remains to be found. More recently, Henderson [7, 8] has generalized the definition of shock wave impedance given by Polachek & Seeger for the refraction of normal shocks.

Experiments with shock waves refracting in gases have been done by Jahn [9], Abd-el-Fattah, Henderson & Lozzi [10], and Abd-el-Fattah & Henderson [2, 11]. More recently, Reichenbach [12] has done experiments with shocks refracting at thermal layers and Haas & Sturtevant [13] have studied refraction by gaseous cylindrical and spherical inhomogeneities. Earlier, Dewey [14] reported on precursor shocks from large scale explosions in the atmosphere. Some multiphase experiments have also been done: Sommerfeld [15] has studied shocks refracting from pure air into air containing dust particles while Gvozdeava *et al.* [16] have experimented with shocks passing from air into a variety of foam plastics.

Some recent numerical work on shock wave refractions include Grove & Menikoff [17] who examined anomalous refraction at interfaces between air and water and Picone *et al.* [18] who studied the Haas & Sturtevant experiments at Air/He and Air/Freon cylindrical and spherical interfaces. Fry & Book [19] have considered refraction at heated layers while Glowacki *et al.* [20] have studied refraction at high speed sound layers and Sugimura, Tokita & Fujiwara [21] have examined refraction in a bubble-liquid system.

2. The Shock Polar Theory.

2.1 A Brief Introduction to the Theory. In this section we present a brief introduction to the theory of regular refraction. This theory is a straightforward extension of von Neumann's theory for regular reflection (von Neumann [22]) and is most easily understood in terms of shock polars. The theory is predicated on the observation that oblique shocks turn the flow. Consider a stationary oblique shock. If we call the angle by which the flow is turned δ (see figure 3), then δ is completely determined by the upstream state (ρ_0, p_0, u_0, v_0) and the shock strength p/p_0 where p denotes the post-shock pressure.

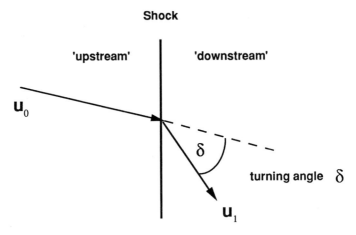

Figure 3 An oblique shock turns the flow velocity towards the shock

For a γ-law gas the equation governing this relation is

$$(2.1) \qquad tan\,(\delta) = \frac{\left(\frac{p}{p_0} - 1\right)\sqrt{\frac{2\gamma M_S^2}{\gamma+1} - \frac{\gamma-1}{\gamma+1} - \frac{p}{p_0}}}{\left(1 + \gamma M_S^2 - \frac{p}{p_0}\right)\sqrt{\frac{\gamma-1}{\gamma+1} + \frac{p}{p_0}}}$$

where M_S is the freestream Mach number upstream of the shock (e.g. see Courant & Friedrichs [23]). If we now allow the shock strength to vary and plot $\log\,(p/p_0)$ versus the turning angle δ we obtain the graph shown in figure 4, commonly referred to as a *shock polar*.

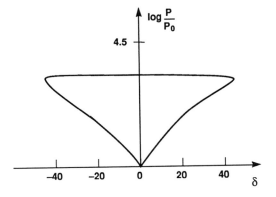

Figure 4 A Shock Polar

Recall that, by definition, in regular refraction the incident, transmitted, and reflected waves all meet at a single point on the interface. We now assume that these waves are locally straight lines in a neighborhood of this point and (for the moment) that the reflected wave is a shock. Each of these shocks will turn the flow by some amount, say δ_i, δ_t, and δ_r respectively (figure 5) and each of these angles will satisfy (2.1) with the appropriate choice of M_S, γ, and p/p_0.

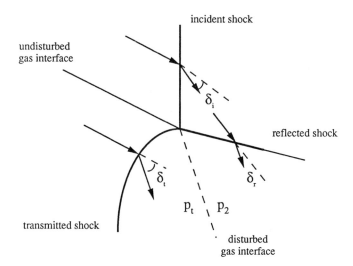

Figure 5 The shock polar theory for regular refraction is based on the fact that the flow must be parallel to the gas interface both above and below the intersection of the shocks. Thus, $\delta_t = \delta_i + \delta_r$. All shocks are assumed to be locally straight in a neighborhood of this intersection.

Furthermore, since the interface is a contact discontinuity we must have

(2.2)
$$p_t = p_2$$

(2.3)
$$\delta_i + \delta_r = \delta_t$$

where the latter condition follows from the fact that the flow is parallel to the interface both upstream and downstream of the intersection of the incident, transmitted and reflected shocks. Note that the interface is, in general, deflected forward downstream of this intersection.

The problem now is as follows. Given the upstream state on both sides of the interface $(\rho_{0i}, p_0, u_{0i}, v_{0i})$ and $(\rho_{0t}, p_0, u_{0t}, v_{0t})$, the inverse incident shock strength ξ_i, and the angle of incidence α_i determine all other states. Let (ρ_1, p_1, u_1, v_1) denote the state downstream of the incident shock (upstream of the reflected shock) and let (ρ_t, p_t, u_t, v_t) and (ρ_2, p_2, u_2, v_2) denote the states downstream of the transmitted and reflected shocks respectively. For certain values of the given data this information is sufficient to completely determine all of the unknown states, although not necessarily uniquely. For example one can derive a 12th degree polynomial in the transmitted shock strength p_t/p_0 from (2.1-3), which for regular reflection has as one root the observed transmitted shock strength (Henderson [8]). The other roots either do not appear in laboratory experiments or are complex, and hence not physically meaningful. Note that knowledge of the transmitted shock strength p_t/p_0 is sufficient to determine all of the other states.

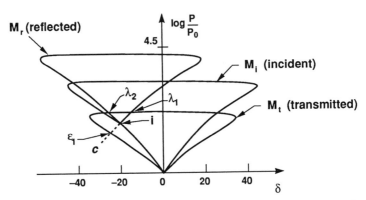

Figure 6 Each intersection of the transmitted shock polar and the reflected shock polar represents a possible wave configuration for regular refraction.

The physically meaningful roots of this polynomial may also be found by plotting the shock polars for the three waves in a common coordinate system. An example is shown in figure 6. Note that we have scaled the reflected shock strength p_2/p_1 by p_1/p_0 and translated δ_r by δ_i. Thus the plot of the reflected shock polar is given by $\log(p_2/p_0) = \log(p_2/p_1) + \log(p_1/p_0)$ versus $\delta_r + \delta_i$. This causes the base of the reflected shock polar ($p_2 = p_1$) to coincide with the map of the incident shock on the incident shock polar (δ_i, p_1/p_0), labeled 'i' in the figure. In this shock polar diagram any intersection of the transmitted and reflected shock polars represents a physically meaningful solution to the problem, i.e. a pair of downstream states (ρ_t, p_t, u_t, v_t) and (ρ_2, p_2, u_2, v_2) such that all of the states satisfy the appropriate shock jump conditions and the boundary conditions (2.2-3). Note that more than one such intersection may exist. For example, in figure 6 there are two, labeled λ_1 and λ_2.

It is also the case that for some values of the initial data $(\rho_{0i}, p_0, u_{0i}, v_{0i})$, $(\rho_{0t}, p_0, u_{0t}, v_{0t})$, ξ_i and α_i, the transmitted and reflected shock polars do not intersect. It is interesting to inquire whether the existence of such an intersection exactly coincides with the occurrence of regular refraction in laboratory experiment. We will discuss this point further below.

We can extend the shock polar theory to include reflected waves which are centered expansions by adjoining to the reflected shock polar the appropriate rarefaction curve for values of $p_2 < p_1$. Let $q = \sqrt{u^2 + v^2}$ denote the magnitude of flow velocity, c the sound speed, and define the Mach angle μ by $\mu = \sin^{-1} 1/M$ where $M = q/c$ is the local Mach number of the flow. Then this rarefaction curve is given by

$$(2.4) \qquad \delta = \pm \int_{p_1}^{p_2} \frac{\cos \mu}{q c \rho} \, dp$$

(see Grove [24]). This curve is sometimes referred to as a *rarefaction polar*. The sign will determine which branch of the shock polar is being extended. In figure 6 the branch corresponding to a negative turning angle δ_r has been plotted with a dotted line and labeled with a c. The intersection of this curve with the transmitted shock polar has been labeled ϵ_1. In some cases there may be two intersections. Each intersection represents a wave system in which the state (ρ_1, p_1, u_1, v_1) is connected to the state (ρ_2, p_2, u_2, v_2) across a centered rarefaction. Such systems are also found to occur in laboratory experiments (e.g. Abd-el-Fattah & Henderson [2]).

2.2 A Shock Polar Sequence. In this section we present the shock polar diagrams for the CO_2/CH_4 refraction with $\xi_i = 0.78$. The data was chosen as specified in Section 1 with only the angle α_i being allowed to vary. In figure 7 we present four shock polar diagrams. These correspond to the two types of regular refraction - namely regular refraction with a reflected expansion (RRE) and regular refraction with a reflected shock (RRR) - the transition between these two states, and the transition between regular and irregular refraction. The polars are labeled M_i, M_t, and M_r, which represent the freestream Mach numbers upstream of the incident, transmitted, and reflected waves respectively. To the right of each shock polar diagram is a small diagram of the wave system in which the initial interface is denoted by an m, and the deflected interface by a D.

In each of the shock polar diagrams the tops of the incident and reflected polars have not been plotted in order to allow us to focus on the intersections which are of interest. As stated above the map of the incident shock on the incident shock polar is labeled i. This point corresponds to the base of the reflected shock polar. The intersection of the incident shock polar with the transmitted shock polar has been labeled A_1.

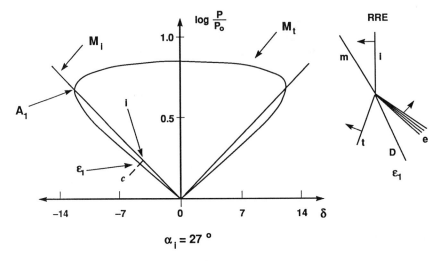

$$\alpha_i = 27\,^\circ$$

Figure 7a

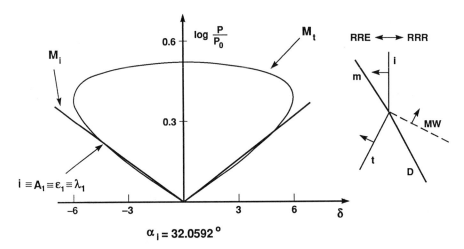

$$\alpha_1 = 32.0592\,^\circ$$

Figure 7b

In figure 7 a) we plot the polars for $\alpha_i = 27^\circ$. Here we have only plotted the reflected rarefaction polar c and its intersection with the transmitted shock polar ϵ_1, not the reflected shock polar. There still exist two solutions λ_1 and λ_2 with a reflected shock but ϵ_1 is the solution observed in the laboratory (Henderson [2]). If we now continuously increase the angle α_i the points i and A_1 move towards each other until they coincide at $\alpha_i \approx 32.0592^\circ$. Here there is no need for a reflected shock or expansion since $\delta_i = \delta_t$. The shock polar diagram for this value of α_i appears in figure 7 b) with the polar for the reflected wave omitted. Note that this is a solution of the problem for which the pressure jump across the reflected wave is vanishingly small, i.e. it is a Mach wave (labeled MW in the small diagram on the right). This is the theoretical transition point between regular reflection with a reflected expansion and regular reflection with a reflected shock, RRE $\rightleftharpoons$ RRR, and at this point the solutions ϵ_1 and λ_1 coincide. At this point we have equality of the wave impedances and hence total transmission of the incident shock into the CH_4. (See Henderson [7, 8].)

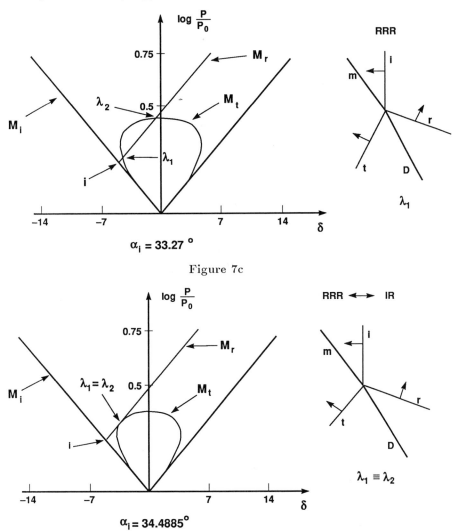

Figure 7c

Figure 7d

Further increase in the angle α_i yields a situation in which the rarefaction curve no longer intersects the transmitted shock polar. However both solutions λ_1 and λ_2 corresponding to reflected shocks still exist. It is the weaker of these solutions, λ_1, which is observed in the laboratory (Henderson [2]). A diagram depicting RRR and the corresponding shock polars appears in figure 7 c). Further continuous increase in α_i results in the transmitted shock polar shrinking relative to the other two polars. This causes the solutions λ_1 and λ_2 to approach each other until they coincide at $\alpha_i \approx 34.4885°$ as shown in figure 7 d). Any further increase in α_i results in a situation where there is no longer an intersection of the reflected and transmitted polars. Thus $\lambda_1 = \lambda_2$ represents the point beyond which regular refraction is (theoretically) impossible. This is denoted RRR $\rightleftharpoons$ IR where IR stands for 'irregular refraction'. Other transition criteria have been proposed. For example, transition could occur when λ_1 coincides with the sonic point, i.e. the value of p_t/p_0 for which the flow speed behind the transmitted shock is sonic. Or transition could occur when λ_1 coincides with the value of p_t/p_0 for which δ_t achieves a maximum. In practice these points often lie so close to each other that it is next to impossible to determine which is the correct criterion from examining schlieren photographs or contour plots. We will discuss this point further after we present the results of our numerical computations in section 4 below.

3. The Numerical Method. We solve the Euler equations for two dimensional, compressible fluid flow. In conservation form these equations are

$$(3.1) \qquad U_t + \nabla \cdot \mathbf{F}(U) = 0$$

where

$$U = (\rho, \rho u, \rho v, \rho E)^T$$

and (u, v) is the velocity, E the total energy per unit mass, and $\mathbf{F} = (F, G)^T$ with

$$F = (\rho u, \rho u^2 + p, \rho u v, \rho u E + u p)^T,$$

$$G = (\rho v, \rho u v, \rho v^2 + p, \rho v E + v p)^T.$$

We solve these equations on a rectangular mesh with grid spacing Δx and Δy and use absorbing boundary conditions on the right hand wall of the computational domain and reflecting boundary conditions on the other three walls.

The following four features of our numerical method are important to the accurate computation of the shock refraction problem.

1) A second order Godunov method for solving the fluid flow equations

2) A local, adaptive gridding strategy

3) A strategy for tracking the fluid interface based on the partial volumes of the fluid components in multifluid cells

4) An algorithm for accurately modeling the disparate thermodynamic properties of the two gases on a subgrid scale.

Currently we use an operator split version of the numerical method. In other words, we solve a succession of one dimensional problems at each time step, alternating the order of the x and y sweeps at alternate time steps. Effective unsplit techniques are available for solving equations (3.1) but our interface tracking algorithm requires operator splitting. We are currently developing an improved interface algorithm that will remove this restriction and we will report on it in a future work.

3.1 The Solution of the Euler equations. We use a second order Godunov method to solve the two dimensional compressible Euler equations. Since these methods have been widely discussed in the literature we refrain from going into detail here. Instead we refer the interested reader to van Leer [25], Colella & Woodward [26], and Colella & Glaz [27]. It should be remarked that in this work we use a piecewise linear approximation to the quantities in each grid cell rather than a piecewise parabolic approximation as discussed in Colella & Woodward [26].

3.2 Adaptive Mesh Refinement. In order to concentrate most of the computational work in regions of physical interest we employ a local adaptive gridding strategy called Adaptive Mesh Refinement (AMR) [28, 29, 30, 31, 32]. The basic idea behind AMR is to estimate the local truncation error at each cell center and tag those cells in which the error is unacceptably large. One then finds a collection of rectangles, all of which are contained in the original grid, in such a way that each of the tagged cells is contained in one of these rectangles. The optimum set of rectangles is determined by minimizing a cost function. So, for example, one large rectangle may be chosen instead of two smaller rectangles with fewer untagged grid points because the large rectangle leads to more optimal vector lengths on a Cray computer. This cost function also takes into account the overhead associated with setting up the boundary conditions for each fine grid.

Each of the new rectangles is then subdivided into smaller cells $1/k$th the size of the original coarse cell (generally $k = 2$ or 4) and the values of the state variables are assigned to each of the new cells in such a way as to conserve all of the appropriate quantities. The equations of motion are then solved on the finer mesh with boundary values obtained from adjacent grids of the same level of refinement or interpolated from the coarser mesh. Note that in order for the CFL condition to be satisfied one must take k times as many time steps on the finer grid, each $1/k$th the size of the coarse grid time step. The value of the state variables in a coarse grid cell which contains fine grid cells is set to the average of the values in the fine grid cells. In order to guarantee conservation at grid boundaries care is taken so that if the boundary of a fine grid abuts a coarse grid (and not another fine grid), then the flux across each coarse cell wall is equal to the sum of the fluxes out of each fine cell which abuts that coarse cell. This adaptive gridding procedure can be recursively extended to obtain multiple levels of refinement.

In figure 8 we show a diagram of a shock wave refraction computation which has been refined in certain important regions. For further details regarding our implementation of the AMR algorithm the reader is referred to Berger & Colella [32].

Adaptive Mesh Refinement

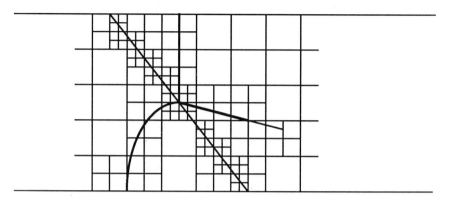

Figure 8 This is how the adaptive gridding algorithm might grid the wave system in figure 2 with two levels or refinement.

3.3 Tracking the Gas Interface.

Partial Volumes

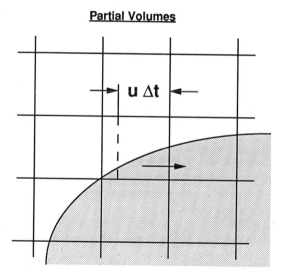

Figure 9 The fraction of dark fluid to the right of the dotted line is advected into the neighboring cell on the right. In this example u is positive.

We employ a partial volumes based approach to the problem of tracking the gas interface. Figure 9 depicts a portion of an interface and its intersection with several grid cells. At the start of the computation in each cell we calculate the ratio f_{ij} of volume occupied by the dark fluid to the total volume of the cell. So $0 \leq f_{ij} \leq 1$ for all cells with $f_{ij} = 0$ if the cell contains all light fluid and $f_{ij} = 1$ if the cell contains all dark fluid. The interface is then advanced in time as follows. At each time step,

1) Given the partial volumes f_{ij} we create an approximation to the interface in each multifluid cell ($0 < f_{ij} < 1$), such that this approximate interface divides the cell into the correct ratio of fluid volumes.

2) For the x-sweep we divide the cell by a vertical line into two rectangles with areas $|u|\Delta t\Delta y$ and $(\Delta x - |u|\Delta t)\Delta y$. We then move that portion of the dark fluid which lies inside the rectangle on the right (if $u > 0$ and on the left if $u < 0$) into the adjacent cell to the right (if $u > 0$ and left if $u < 0$). A cartoon depicting an example of this procedure when $u > 0$ is shown in figure 9. An identical procedure is performed for the y-sweep with u replaced by v, Δy replaced by Δx, etc.

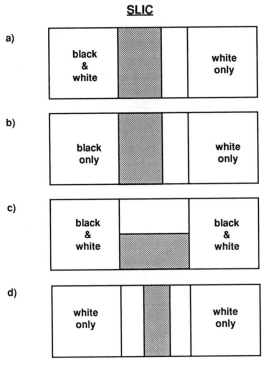

Figure 10 Given the states in the adjacent cells this is how SLIC will draw the interface for a pass in the x-direction. There are five other cases obtained by interchanging black and white and/or left and right.

It remains for us to specify how one recreates the interface given the partial volumes f_{ij}. Here we employ the SLIC (Simple Line Interface Calculation) algorithm created by Noh and Woodward [33]. In determining the interface in the i,jth cell for an x-sweep SLIC considers only the ratio f_{ij} in that cell and the presence or absence of light and dark fluids in the two adjoining (in the x-direction) cells. Figure 10 depicts how the interface is drawn in four of the nine possible cases. The other five cases can be found by reversing the roles of the light and dark fluids and/or by reversing left and right. Figure 11 contains an example of how the SLIC algorithm would reconstruct the interface in figure 9 for a sweep in the x-direction. The interface is reconstructed in an analogous manner for a sweep in the y-direction.

SLIC

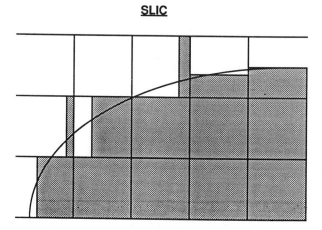

Figure 11 This is how SLIC would recreate the interface in figure 9 for a sweep in the x-direction.

It should be emphasized that the only feature of the flow which we are tracking is the actual gas interface. All of the shocks and other discontinuities in the flow are *captured* by the underlying numerical solution of the equations of gas dynamics.

3.4 Subgrid Modeling of the Multifluid Components. We employ a recent innovation for modeling the thermodynamic properties of distinct fluid components which occupy the same grid cell. The principle goal of this algorithm is to ensure that fluid components of different densities will undergo the correct relative compressions or expansions when the cell they occupy is subjected to pressure forces. This algorithm is based on the assumption that the various fluid components in each cell are in pressure equilibrium with one another and that each cell has a single velocity. From a physical point of view the assumption of pressure equilibrium is not unreasonable since pressure is continuous across a contact discontinuity. The requirement that the cell have a single velocity is not appropriate in more than one dimension since slip will be generated at a fluid interface. Thus we track the jump in thermodynamic variables across the interface, while capturing the jump in tangential velocity using the underlying conservative finite difference method. This algorithm is applicable to any number of fluid components. We refer the reader to Colella, Glaz & Ferguson [34] for a detailed description of this algorithm.

4. The Computational Results. We used the computational method described above to model a weak shock ($\xi_i = 0.78$) in CO_2 striking an plane interface with CH_4. In this context the word 'weak' means that the flow downstream of the incident shock remains supersonic and hence it is possible for the reflected wave to be a shock. We remark that Abd-el-Fattah & Henderson [2] (who refer to the case with $\xi_i = 0.78$ as being a 'very weak' incident shock) examined the effect changing the incident shock strength has on the structure of these wave systems.

In figure 12 we reproduce contour plots of $\log p$ for a sequence of α_i with $27° \leq \alpha_i \leq 65°$. Next to each contour plot we show an enlargement of the intersection of the incident, transmitted, and reflected waves with the gas interface. In each of these contour plots there is a straight line running diagonally from upper left to lower right. This line represents the initial gas interface before being struck by the shock. It is simply a line drawn for easy reference and is *not* a contour of $\log p$.

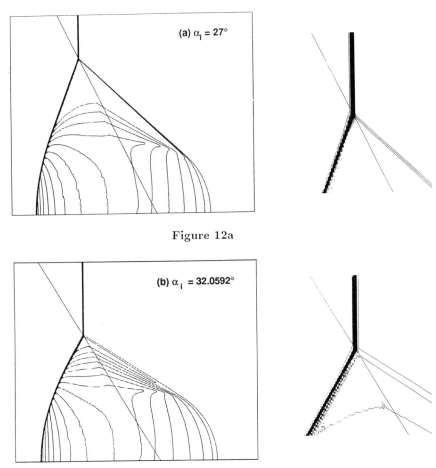

Figure 12a

Figure 12b

In figure 12 a) $\alpha_i = 27°$ and the resulting wave system is a regular reflection with a reflected expansion corresponding to figure 7 a). In figure 12 b) we have increased α_i to 32.0592°. This is the theoretical boundary between a reflected expansion and a reflected shock and corresponds to figure 7 b). It is apparent from the enlargement in figure 12 b) that the reflected wave is very nearly a Mach wave. This can be seen by noting that the reflected wave consists of two contours. The first is a continuation of the last contour in the incident shock and hence we know that the pressure decreases as we move across it from the state (ρ_1, p_1, u_1, v_1) towards the state (ρ_2, p_2, u_2, v_2). Similarly, since the second contour is a continuation of the last contour in the transmitted shock, the pressure must rise again as we cross it. The plotting program plots values of the pressure at fixed increments and hence the pressure value must be the same on these two contours. In other words, if these two contours were coincident then we would have a Mach wave. The discrepancy between figure 12 b) and an actual Mach wave is well within the range of acceptable numerical error given the level of refinement of the computation.

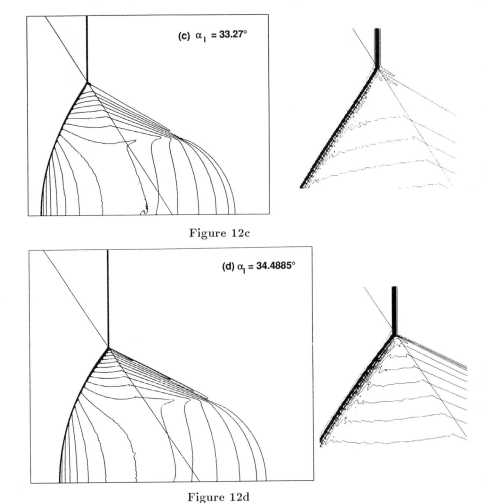

Figure 12c

Figure 12d

If we increase α_i to 33.27° we find that the reflected wave has now become a shock. The contour plot appears in figure 12 c) and corresponds to the solution λ_1 shown in figure 7 c). Figure 12 d) contains the contour plot of $\log p$ for $\alpha_i = 34.4885°$. This is the case $\lambda_1 = \lambda_2$ shown in figure 7d) and is the theoretical limit of regular refraction. Note that as α_i increases from 27° to 34.4885° the pressure contours emanating from the transmitted shock swing around until they are parallel and nearly coincident with the reflected shock. The relation (if any) of this phenomenon to the onset of irregular refraction and/or a precursor wave system remains to be investigated.

Figures 12 e) - h) contain a sequence of contours for α_i in the range beyond the theoretical limit of regular refraction. In figure 12 e) $\alpha_i = 38°$ which is several degrees beyond the theoretical limit of regular refraction. However it is very difficult to determine from the contour plot if this is still a regular refraction or not. This is a problem confronting anyone who attempts to interpret either numerical or experimental results in the region close to the transition. Similar problems are encountered when one attempts to determine the proper transition criterion for shock reflection. We remark that for the shock reflection problem it has been observed that for certain incident shock strengths regular reflection persists beyond the theoretical limit. See Colella & Henderson [35] for further details.

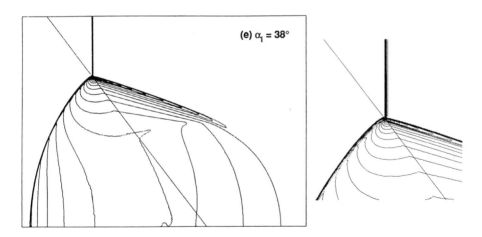

Figure 12e

278

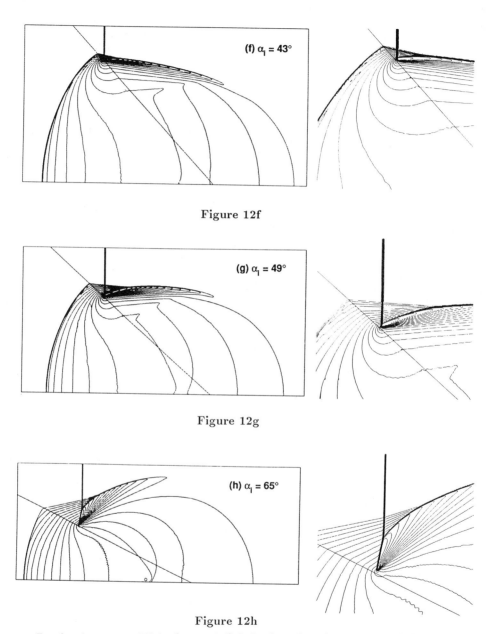

Figure 12f

Figure 12g

Figure 12h

By the time $\alpha_i = 43°$ in figure 12 f) it is clear that the transmitted shock has moved ahead of the incident shock and formed a precursor. Note that the precursor itself refracts back into the incident gas producing a small 'side shock' which interacts with the bottom of the incident shock. There is some indication from the enlargement that near the interface the precursor and side shocks are beginning to break up into a band of compression waves. In figures 12 g) & h) this effect is much more pronounced. Based on schlieren photographs of shock tube experiments it had previously been thought that the transmitted and side waves remained shocks for all values of α_i in this range.

6. Conclusions. We have used a second order conservative finite difference method for solving the 2d compressible Euler equations to model the refraction of a weak shock at a CO_2/CH_4 interface. We modeled the gas interface with a simple interface tracking algorithm based on volume fractions. A new feature of this method is that the thermodynamic properties of distinct fluid components sharing a common multifluid cell are modeled separately. This allows components with disparate thermodynamic properties to undergo the appropriate expansions or compressions on a subgrid scale. We computed a sequence of refractions with fixed incident shock strength, varying only the angle of incidence α_i. For values of this angle lying in the theoretical range of regular refraction the computational results were in complete agreement with the shock polar theory. For values of α_i outside this range the contour plots reveal a detailed picture of the development of an irregular wave system with a precursor shock in the CH_4. Furthermore our results indicate that for the larger values of α_i in the irregular range the precursor degenerates into a band of compression waves. This is a new observation based on the greater detail available from the computational results as compared to earlier schlieren photographs.

Acknowledgements. The author would like to graciously acknowledge many fruitful discussions with Phil Colella and Roy Henderson.

REFERENCES

[1]   P. Colella, L. F. Henderson and E. G. Puckett, *A Numerical Study of Shock Wave Refractions at a Gas Interface*, Proceedings of the AIAA 9th Computational Fluid Dynamics Conference (June 13-15,1989), pp. 426–439, Buffalo, New York.

[2]   A. M. Abd-el-Fattah and L. F. Henderson, *Shock waves at a slow-fast gas interface*, J. Fluid Mech., 89 (1978), pp. 79–95.

[3]   L. F. Henderson, P. Colella, and E. G. Puckett, *On the refraction of shock waves at a slow-fast gas interface*, submitted to the Journal of Fluid Mechanics.

[4]   A. H. Taub, *Refraction of Plane Shock Waves*, Phys. Rev., 72 (1947), pp. 51–60.

[5]   H. Polachek and R. J. Seeger, *On Shock Wave Phenomenon: Refraction of Shock Waves at a Gaseous Interface*, Phys. Rev., 84 (1951), pp. 922–929.

[6]   L. F. Henderson, *The refraction of a plane shock wave at a gas interface*, J. Fluid Mech., 26 (1966), pp. 607–637.

[7]   L. F. Henderson, *On shock impedance*, J. Fluid Mech., 40 (1970), pp. 719–735.

[8]   L. F. Henderson, *On the refraction of shock waves*, J. Fluid Mech., 198 (1989), pp. 365–386.

[9]   R. G. Jahn, *The refraction of shock waves at a gaseous interface*, J. Fluid Mech., 1 (1956), pp. 457–489.

[10] A. M. Abd-el-Fattah, L. F. Henderson and A. Lozzi, *Precursor shock waves at a slow-fast gas interface*, J. Fluid Mech., 76 (1976), pp. 157–176.

[11] A. M. Abd-el-Fattah and L. F. Henderson, *Shock waves at a fast-slow gas interface*, J. Fluid Mech., 86 (1978), pp. 15–32.

[12] H. Reichenbach, *Roughness and heated layer effects on shock wave propagation and reflection-experimental results*, E-Mach Inst. Rep. E24/85, Frieberg: Fraunhofer-Gesellschaft.

[13] J.-F. Haas and B. Sturtevant, *Interaction of weak shock waves with cylindrical and spherical gas inhomogeneities*, J. Fluid Mech., 181 (1987), pp. 41–76.

[14] J. M. Dewey, *Precursor Shocks Produced by a Large Yield Chemical Explosion*, Nature, 205 (1965), pp. 1306 only.

[15] M. Sommerfeld, *The unsteadiness of shock waves propagating through gas-particle mixtures*, Experiments in Fluids, 3 (1985), pp. 197–206.

[16] L. G. Gvozdeava, X. Faresov, M. Yu, J. Brossard, and N. Charpentier, *Normal shock wave reflexion on porous compressible material*, Prog. Astron. Aeron., 106 (1986), pp. 155-165.

[17] J. W. Grove and R. W. Menikoff, *Anomalous Refraction of a Shock Wave at a Fluid Interface*, Los Alamos Report LA-UR-89-778.

[18] J. M. Picone, J. P. Boris, E. S. Oran, and R. Ahearne, *Rotational Motion Generated by Shock Propagation Through a Nonuniform Gas*, in Proc. 15th Intl. Symp. on Shock Waves and Shock Tubes, ed. D. Bershader and R. Hanson, Stanford University Press.

[19] M. A. Fry and D. L. Book, *Shock Dynamics in Heated Layers*, in Proc. 15th Intl. Symp. on Shock Waves and Shock Tubes, ed. D. Bershader and R. Hanson, Stanford University Press.

[20] W. J. Glowacki, A. L. Kuhl, H. M. Glaz, and R. E. Ferguson, *Shock Wave Interaction with High Speed Sound Layers*, in Proc. 15th Intl. Symp. on Shock Waves and Shock Tubes, ed. D. Bershader and R. Hanson, Stanford University Press.

[21] T. Sugimura, K. Tokita, and T. Fujiwara, *Nonsteady Shock Wave Propagating in a Bubble-Liquid System*, AIAA J., 94 (1984), pp. 320-331.

[22] J. von Neumann, *Oblique Reflection of Shocks*, in Collected Works 6, ed. A. H. Taub Pergamon Press, New York (1963), pp. 238–299.

[23] R. Courant and K. O. Friedrichs, *Supersonic Flow and Shock Waves*, Springer Verlag, New York, 1976.

[24] J. Grove, *The Interaction of Shock Waves with Fluid Interfaces*, Advances in Applied Mathematics, 10 (1989), pp. 201–227.

[25] B. van Leer, *Towards the ultimate conservative difference scheme, a second order sequel to Godunov methods*, J. Comp. Phys., 32 (1979), pp. 101–136.

[26] P. Colella and P. Woodward, *The piecewise parabolic method (PPM) for gas dynamical simulations*, J. Comp. Phys., 54 (1984), pp. 174–201.

[27] P. Colella and H. M. Glaz, *Efficient solution algorithms for the Riemann problem for real gases*, J. Comp. Phys., 59 (1985), pp. 264–289.

[28] M. J. Berger and J. Oliger, *Adaptive mesh refinement for hyperbolic partial differential equations*, J. Comp. Phys., 53 (1984), pp. 482-512.

[29] M. J. Berger and A. Jameson, *Automatic adaptive grid refinement for the Euler equations*, AIAA J., 23 (1985), pp. 561–568.

[30] M. J. Berger, *Data structures for adaptive grid generation*, SIAM J. Sci. Stat. Comp., 7 (1986), pp. 904–916.

[31] M. J. Berger, *On conservation at grid interfaces*, SIAM J. Num. Anal. (1987).

[32] M. J. Berger and P. Colella, *Local adaptive mesh refinement for shock hydrodynamics*, J. Comp. Phys., 82 (1989), pp. 64–84.

[33] W. F. Noh and P. Woodward, *SLIC (Simple Line Interface Calculation)*, UCRL preprint 97196, Lawrence Livermore National Laboratory.

[34] P. Colella, H. M. Glaz, and R. E. Ferguson, *Multifluid algorithms for Eulerian finite difference methods*, manuscript.

[35] P. Colella and L. F. Henderson, *The von Neumann paradox for the diffraction of weak shock waves*, J. Fluid Mech. (to appear).

AN INTRODUCTION TO WEAKLY NONLINEAR
GEOMETRICAL OPTICS*

RODOLFO R. ROSALES†

Abstract. Many natural phenomenae are governed by systems of nonlinear conservation laws that are – in a first approximation – hyperbolic. In this context, the understanding of the laws governing the propagation and interaction of small but finite amplitude high frequency waves in hyperbolic P.D.E.'s, and their interactions with "large scale" phenomenae (mean flows, shear layers, shock and detonation waves, etc.) is very important. Weakly Nonlinear Geometrical Optics (W.N.G.O.) is an asymptotic formal theory whose objective is precisely to do this.

In this paper we review the theory of W.N.G.O. as it stands currently. An important point is that W.N.G.O. is not (yet?) a complete theory. It fails at caustics, singular rays, etc. Since these are frequently regions of physical interest, a good theory for what happens there is important – and mostly nonexistent. The current status of these important open problems will also be (briefly) reviewed.

1. Introduction. Waves are ubiquitous in the physical world, thus the study of the general laws that govern their propagation is very important. In particular, high frequency-short waves are frequently encountered and, fortunately, the behavior of waves is much simpler in this limit.

It is actually easy to understand why high frequency waves[1] have simpler behaviors: locally, to any such wave, the media over which it propagates looks homogeneous and the wave behaves, locally, as a plane wave. Thus, to these waves, one can often associate a definite direction of propagation along which the waves (together with important physical quantities such as energy, information, etc.) propagate. This is why many of the theories (see below) developed to deal with high frequency waves have a very strong "particle like" flavor. In fact, this phenomenon is at the heart of the particle-wave duality of quantum mechanics.

Exploiting the idea described in the prior paragraph, many approximation theories have been developed to deal with waves in different contexts: Modulation Theory for dispersive waves (see [W1]), e.g.: water waves; W.K.B.J. for the classical limit of quantum mechanics (see [B]); L.G.O. (Linear Geometrical Optics) for the propagation of light and other electromagnetic waves (see [K] and [W1]); etc. All these theories have strong similarities at the formal level[2], which reflects the same

---

*This work was partially supported by grant NSF #DMS-8702625 and was performed, in part, while the author visited the IMA during April 1989.

†Department of Mathematics, Room 2-337, Massachusetts Institute of Technology, Cambridge, MA 02139, USA.

[1]To be precise: what we mean here by high frequency-short waves is "waves whose period and wavelength are much shorter than other typical time and length scales in the problem, including those over which the parameters describing the wave (frequency, wavelength, amplitude, etc.) may vary."

[2]Reference [H2] illustrates the point by using a W.N.G.O. formalism to study general high frequency, small dispersion, small amplitude solutions for dominantly hyperbolic systems.The reader may also want to compare the expansions introduced later in this paper with those reviewed in Reference [R3].

core idea at the level of the approximations involved. One must, however, be careful not to push the similarities too much as the physical content – and consequently the properties of the mathematical equations involved – can be quite different.

Linear Geometrical Optics (L.G.O.) actually applies to general linear hyperbolic systems of equations; and Weakly Nonlinear Geometrical Optics (W.N.G.O.) is a (nontrivial) modification of L.G.O. to incorporate the effects of a small (weak) nonlinearity into the approximation. Although the amount of nonlinearity is small, its effects are important and drastic modifications occur in the asymptotic governing equations. New phenomenae arise, such as: (i) wave breaking and shock formation, (ii) resonant wave interaction, (iii) new wave generation, etc. The challenge of W.N.G.O. is to understand these phenomenae and to describe and formulate their governing laws in practical and useful ways. Thus extremely complex situations are reduced to simpler but often non-trivial problems which are more readily understood. In fact, as we will see, weakly nonlinear hyperbolic wave motions in 1-D and multi-D are often well approximated through solutions of much simpler equations – such as the inviscid Burgers' equation.

Many of the basic ideas of how to deal with weak nonlinearities in hyperbolic problems date back to the pioneering work of Landau, Lighthill and Whitham in the 1940's and 1950's (see [Lan], [Li] and [W2]). A formal version (using the method of multiple scales) of the W.N.G.O. theory in its simplest form was developed in [CB]. Recently the theory has been extended and generalized in [HK1], [HMR] and [MR2]. It should be clear that the main driving force behind the development of this theory has been in its applications – mainly to compressible fluid mechanics – as a quick glance at the bibliography of this paper will show.

The plan of the paper is as follows:

In §2 we introduce general notation for the basic systems of hyperbolic conservation laws to which W.N.G.O. can be applied. The characteristic values and characteristic vectors are introduced with the proper normalizations and Taylor expansions for the densities, fluxes and eigenvalues are exhibited. The example of gas dynamics is introduced.

In §3 we develop the most basic expansion of W.N.G.O. (corresponding to a single high frequency mode) in some detail and generality. *We work in conservation form to emphasize that the expansion remains valid after shocks form.* The asymptotic equations can be found in (3.10), (3.19) and (3.20). Although the overall motion of the waves is governed by the same *Eikonal Equation* that appears in L.G.O. (Linear Geometrical Optics), the wave amplitude equation (3.20) has nonlinear corrections in it, which are crucial for shock formation. The effects of wave amplification and damping due to geometrical focusing - defocusing and due to interactions with a nonhomogeneous media are studied in some detail. In the symmetric case an energy conservation equation (3.31) is derived and an example from gas dynamic is given.

In §4 we consider an extension of W.N.G.O. describing the propagation of a single (sharp) wavefront zone, rather than the situation where an extended high frequency region exists – as in §3. Three cases are considered. In the more interesting one, Case 3, the expansion is nonuniform and a brief discussion is included as to the reason for this nonuniformity and its possible relationship to the special behavior of shocks at caustics. This Case 3 requires matching to complete the expansion and the asymptotic governing equations are then (3.10) and (4.5) through (4.11).

In §5 the effects of adding a small amount of diffusive dissipation to the governing equations is considered. The modifications introduced in the asymptotic expansions are quite simple and easy to follow through. In some problems consideration of these effects is very important, e.g.: Example 9.1 would not make any sense without diffusive effects and in [RM] the reader can find some discussion of the interesting effects diffusion can have on the behavior of the model of Example 9.2.

In §6 a brief outline of the situation for multiple wave interactions in W.N.G.O. is presented. Only limited results exist in this case and the expansions become quite involved and nontrivial. Since these multiple wave situations arise in many problems of interest in the applications, the open problems that remain in this area are important.

In §7 the question of the "zones of failure" of the W.N.G.O. is reviewed. This is a very important open problem and very little is known about its resolution. For L.G.O. (Linear Geometrical Optics) one knows what to do, but the same techniques that work there, when generalized to W.N.G.O., either do not work or provide answers too complex to be satisfactory.

In §8 a generalization of the basic expansion of §3 to situations where the equations have (small) lower order terms is developed and an important example from combustion theory is presented. The example of the general semilinear system of equations is also considered.

In §9, we consider the same situation as in §8, but generalizing the expansions in §4. Two examples from combustion theory are provided: one a simple model for flames and the other a simple model for detonations.

Finally: the unaware reader may get the impression, after reading this paper, that the use of the methods presented here – once developed – is pretty mechanical. This is not so. In addition to the mathematical difficulties that one may encounter (eg.: see Remark 4.3, §6 and §7) we would like to point out that:

(i) In the applications the equations (even when there is a well defined set of equations almost everybody agrees on) do not come in the "clean" forms used in this paper.In fact, they usually involve dimensions and must be properly nondimensionalized so that the proper nondimensional parameters can be identified. There is usually no unique way of doing this nondimensionalization, and some ways may be "better" than others.

(ii) Once the equations have been nondimensionalized and the proper parameters have been identified (there is usually several of them) one must decide what is small and what is large, and in what fashion. Although in this paper we will regularly deal with limits where $\epsilon \to 0$, in practice ϵ is small, but never zero and sometimes not even "very small".

Now, to illustrate the point in (ii) above, if we have a parameter whose value is $P = \frac{1}{6}$ and a second parameter whose value is $Q = \frac{1}{45}$, then: should we set $Q = \epsilon$ and $P = O(1)$? or perhaps we should set $Q = \epsilon$ and $P = O(\sqrt{\epsilon})$? The difference between the two choices can be substantial, since various physical processes will be given different weights – depending on the choice. Which one, if any, is the correct one that gives the *correct qualitative behavior* [3] depends on having made the "right" physical choice. In the end the applied mathematician, like all scientists, must rely on his intuition, comparison with numerical calculations and (the final judge off all truths) comparison with experimental and observational measurements.

Acknowledgements. I would like to thank my scientific collaborators and personal friends, J.K. Hunter and A.J. Majda, with whom I have been exploring the W.N.G.O. "jungle", where I got lost in a moment of inattention in late 1979. My survival would have been dubious without them. I would also like to thank my student, M. Engel, whose help was critical for the completion of this paper.

2. The equations and properties. The purpose of this section is to present and introduce appropriate notation, which is then used throughout the rest of the paper, for the basic sets of hyperbolic conservation laws to which the W.N.G.O. approximation can be applied. Small modifications and generalizations will be introduced in later sections, as the W.N.G.O theory is developed. We include here some short general considerations regarding shocks in the unperturbed base state, strict hyperbolicity, conservation forms and validity of the expansions past the time of shock formation (see the remarks). Also, at the end, the example of gas dynamics is introduced.

We consider a system of n $(n = 1, 2, 3, \ldots,)$ conservation laws in n unknowns $\mathbf{u} = \mathbf{u}(\mathbf{x}, t) = (u_1, \ldots, u_n)^T$ in one time t and N $(N = 1, 2, 3, \ldots,)$ space dimensions $\mathbf{x} = (x_1, \ldots, x_N)^T$

$$(2.1) \qquad \hat{N}[\mathbf{u}] = \sum_{0}^{N} \frac{\partial}{\partial x_j} \mathbf{F}_j(\mathbf{x}, t, \mathbf{u}) = 0,$$

where $\hat{N}$ is defined by the formula, we have introduced the *notation* $x_0 = t$ and the n-vectors $\mathbf{F}_j$ of densities (for $j = 0$) and fluxes (for $1 \leq j \leq N$) are presumed to depend smoothly on their arguments. Actually, for our purposes in this paper, continuous second derivatives are enough.

---

[3] In this modern age of fast computers, asymptotic expansions are seldom intented for computational uses. What one aims at is *understanding*. In fact, asymptotic models can often be used as testing grounds for numerical algorithms and, furthermore, the qualitative understanding they provide can also help in the design of reasonable numerical algorithms – less likely to produce "fake" answers.

Now let $\mathbf{u} = \mathbf{u}_0(\mathbf{x}, t)$ be a smooth solution of (2.1) (again, we do not really need smoothness). We will call $\mathbf{u}_0$ the *base state*. Without loss of generality (redefine the densities and fluxes if necessary) we may assume $\mathbf{u}_0 \equiv \mathbf{0}$, so that

$$(2.2) \qquad \sum_0^N \frac{\partial}{\partial x_j} \mathbf{F}_j^0(\mathbf{x}, t) = 0,$$

where the *superscript 0 denotes evaluation of a function of* $\mathbf{u}$ *at* $\mathbf{u} = \mathbf{0}$. We will use this *notation* consistently throughout the paper.

REMARK 2.1. Although in this paper we will not consider situations where $\mathbf{u}_0$ is allowed to have discontinuities (shocks, vortex lines, etc.) and other singularities, such situations can be considered. The discontinuities must then be handled separately by special methods that go beyond the limited aims of the present paper. The interested reader can consult, for example, references [AM], [M], [MA], [MR1] and [R1].

Assume now that the system (2.1) is, at least for $\|\mathbf{u}\|$ small, strictly hyperbolic with $t = x_0$ a time-like direction. That is, if the $n \times n$ matrices $A_j = A_j(\mathbf{x}, t, \mathbf{u})$ are the gradients with respect to $\mathbf{u}$ of the $\mathbf{F}_j, 0 \le j \le N$, we have:

$$(2.3) \qquad \det A_0 \ne 0.$$

For any $\boldsymbol{\xi} = (\xi_1, \ldots, \xi_N) \in R^N$ nonvanishing, the generalized eigenvalue problem

$$(2.4) \qquad \left(\lambda A_0 - \sum_1^N \xi_j A_j\right) \cdot \mathbf{r} = 0, \quad \boldsymbol{\ell} \cdot \left(\lambda A_0 - \sum_1^N \xi_j A_j\right) = 0$$

has n *distinct* eigenvalues $\lambda_s = \lambda_s(\mathbf{x}, t, \mathbf{u}, \boldsymbol{\xi}), 1 \le s \le n$, with corresponding left $\boldsymbol{\ell}_s = \boldsymbol{\ell}_s(\mathbf{x}, t, \mathbf{u}, \boldsymbol{\xi})$ and right $\mathbf{r}_s = \mathbf{r}_s(\mathbf{x}, t, \mathbf{u}, \boldsymbol{\xi})$ eigenvectors (all depending smoothly on their arguments) which we normalize by

$$(2.5) \qquad \boldsymbol{\ell}_p \cdot A_0 \cdot \mathbf{r}_q = \delta_{pq}, \quad 1 \le p, \ q \le n,$$

where δ_{pq} denotes Kronecker's delta. To determine the eigenvectors uniquely, one further normalization is necessary. Particular problems may dictate their own choices, but, for the moment we take

$$(2.6) \qquad \|A_0 \cdot \mathbf{r}_s\|_2 = 1, \quad 1 \le s \le n.$$

Then the eigenvector pairs $(\boldsymbol{\ell}_s, \mathbf{r}_s)$ are determined up to signs.

REMARK 2.2. We have restricted the system (2.1) to be strictly hyperbolic. However, for the development of the W.N.G.O. theory strict hyperbolicity is not always necessary (but hyperbolicity is). Most of the theory presented in this paper has obvious generalizations to situations where the eigenvalues are not simple, and we will point out some of them as we go along. This is not a matter of empty generalization: one of the most important practical problems where W.N.G.O. can be used (and which in fact has spurred its development to a great extent) is that of gas dynamics;[4] there the particle path characteristics are multiple (even when $N = 1$ for reacting gas dynamics). In the context of this latter example, because of the many symmetries involved, the rather complicated asymptotic equations that would arise in a general W.N.G.O. treatment involving multiple eigenvalues, simplify enormously – with many of the general coefficients vanishing. Thus, it seems pointless to unnecessarily complicate the presentation of the theory here by introducing non-strict hyperbolicity in the general development.

REMARK 2.3. The theory of W.N.G.O. can be used for *general* hyperbolic systems, not necessarily in conservation form. However, this seems pointless, unless one can be assured that no wave breaking will occur (as in the linear and semilinear cases, see Example 8.1). In fact, one of the more attractive features of W.N.G.O. is that *it can deal with wave breaking and shock formation.* The expansion does not depend on smoothness and we will emphasize this aspect by carrying out the expansions in conservation form. A very general discussion of these issues, for W.N.G.O. in multiple wave contexts, can be found in [CR]. Furthermore, we point out that some rigorous proofs of validity for the W.N.G.O. approximations (including shocks) are available – see [DM].

Finally, following up on the assumption of smoothness of the fluxes and densities $\mathbf{F}_j$, we record here *notation* for their Taylor expansions near $\mathbf{u} = 0$ which will be used further on

(2.7) $\quad \mathbf{F}_j(\mathbf{x}, t, \mathbf{u}) = \mathbf{F}_j^o(\mathbf{x}, t) + A_j^o(\mathbf{x}, t) \cdot \mathbf{u} + \frac{1}{2} \mathbf{B}_j^o(\mathbf{x}, t)[\mathbf{u}, \mathbf{u}] + O(\|\mathbf{u}\|^3), \quad 0 \le j \le N,$

where the $\mathbf{B}_j^o(\mathbf{x}, t)[\cdot, \cdot]$ are n-vector valued symmetric bilinear forms and we continue using the superscript 0 notation to denote functions of $\mathbf{u}$ evaluated at $\mathbf{u} = \mathbf{u}_0 = 0$.

THEOREM 2.1. *The following identities apply (easy to prove):*

(2.8) $\qquad\qquad \dfrac{\partial}{\partial \xi_j} \lambda_s = \boldsymbol{\ell}_s \cdot A_j \cdot \mathbf{r}_s, \quad 1 \le j \le N, \quad 1 \le s \le n.$

(2.9) $\qquad\qquad \boldsymbol{\xi} \cdot \mathbf{grad}_\xi \lambda_s = \lambda_s, \quad 1 \le s \le n.$

(2.10) $\quad \lambda_s = \lambda_s^o + \left\{ \displaystyle\sum_1^N \xi_j \boldsymbol{\ell}_s^o \cdot \mathbf{B}_j^o[\mathbf{u}, \mathbf{r}_s^o] - \lambda_s^o \boldsymbol{\ell}_s^o \cdot \mathbf{B}_0^o[\mathbf{u}, \mathbf{r}_s^o] \right\} + O(\|\mathbf{u}\|^2), \quad 1 \le s \le n.$

EXAMPLE 2.1. The Euler Equations of Gas-Dynamics can be written in the form (see [W1])

$$\rho_t + \operatorname{div}(\rho \mathbf{v}) = 0,$$

(2.11) $\qquad\qquad (\rho \mathbf{v})_t + \operatorname{div}(\rho \mathbf{v} \otimes \mathbf{v} + pI) = 0,$

$$(\rho E)_t + \operatorname{div}(\rho E \mathbf{v} + p \mathbf{v}) = 0,$$

---

[4]See Examples 2.1, 3.1, 6.1, 8.2, 9.1 and 9.2.

with (equations of state)

(2.12) $$E = \frac{1}{2}\mathbf{v} \cdot \mathbf{v} + e, \quad e = e(\rho, S), \quad p = p(\rho, S),$$

where ρ is the density, $\mathbf{v}$ is the flow velocity N-vector, p is the pressure, I is the rank 2 identity tensor, E is the total energy per unit mass, e is the internal energy per unit mass and S is the entropy. The absolute temperature T is related to the other variables by the thermodynamic identity

(2.13) $$T\,dS = de + p\,d\tau,$$

where $\tau = 1/\rho$ is the specific volume. Introducing $\mathbf{u} = (\rho, \mathbf{v}, S)^T$, the system is seen to have the form (2.1) with $n = N+2$, $N = 1, 2$ or 3. These equations are hyperbolic, but not strictly hyperbolic (except for $N = 1$). In fact, if in (2.4) $\mathbf{r} = (\rho', \mathbf{v}', S')^T$, the eigenvalues and right eigenvectors are

(2.14) $$\lambda = \boldsymbol{\xi} \cdot \mathbf{v} \quad \text{with} \quad \boldsymbol{\xi} \cdot \mathbf{v}' = 0 \quad \text{and} \quad c^2 \rho' + p_S S' = 0,$$

for the particle paths and

(2.15) $$\lambda = \boldsymbol{\xi} \cdot \mathbf{v} \pm c\xi \quad \text{with} \quad \mathbf{v}'_p = 0, \quad S' = 0 \quad \text{and} \quad \rho \boldsymbol{\xi} \cdot \mathbf{v}' \mp c\xi \rho' = 0,$$

for the two acoustical modes. Here $c = \sqrt{p_\rho}$ is the sound speed, $\xi = ||\boldsymbol{\xi}||_2$ and $\mathbf{v}'_p$ are the components of $\mathbf{v}'$ perpendicular to $\boldsymbol{\xi}$. We note that the particle path eigenvalue has multiplicity N while the acoustical modes have multiplicity 1.

3. The basic expansion (single high frequency mode). In this section we will consider the most basic situation in W.N.G.O.; that of a single high frequency small amplitude simple wave oscillatory mode at leading order, interacting with the base state and (possibly) with a mean field–also of small amplitude. The asymptotic equations are displayed in (3.10), (3.19) and (3.20).

Let $0 < \epsilon << 1$ be a measure of the amplitude of the waves. Then the W.N.G.O. ansatz for the situation just described is the following (locally plane and monochromatic) formal expansion for the solution of (2.1)

(3.1) $$\mathbf{u}_\epsilon = \mathbf{u}_\epsilon(\mathbf{x}, t, \theta) = \epsilon \mathbf{u}_1(\mathbf{x}, t, \theta) + \epsilon^2 \mathbf{u}_2(\mathbf{x}, t, \theta) + \ldots,$$

where $\theta = \frac{1}{\epsilon}\varphi(\mathbf{x}, t)$ is the *phase,* and it is assumed that the wave number vector $\mathbf{k}$ does not vanish.[5] The *wave number* and *wave frequency* are defined by (φ is a scalar)

(3.2) $$\mathbf{k} = (k_1, ..., k_N)^T = \mathbf{grad}\ \varphi \quad \text{and} \quad \omega = -\varphi_t.$$

It is presumed that the only dependence on ϵ of the solution is as explicitly stated in (3.1) – up to the order displayed.

In the usual fashion of the method of multiple scales (see [KC]), once (3.1) is substituted into the equations and the appropriate substitutions are made for the partial derivation operators

(3.3) $$\frac{\partial}{\partial t} \longrightarrow -\frac{\omega}{\epsilon}\frac{\partial}{\partial \theta} + \frac{\partial}{\partial t}, \quad \frac{\partial}{\partial x_j} \longrightarrow \frac{1}{\epsilon}k_j\frac{\partial}{\partial \theta} + \frac{\partial}{\partial x_j}, \quad 1 \le j \le N,$$

---

[5] The expansion generally fails when this happens.

the equations are solved sequentially, by collecting equal orders of ϵ and treating θ as a new independent variable and φ as a new dependent variable. Of course, having introduced two new variables into the equations, a certain amount of ambiguity arises. This is the key to the whole process, for it allows us to require the dependence on the "fast" variable θ to have some desirable properties so that the ambiguities disappear. We will require here (other possibility is considered later in §4) that:

(i) *The dependence of* $\mathbf{u}_1$ *and its derivatives on* θ *is oscillatory,* e.g.: periodic, quasi-periodic, almost periodic, etc., in θ (see below, (3.6)). In particular we define the mean of any oscillatory function $f = f(\theta)$ by

$$(3.4) \qquad \overline{f} = \lim_{T \to \infty} \frac{1}{2T} \int_{-T}^{T} f(\theta)\, d\theta$$

and require that such means exist and that in fact

$$(3.5) \qquad \overline{f} = \lim_{T \to \infty} \frac{1}{2T} \int_{-T}^{T} f(\theta + \eta)\, d\theta$$

uniformly in $\infty < \eta < \infty$. Such an assumption is always satisfied when f can be decomposed as

$$(3.6) \qquad f = f_i + f_a\,,$$

where f_a is almost periodic and f_i is bounded and vanishes as $|\theta| \to \infty$.

(ii) *The dependence of* $\mathbf{u}_2$ *on* θ *is sublineal, that is:*

$$(3.7) \qquad \lim_{|\theta| \to \infty} \frac{1}{\theta} \mathbf{u}_2 = 0\,.$$

This guarantees the formal validity of the expansion on at least $O(1)$ regions of $(\mathbf{x}, t)$ space because there[6] $\|\epsilon \mathbf{u}_2\| << \|\mathbf{u}_1\|$.

When (3.7) applies we will say that the expansion is *semi-uniform* while if $\mathbf{u}_2$ is *bounded* then we will say that it is *uniform*.

REMARK 3.1. This is perhaps a convenient place to point out a fundamental difference between L.G.O. (Linear Geometrical Optics) and W.N.G.O. In L.G.O. we know, given the oscillatory assumptions above, that the solutions can be written as sums of exponentials in θ. Thus we can concentrate on a single exponential and replace (3.1) by (see [W1] and [K])

$$(3.8) \qquad \mathbf{u}_\epsilon = \Sigma \epsilon^m \mathbf{a}_m(\mathbf{x}, t) e^{i\theta},$$

with the more general solution obtained by linear superposition. In W.N.G.O. the nonlinearity generates harmonics that interact between them, so that the precise form of the solution is not known in advance and must be resolved as part of the problem;

---

[6]For example, if $\|\mathbf{u}_2\| = O(\sqrt{|\theta|})$ then $\|\epsilon \mathbf{u}_2\| << \|\mathbf{u}_1\|$ for $|\theta| << \epsilon^{-2}$ i.e., $|\varphi| << \epsilon^{-1}$.

thus (3.1). Furthermore, in (3.8) the summation can start at any m, there is no limitation to small amplitudes, as in W.N.G.O., for L.G.O.

REMARK 3.2. The ansatz (3.1) implies a distinguished limit, where both the amplitude and wavelength are $O(\epsilon)$. This is, generally, the most important high frequency - small amplitude limit for if:

(a) amplitude $<<$ wavelength, then the limit behavior is linear.

(b) amplitude $>>$ wavelength, then (generally) shock formation and the associated decay will rapidly reduce the amplitude down to the situation here.

Of course, there are special situations where (b) may not apply – failure of genuine nonlinearity being one (see Remark 3.10) – and then variations of (3.1) will be necessary, but we will not consider such situations in this paper.

REMARK 3.3. In no way does (3.1) cover the most general high frequency and small amplitude situation (exclusive of Remark 3.2), e.g.: multiple mode (several phases) solutions are also possible (see §6). Thus for the validity of the expansions here appropriate restrictions on the initial and boundary conditions are required. In this review we will not consider W.N.G.O. multiple phase expansions in any detail, the interested reader can consult, for example, references [CR], [HK1], [HMR] and [MR2].

Next we carry out the details of the expansion. As pointed out before, in Remark 2.3, we carry the *analysis in conservation form* to emphasize that *it does not depend on smoothness* and that *it remains valid when shocks arise*. Substituting (3.1) into (2.1) and collecting equal orders of ϵ, we find (using the results and notation introduced in §2) at $O(1)$ the equation

$$(3.9) \qquad \frac{\partial}{\partial \theta} \left\{ (\sum_{1}^{N} k_j A_j^o - \omega A_0^o) \cdot \mathbf{u}_1 \right\} = 0.$$

Since, obviously, we do not want $\mathbf{u}_1$ to be trivial in θ, we conclude that φ must satisfy the

Eikonal Equation

$$(3.10) \qquad \det(\omega A_0^o - \sum_{1}^{N} k_j A_j^o) = 0.$$

That is, for some p ($1 \leq p \leq n - p$ *fixed from now on*),

$$(3.11) \qquad \omega = \lambda_p^o(\mathbf{x}, t, \mathbf{k}).$$

Furthermore

(3.12)
$$\mathbf{u}_1 = \mathbf{v}(\mathbf{x}, t) + \sigma(\mathbf{x}, t, \theta)\mathbf{r}_p^o(\mathbf{x}, t, \mathbf{k}),$$

where σ is a scalar oscillatory function of θ and $\mathbf{v} = \overline{\mathbf{u}_1}$ so that $\overline{\sigma} = 0$ (see (3.4) – (3.6) above). This insures the conservation form (3.9): $(\Sigma k_j A_j^o - \omega A_0^o)\mathbf{u}_1 = $ function of $(\mathbf{x}, t)$ alone.

REMARK 3.4. We introduce the *notation*, to be used here and in the rest of the paper,
(3.13)
$$\lambda^o = \lambda^o(\mathbf{x}, t) = \lambda_p^o(\mathbf{x}, t, \mathbf{k}), \ \ \mathbf{r}^o = \mathbf{r}^o(\mathbf{x}, t) = \mathbf{r}_p^o(\mathbf{x}, t, \mathbf{k}) \text{ and } \boldsymbol{\ell}^o = \boldsymbol{\ell}^o(\mathbf{x}, t) = \boldsymbol{\ell}_p^o(\mathbf{x}, t, \mathbf{k}).$$

It is now convenient to introduce the

Characteristic Form of the Eikonal Equation

Along the *rays*, defined by (use (2.8))

(3.14)
$$\frac{d}{dt}\mathbf{x} = (\mathbf{grad}_\xi \lambda_p^o)(\mathbf{x}, t, \mathbf{k}) = \{\boldsymbol{\ell}^o \cdot A_j^o \cdot \mathbf{r}^o\}_{j=1}^N,$$

we have

(3.15)
$$\frac{d}{dt}\varphi = 0,$$

and in fact also

(3.16)
$$\frac{d}{dt}\omega = (\lambda_p^o)_t \Big|_{\boldsymbol{\xi}=\mathbf{k}} \quad and \quad \frac{d}{dt}\mathbf{k} = -(\mathbf{grad}_x \lambda_p^o) \Big|_{\boldsymbol{\xi}=\mathbf{k}}.$$

Thus, the wavefronts $\varphi = $ const. propagate along the rays at *ray speed velocity:* $\mathbf{grad}_\xi \lambda_p^o$. From now on we will use the *notation* $\frac{d}{dt}$ whenever we want to indicate differentiation along rays.

REMARK 3.5. In the special case of an *homogeneous, time independent base state* ((2.1) constant coefficients), λ_p^o is independent of $(\mathbf{x}, t)$ and $\mathbf{k}, \omega$ are *constant* along rays.

REMARK 3.6. For an *isotropic base state* (where $\lambda_p^o = \lambda_p^o(\mathbf{x}, t, \xi)$, $\xi = ||\boldsymbol{\xi}||_2$) it is easy to see that (use (2.9)) for some $c_p^o = c_p^o(\mathbf{x}, t)$ – the local "wave speed" – we have

(3.17)
$$\lambda_p^o = c_p^o \xi \quad so \ that \quad \frac{d}{dt}\mathbf{x} = c_p^o \hat{\mathbf{k}} = c_p^o \mathbf{k}/||\mathbf{k}||_2,$$

where $\hat{\mathbf{k}}$ is defined by the equation. Thus at each point in space the *wavefronts propagate normal to themselves at speed c_p^o and the rays are normal to the wavefronts.*

REMARK 3.7. In the case of *multiple eigenvalues*, the only difference at this stage is that in (3.12) one must introduce a different scalar σ for each eigenvector corresponding to λ_p^o. On the other hand, if (2.1) is hyperbolic – but not strictly hyperbolic – and the particular wave mode excited (p^{th}) is simple, then everything in this section applies without change.

REMARK 3.8. Given initial values for φ, it is clear that, after some time we should generally expect the rays to start crossing, with the wavefronts attempting to fold and cross themselves. When this happens the solution of the *Eikonal Equation* becomes multiple valued, the expansion loses validity and – as we will see in (3.30) – infinities appear. This is the problem of *caustics*. Problems also arise in the presence of obstacles; the present expansion cannot account for diffraction effects and predicts sharp transitions between "illuminated" and "shadow" zones where no rays reach. At the boundaries of the "illuminated" zone ("*singular rays*") the expansion fails. We come back again to these problems later on in §7.

Having completed the problem at leading order, we now move on to the next order, $O(\epsilon)$. We assume φ smooth in the region of space where we carry the expansion. At $O(\epsilon)$ we find

(3.18)
$$\frac{\partial}{\partial \theta}\left\{\left(\textstyle\sum_1^N k_j A_j^o - \omega A_0^o\right) \cdot \mathbf{u}_2\right\} = \; -\frac{\partial}{\partial t}\left\{A_0^o \cdot \mathbf{v} + \sigma A_0^o \cdot \mathbf{r}^o\right\}$$
$$-\textstyle\sum_1^N \frac{\partial}{\partial x_j}\left\{A_j^o \cdot \mathbf{v} + \sigma A_j^o \cdot \mathbf{r}^o\right\}$$
$$-\frac{\partial}{\partial \theta}\left\{\left(\textstyle\sum_1^N k_j \mathbf{B}_j^o[\mathbf{v}, \mathbf{r}^o] - \omega \mathbf{B}_0^o[\mathbf{v}, \mathbf{r}^o]\right)\sigma\right.$$
$$\left.+\tfrac{1}{2}\left(\textstyle\sum_1^N k_j \mathbf{B}_j^o[\mathbf{r}^o, \mathbf{r}^o] - \omega \mathbf{B}_0^o[\mathbf{r}^o, \mathbf{r}^o]\right)\sigma^2\right\} \; .$$

Now, because $\mathbf{u}_2$ is sublinear in θ (i.e. (3.7)) and σ is oscillatory of mean zero, we must have

Mean Field Equations

(3.19)
$$\frac{\partial}{\partial t}(A_0^o \cdot \mathbf{v}) + \sum_1^N \frac{\partial}{\partial x_j}(A_j^o \cdot \mathbf{v}) = 0.$$

Multiplying (3.18) on the left by $\boldsymbol{\ell}^o$ and using the fact that it is a left eigenvector, we obtain[7]

High Frequency Equation

(3.20)
$$\frac{d}{dt}\sigma + \Delta\sigma + \frac{\partial}{\partial \theta}(a\sigma + \tfrac{1}{2}b\sigma^2) = 0,$$

---

[7]Since $\boldsymbol{\ell}_0$ is smooth, this does not violate the conservation form and poses no problems across shocks.

where $\frac{d}{dt}$ is as in (3.14) and

$$\Delta = \Delta(\mathbf{x}, t) = \sum_0^N \boldsymbol{\ell}^o \cdot \frac{\partial}{\partial x_j}(A_j^o \cdot \mathbf{r}^o),$$

(3.21)
$$a = a(\mathbf{x}, t, \mathbf{v}) = \sum_1^N k_j \boldsymbol{\ell}^o \cdot \mathbf{B}_j^o[\mathbf{v}, \mathbf{r}^o] - \omega \boldsymbol{\ell}^o \cdot \mathbf{B}_0^o[\mathbf{v}, \mathbf{r}^o],$$

$$b = b(\mathbf{x}, t) = a(\mathbf{x}, t, \mathbf{r}^o),$$

where, obviously, the $(\mathbf{x}, t)$ dependence involves φ. This *equation is in conservation form and remains valid after shocks form*, which will inevitably occur if $b \neq 0$.

It is now easy to see that (3.19) and (3.20) guarantee the sublinearity condition (3.7). In fact, writing

(3.22)
$$\mathbf{u}_2 = \sum_1^n m_j(\mathbf{x}, t, \theta) \mathbf{r}_j^o(\mathbf{x}, t, \mathbf{k})$$

and substituting into (3.18), one can easily write explicit solutions for the m_j, $1 \leq j \leq n$, and verify the sublinearity.

REMARK 3.9. The *High Frequency Equation* (3.20), is (along each ray) the *inviscid Burgers' equation* with time dependent coefficients and a forcing term $\Delta \sigma$ added. In fact, *along any given ray*, where $\Delta = \Delta(t), a = a(t)$ and $b = b(t)$ – we assume $\mathbf{v}$ known – the transformation

(3.23)
$$\sigma(\theta, t) = \alpha(t) M(\xi, t), \quad \alpha = e^{-\int^t \Delta}, \quad \xi = \theta - \int^t a$$

yields the equation for M

(3.24)
$$\frac{d}{dt} M + \frac{\partial}{\partial \xi}(\frac{1}{2} \alpha b M^2) = 0.$$

REMARK 3.10. Using (2.10) it is clear that we have

(3.25)
$$\lambda_p(\mathbf{x}, t, \mathbf{u}_\epsilon, \mathbf{k}) = \omega + \epsilon(a + b\sigma) + O(\epsilon^2).$$

Thus we see that a and b above in (3.20) provide (nonlinear) corrections for the dominant characteristic speed due to the effect of the mean field $\mathbf{v}$ and the wave amplitude σ. In fact b *is the genuine nonlinearity coefficient* of Lax (see [Lax]) as it can also be written in the form

(3.26)
$$b = (\mathbf{r} \cdot \mathbf{grad}_u \lambda_p)^o \Big|_{\xi=\mathbf{k}}.$$

It should be pointed out that, in the exceptional cases where b vanishes (so that (3.20) is linear in σ) but the problem is still not linear, then the scaling implied by (3.1) is incorrect and the waves should be taken of larger amplitude than ϵ for nonlinear effects to appear (see Remark 3.2; and reference [R3] for the analogous situation in nonlinear long dispersive waves).

REMARK 3.11. Note that in the *periodic case,* i.e. σ periodic in θ, $\mathbf{u}_2$ is also periodic (thus bounded) and the *expansion is uniform.*

REMARK 3.12. When the *system (2.1) is linear,* the $\mathbf{B}_j$'s in (2.7) vanish, thus a and b also vanish in (3.21) and the *High Frequency Equation* for σ reduces to

$$(3.27) \qquad \frac{d}{dt}\sigma + \Delta\sigma = 0 \,.$$

Separating $\sigma = \text{Re}(\tilde{a}(\mathbf{x},t)e^{i\theta})$, we get the usual Linear Geometrical Optics equation for the amplitude $\tilde{a}$ (see [W1] and [K]).

REMARK 3.13. Finally we examine the *role of the term $\Delta\sigma$ in* (3.20) and (3.27). It is clear that this term accounts for the increases and decreases in the wave amplitude σ due to: (i) wave interactions with the underlying base state, and (ii) geometrical focusing and defocusing of the waves as the fronts move along the rays. In some special but very important cases these interactions can be formulated in a very simple fashion, as follows from the following easy to prove theorem – whose hypothesis apply in the gas dynamics case, provided we choose the variables properly (see Example 3.1 below).

THEOREM 3.1. Let the A_j^o be symmetric such that $\frac{\partial}{\partial x_j}A_j^o = 0$, $0 \le j \le N$, and normalize [8] the $\mathbf{r}_s^o$, $1 \le s \le n$, by requiring $|(\mathbf{r}_s^o)^T \cdot A_0^o \cdot \mathbf{r}_s^o| = 1$.

Then

$$(3.28) \qquad \Delta = \frac{1}{2}\sum_1^N \frac{\partial}{\partial x_j}(\boldsymbol{\ell}^o \cdot A_j^o \cdot \mathbf{r}^o) \,,$$

so that Δ is one half the divergency of the ray speed vector in N-space (see (3.14)). Furthermore, in the isotropic case (when (3.17)) applies

$$(3.29) \qquad \Delta = \frac{1}{2}c_p^o\Sigma\kappa_i + \frac{1}{2}\hat{\mathbf{k}}\cdot\mathbf{grad}c_p^o \,,$$

where the κ_i's, $1 \le i \le N-1$, are the principal curvatures of the wavefronts in N-space[9].

In (3.29) the contributions to Δ from focusing - defocusing terms and the inhomogeneities in the medium/base state are clearly separated.Furthermore, equation (3.28) implies that

$$(3.30) \qquad \frac{d}{dt}\sigma + \Delta\sigma = \frac{1}{2\sigma}\left\{\frac{\partial}{\partial t}(\sigma^2) + \sum_1^N \frac{\partial}{\partial x_j}(\sigma^2\boldsymbol{\ell}^o \cdot A_j^o \cdot \mathbf{r}^o)\right\}$$

where $\boldsymbol{\ell}^o \cdot A_j^o \cdot \mathbf{r}^o = c_p^o\hat{k}_j$ in the isotropic case. This shows, very clearly, that when the ray tubes collapse (caustics) σ blows up and the expansion fails (see Remark 3.8.).

---

[8]Note this normalization is different from (2.6). It guarantees that $\boldsymbol{\ell}_s^o = \pm(\mathbf{r}_s^o)^T$.

[9]$\kappa_i > 0$ if the front is curved away from the direction of propagation $\hat{\mathbf{k}}$; i.e. rays diverging.

Equation (3.30) can be *interpreted in terms of energy conservation,* since together with (3.20) it yields

$$(3.31) \qquad \frac{\partial}{\partial t}\left(\frac{1}{2}\sigma^2\right) + \sum_1^N \frac{\partial}{\partial x_j}\left(\frac{1}{2}\sigma^2 \ell^o \cdot A_j^o \cdot r^o\right) + \frac{\partial}{\partial \theta}\left(\frac{1}{2}a\sigma^2 + \frac{1}{3}b\sigma^3\right) + \mathcal{E} = 0,$$

where $\mathcal{E} \geq 0$ is the energy dissipated at the shocks. More precisely, along each ray, if the i-th shock is located at $\theta = \theta_i(t)$ with jump in σ across the shock given by $[\sigma]_i = \sigma\Big|_{\theta_i+0}^{\theta_i-0}$, we have

$$(3.32) \qquad \mathcal{E} = \frac{1}{12}b\sum_i([\sigma]_i)^3\delta(\theta - \theta_i)$$

along the ray, where $\delta(\cdot)$ denotes Dirac's delta function. In the linear case (see (3.27)) equation (3.31) reduces to the usual formula for conservation of $|\tilde{a}|^2$ along ray tubes (see [W1] and [K]).

EXAMPLE 3.1. Consider the example of an isentropic gas with an applied force $\mathbf{F} = \mathbf{F}(\mathbf{x})$. In mass-Lagrangian coordinates the equations can be written in the form

$$(3.33) \qquad \frac{\partial}{\partial t}\tau - \operatorname{div}\mathbf{v} = 0 \text{ and } \frac{\partial}{\partial t}\mathbf{v} + \operatorname{grad}p = \mathbf{F},$$

where p is the pressure, $\mathbf{v}$ is the flow velocity and $\tau = V(p)$ is the specific volume with $c = \sqrt{-\tau^2/\frac{dV}{dp}}$ the sound speed. Then if $\mathbf{F} = \operatorname{grad}\phi$ is a gradient we can take the base solution as $p_0 = p_0(\mathbf{x}) = \phi$, $\tau_0 = \tau_0(\mathbf{x}) = V(\phi)$ and $\mathbf{v}_0 = 0$. Introducing $\mathbf{u} = (p - p_0, \mathbf{v})^T$ as the set of variables, the hypothesis of Theorem 3.1 are satisfied and the system is isotropic. Actually, strict hyperbolicity fails except when $N = 1$, but the acoustical modes are simple with $c_p^o = \pm c_0$. See Remark 3.7 and Example 2.1.

4. The basic expansion (single wavefront case).

In this section we will consider a small variation of the basic expansion in §3 that allows us to deal with the propagation of single wavefronts, in particular weak shocks. Basically, the only difference is at the level of: *which requirements we enforce on the θ dependence of the expansion.* Here we remove the oscillatory condition and replace it by the requirement that *definite limits should be achieved as $\theta \to \pm\infty$*[10]. Unfortunately, in this case it is generally not possible to impose a sublinearity condition, as in (3.7), on $\mathbf{u}_2$ and the *expansion is generally non-uniform,* which (as we will see) forces the need to use matched asymptotic expansion techniques (see [KC]) to complete the expansion. For a discussion of the meaning of this non-uniformity, see Remark 4.3.

We start again with the ansatz (3.1) and up to equation (3.18) everything is the same as in §3. The only difference is as to the interpretation of equation (3.12). Since the condition $\overline{\sigma} = 0$ is now meaningless, *the splitting into $\mathbf{v}$ and σ is now made unique*

---

[10]Obviously, this is the natural condition on a situation where a narrow transition layer (shock region) connects two different states.

by requiring (see Remark 3.4 for notation) $\boldsymbol{\ell}^o \cdot A_0^o \cdot \mathbf{v} = 0$. Thus $\sigma \mathbf{r}^o$ includes *all* the $\mathbf{r}^o$ component of $\mathbf{u}_1$, not just the oscillatory part as in (3.12). We also introduce

$$(4.1) \qquad \mathbf{v}_\pm = \mathbf{v} \pm \sigma_\pm \mathbf{r}^o, \quad where \quad \sigma_\pm = \lim_{\theta \to \pm\infty} \sigma,$$

so that $+$ and $-$ indicate limits ahead $(\theta > 0)$ and behind $(\theta < 0)$ the wave.

Now, clearly, the arguments leading from (3.18) to (3.19) and so on do not apply. We consider three cases:

CASE 1. $\sigma_+ = \sigma_- = \overline{\sigma}$.

This case is actually a particular instance of the situation considered in §3 as we can see by redefining σ and $\mathbf{v}$ as follows:

$$(4.2) \qquad \sigma_{\text{new}} = \sigma_{\text{old}} - \overline{\sigma}, \quad \mathbf{v}_{\text{new}} = \mathbf{v}_{\text{old}} + \overline{\sigma} \mathbf{r}^o.$$

Then we are in the situation of §3 when in (3.6) $f_a = 0$. In this case the expansion will be *uniform* ($\mathbf{u}_2$ bounded) provided σ_{new} is integrable, otherwise it will be *semi-uniform* with $\mathbf{u}_2$ merely sublinear in θ.

CASE 2. $\sigma_+ \neq \sigma_-$ but, if $\sigma_\Delta = \sigma_- - \sigma_+$,

$$(4.3) \qquad 0 = \sum_0^N \frac{\partial}{\partial x_j}(\sigma_\Delta A_j^o \cdot \mathbf{r}^o).$$

This is a rather exceptional situation that can occur, for example, if all the A_j^o, $0 \leq j \leq N$, are constant and $\varphi = \mathbf{k} \cdot \mathbf{x} - \omega t$ is a plane wave with $\mathbf{k}$ and ω constants. Then $\mathbf{r}^o$ is also constant and (4.3) will apply if, for example, σ_Δ is constant. Case 1 follows when $\sigma_\Delta = 0$.

In this case the expansion can also be made *uniform* ($\mathbf{u}_2$ bounded in θ), or *semi-uniform* ($\mathbf{u}_2$ semilinear in θ), depending on the rate of convergency of σ to $\sigma_\pm$ as $\theta \to \pm\infty$. In any event, it is quite clear that the "solvability" conditions in (3.18) that guarantee a solution $\mathbf{u}_2$ with the proper behavior are now

Limit Equations

$$(4.4) \qquad \frac{\partial}{\partial t}(A_0^o \cdot \mathbf{v}_\pm) + \sum_1^N \frac{\partial}{\partial x_j}\left(A_j^o \cdot \mathbf{v}_\pm\right) = 0,$$

where $\mathbf{v}_\pm$ are defined in (4.1) and these two systems are consistent because of (4.3).

Layer Equation

(4.5)
$$\frac{d}{dt}\sigma + \Delta\sigma + \frac{\partial}{\partial\theta}(a\sigma + \frac{1}{2}b\sigma^2) = \Gamma,$$

where $\Gamma = \Gamma(\mathbf{x}, t; \mathbf{v}) = -\boldsymbol{\ell}^o \cdot \sum_0^N \frac{\partial}{\partial x_j}(A_j^o \cdot \mathbf{v})$ and the rest is as in (3.21). Thus, with very small obvious modifications, all the considerations in §3, from Remark 3.9 to the end, apply here. Obviously, as $\theta \to \pm\infty$, (4.5) yields

(4.6)
$$\frac{d}{dt}\sigma_\pm + \Delta\sigma_\pm = \Gamma,$$

which also follows from (4.4) upon multiplication by $\boldsymbol{\ell}^o$ and use of (4.1).

CASE 3. $\sigma_+ \neq \sigma_-$ and (4.3) cannot be enforced.

Clearly, from (3.18), (4.5) and thus (4.6) must still apply. On the other hand, (4.4) cannot be enforced – at least not on both the (+) and (–) sides. No other condition can be imposed on (3.18) and we are faced with the fact that $\|\mathbf{u}_2\|$ will grow secularly as $|\theta|$ when $|\theta| \to \infty$. Thus the *expansion is nonuniform* and is formally valid only in a narrow zone $|\epsilon\theta| = |\varphi| << 1$ (the condition for $\epsilon\|\mathbf{u}_2\| << \|\mathbf{u}_1\|$). Furthermore, in order to determine $\mathbf{v}$ and complete the expansion, we must resort to *a matching technique approach* (see [KC]), namely:

On each side of the "thin" layer $|\theta| << \epsilon^{-1}$ where the W.N.G.O. expansion is valid, we propose regular expansions

(4.7)
$$(\mathbf{u}_\epsilon)_\pm = \epsilon(\mathbf{u}_1)_\pm(\mathbf{x}, t) + \epsilon^2(\mathbf{u}_2)_\pm(\mathbf{x}, t) + \dots$$

for the solution of (2.1). Then both $(\mathbf{u}_1)\pm$ satisfy the linearized equations (4.4) – or (3.19) for that matter – with appropriate Boundary Conditions (see below) along the characteristic surface $\varphi = 0$, relating $(\mathbf{u}_1)_+$ valid on $\varphi > 0$ to $(\mathbf{u}_1)_-$ valid on $\varphi < 0$.

The matching conditions that determine both $\mathbf{v}$ and the B.C. along $\varphi = 0$ for $(\mathbf{u}_1)_\pm$ follow easily now. But first we note that since the W.N.G.O. expansion is valid only for $|\varphi| \ll 1$, it is meaningless to insist in having $\mathbf{v}$ and σ in (3.12) as functions of $(\mathbf{x}, t)$ with $\mathbf{x}$ and t independent. Rather:

(4.8) *The $(\mathbf{x}, t)$ dependence that appears in (3.12) should be interpreted as meaning that σ and $\mathbf{v}$ are functions defined on the hypersurface $\varphi = 0$.*

This said, the matching conditions determining $\mathbf{v}$ are then

(4.9)
$$\mathbf{v}_+ = (\mathbf{u}_1)_+ \quad \text{and} \quad \mathbf{v}_- = (\mathbf{u}_1)_- \quad \text{along} \ \varphi = 0,$$

where we use (4.1). This gives us the Boundary Conditions for $(\mathbf{u}_1)_\pm$

(4.10)
$$\boldsymbol{\ell}_s^o(\mathbf{x}, t, \mathbf{k}) \cdot A_0^o(\mathbf{x}, t) \cdot (\mathbf{u}_1)_-(\mathbf{x}, t) = \boldsymbol{\ell}_s^o(\mathbf{x}, t, \mathbf{k}) \cdot A_0^o(\mathbf{x}, t) \cdot (\mathbf{u}_1)_+(\mathbf{x}, t),$$

valid for every $1 \leq s \leq n$, $s \neq p$, along the hypersurface $\varphi = 0$. Finally, $\sigma_\pm$ must satisfy

(4.11)
$$\sigma_\pm = \boldsymbol{\ell}^o \cdot A_0^o \cdot (\mathbf{u}_1)_\pm,$$

where (4.6) is the consistency condition for B.C. along characteristics.

REMARK 4.1. We note that if $\omega \neq 0$, then we can parameterize the hypersurface $\varphi = 0$ as $t = S(\mathbf{x})$[11] and consider $\mathbf{v}$ and σ in (3.12) as functions of $\mathbf{x}$ alone. In particular, for time independent media, this can be made very explicit as one may consider special solutions of the *Eikonal Equation* (3.10)–(3.11) with $\omega = $ constant (as follows from the characteristic equations (3.16)) with $\varphi = (S - t)\omega$.

REMARK 4.2. Consider the n eigenvalues $\lambda_s^o(\mathbf{x}, t, \mathbf{k})$, $1 \leq s \leq n$, and at each point along the wavefront $\varphi = 0$ split the eigenvalues for which $s \neq p$ in two groups

(4.12) $$G_- = \left\{ \lambda_s^o : \lambda_s^o > \lambda_p^o \right\}, \quad G_+ = \left\{ \lambda_s^o : \lambda_s^o < \lambda_p^o \right\},$$

where G_- corresponds to the signals propagating faster than the wavefront $\varphi = 0$ and arriving at it from behind, i.e., $\theta = -\infty$, while G_+ corresponds to the signals propagating slower than the main wave. If either G_+ or G_- is empty then the expansion can be made uniform (or semiuniform)[12] on the other side, as then we can take $\mathbf{v}_- = (\mathbf{u}_1)_-$ or $\mathbf{v}_+ = (\mathbf{u}_1)_+$, respectively.

REMARK 4.3. There is no good understanding of why this nonuniformity arises here and not on the oscillatory case, nor is it known whether there is a way to "uniformize" the expansion. Because of this nonuniformity it is not possible to extend the expansion in this section to the multiple-wave case (see §6). One may conjecture that the nonuniformity is related to the special behavior of weak shocks at caustics (see §7). We mention the following points that may help the reader understand the nature of the difficulty:

(i) For oscillatory waves there is a well-defined mean velocity of propagation $\omega + \epsilon a$ (see (3.25)) which is independent of the wave amplitude σ, at least to the order considered. Thus, it makes sense to have an equation for the overall wave propagation – the *Eikonal*, (3.10) / (3.11) – that is independent of σ.

(ii) For the non-oscillatory case here, where there is no well-defined mean, it does not make sense (apparently) to talk of an overall velocity of propagation. The errors made by moving the waves macroscopically using the *Eikonal Equation* – which ignores amplitude dependent σ contributions – can accumulate systematically leading to deviations important at leading order. Thus the non-uniformity

(iii) In view of the above, it would seem necessary to incorporate wave amplitude corrections into the *Eikonal Equation* to eliminate the non-uniformity. It is not clear how to do this, but perhaps Whitham's *Geometrical Shock Dynamics* theory (see [W1]) is a good model to look at for inspiration.

---

[11] Note $\frac{d}{dt}S = 1$ along the rays.

[12] Depending on the rate of convergency of σ to σ_- or to σ_+.

All of this points to the fact that this problem of *lack of uniformity of the expansion for the single wavefront case may be a very important problem to investigate,* in spite of its apparent dryness and technical nature.

5. Diffusion. It is very easy and straightforward to incorporate the effects of a small amount of diffusion into the equations of W.N.G.O., as follows:

Assume that the basic system under consideration, i.e. (2.1), is modified by adding to it a small amount of diffusive dissipation, so that the new system is

$$
(5.1) \qquad \hat{N}[\mathbf{u}] = \epsilon^2 \sum_{i,j=1}^{N} \frac{\partial}{\partial x_i} \left\{ J_{ij} \frac{\partial}{\partial x_j} \mathbf{u} \right\},
$$

where the N^2 $n \times n$ "viscosity matrices" $J_{ij} = J_{ij}(\mathbf{x}, t, \mathbf{u})$ satisfy the strict stability condition (see [MP])

$$
(5.2) \qquad \sum_{i,j=1}^{N} \xi_i \xi_j \boldsymbol{\ell}_s \cdot J_{ij} \cdot \mathbf{r}_s \geq 0, \quad \forall \, 1 \leq s \leq n \text{ and } \boldsymbol{\xi} \in R^N.
$$

REMARK 5.1. Again, there is here a distinguished limit involved (see Remark 3.2). If the "size" of the added diffusion is very different from $O(\epsilon^2)$, as in (5.1), then either diffusion will dominate to the exclusion of all other effects or it will not be important at leading order.

The new term in (5.1) does not affect the expansion up to the level of equation (3.18), where the term

$$
(5.3) \qquad \frac{\partial^2}{\partial \theta^2} \left[\left(\sum_{i,j=1}^{N} k_i k_j J_{ij}^o \cdot \mathbf{r}^o \right) \sigma \right]
$$

added to the righthand side is the only change. Clearly then

(i) The *Mean Field Equations* (3.19) are not affected.

(ii) The *High Frequency Equation* (3.20) must be modified to the (viscous) Burgers' equation

$$
(5.4) \qquad \frac{d}{dt}\sigma + \Delta\sigma + \frac{\partial}{\partial \theta}\left(a\sigma + \frac{1}{2}b\sigma^2\right) = \frac{\partial}{\partial \theta^2}(\mu\sigma),
$$

where

$$
(5.5) \qquad \mu = \sum_{i,j=1}^{N} k_i k_j \boldsymbol{\ell}^o \cdot J_{ij}^o \cdot \mathbf{r}^o = \mu(\mathbf{x}, t) \geq 0.
$$

Similarly, (4.4) and (4.6) experience no changes, while in (4.5) a term $\mu\sigma_{\theta\theta}$ appears in the right hand side. Similar changes occur in the expansions of the following sections.

6. Multiple Waves. In this section we review very briefly the extension of the W.N.G.O. expansion in §3, to the situation where, at leading order, more than one characteristic mode is excited. That is, in (3.1), more than one fast phase θ appears. For more details the reader can consult [CR], [HK1], [HMR] and [MR2]. As we will see, many open problems remain in this area.

The need to study multiple waves is obvious. The requirement of a mostly "monochromatic" wave in (3.1) is too restrictive. In many problems of interest (in fact in most of them) more than one phase is needed. Multiple phases can arise in many ways, for example: (i) they are already there in the initial conditions, (ii) an originally single phase wave reflects of an object or folds on itself along a caustic, (iii) the initial conditions have different waves in different regions of space, which as time goes by, end up coalescing in the same region, (iv) when a wave interacts with a shock, or a contact discontinuity, or a shear wave, new waves are generated.

REMARK 6.1. of course, in the *linear case,* where the different waves do not interact, there is not much to be said. The Linear Geometrical Optics multiple wave expansion is simply a sum of expansions, each exactly the same as the single wave expansion, with different phases.

The differences between the situation considered in this section and the expansion in §3 start at the leading order, where they may seem harmless (but are not). Basically now (3.12) is replaced by a formula of the form

$$(6.1) \qquad \mathbf{u}_1 = \mathbf{v}(\mathbf{x}, t) + \sum_m \sigma_m(\mathbf{x}, t, \theta_m) \mathbf{R}_m(\mathbf{x}, t),$$

where $\theta_m = \frac{1}{\epsilon} \varphi_m(\mathbf{x}, t)$, $1 \le m \le M$ (some M integer), is the set of phases; each one of them satisfying the *Eikonal Equation*

$$(6.2) \qquad \det \left(A_0^o \frac{\partial}{\partial t} \varphi_m + \sum_1^N A_j^o \frac{\partial}{\partial x_j} \varphi_m \right) = 0,$$

with the $\mathbf{R}_m$ the corresponding right eigenvectors

$$(6.3) \qquad \left(A_0^o \frac{\partial}{\partial t} \varphi_m + \sum_1^N A_j^o \frac{\partial}{\partial x_j} \varphi_m \right) \cdot \mathbf{R}_m = 0.$$

The difficulties now arise because of the nonlinearity: different waves interact and produce new waves. If these new waves "resonate" with the linear part of the equation, then they cannot be ignored and must be taken care of at the leading order (6.1) of the expansion. This means that the set of phases in (6.1) cannot be arbitrary.

We explain now, in as simple a fashion as possible, the phenomenon of wave resonance mentioned above. Consider two phases θ_i and θ_j, whose amplitudes σ_i and σ_j have energy in the Fourier components $\exp(i\lambda_i\theta_i)$ and $\exp(i\lambda_j\theta_j)$ – where λ_i and λ_j can be quite arbitrary. Then the nonlinear interactions generally draw energy from these components and put it into producing the modes

$$(6.4) \qquad \exp\left[i(m_i\lambda_i\theta_i + m_j\lambda_j\theta_j) \right], \quad \forall m_i, m_j \text{ integers.}$$

Now, if for some μ_i and μ_j, $\mu_i\varphi_i + \mu_j\varphi_j$ also solves the *Eikonal Equation* (6.2), then the nonlinearly produced waves in (6.4) – for $\mu_i = m_i\lambda_i$ and $\mu_j = m_j\lambda_j$ – will accumulate and a phase corresponding to $\mu_i\varphi_i + \mu_j\varphi_j$ will have to be included in (6.1). This is, basically, the same phenomena of three wave resonance, well known in dispersive wave theory (see [AS] and [Cr] for example). Here, however, the situation is complicated by the homogeneity of the "dispersion relation" (6.2) for the hyperbolic system (2.1): that is, any (constant) multiple of a solution of the *Eikonal Equation* is also a solution. This means that one must consider the resonant "triads" and all their harmonics (which are also resonant) at once. It is not possible to concentrate on a single Fourier component resonant triad – as is usually done in dispersive wave theory.

In [HMR] the following restrictions are imposed on the set of phases, which are sufficient (but perhaps not necessary) to obtain a tractable expansion[13]

(i) The set of phases is finite (obvious restriction). Thus (see also (iii) below) no infinite chain of resonances are allowed, where two waves produce a third, which then resonates with a fourth to produce a fifth, and so on.

(ii) The set of phases is coherent. Partial resonances are not allowed. This means that when two phases resonate, they do so coherently: that is everywhere in space the same way – the μ's mentioned after (6.4) are constant. Basically: if the phase being produced by the resonances in one part of space is not the same as that produced elsewhere, the situation becomes quite involved.

(iii) The set of phases is closed: no new phases are generated not already in the system.

(iv) The set of phases is nondegenerate. This is a technical condition which, in between other things, avoids having repeated phases in the set. But mainly this condition arises from the study of the solvability condition at second order (see below).

At the next level in the expansion, the situation becomes quite a bit more complicated than (6.1). The analog of (3.18) is now a linear p.d.e. in $\mathbf{u}_2$, involving its partial derivatives with respect to all the θ_m's, being forced by complicated expressions depending on the lower order term $\mathbf{u}_1$. The solvability[14] conditions are now rather involved. In [HMR] this problem is solved under the hypothesis (i)-(iv) above on the set of phases. The solvability conditions then give a set of M equations[15] for the σ_m's. These equations for the wave amplitudes are a set of inviscid Burgers' equations (one for each σ_m) coupled through nonlinear integro-differential terms, as the following example illustrates.

---

[13][HMR] deals only with the constant coefficient case: the fluxes and densities in (2.1) are independent of space and time.

[14]Necessary and sufficient conditions on the r.h.s. of (3.18) that guarantee the sublinearity of $\mathbf{u}_2$.

[15]in [HMR] $\mathbf{v}$ in (6.1) is set to zero.

EXAMPLE 6.1. Consider the situation in 2-D isentropic gas dynamics where we have three plane waves: one a vorticity/shear flow wave and the other two are an incident (on the vorticity wave) and a reflected acoustical wave. This leads to a situation where the results of [HMR] can be used. Alternatively, consider 1-D flow for "full" gas dynamics, where the waves are: entropy and the two (right and left moving) sound waves (see[MR2]). In either case, after some manipulations, the equations can be written in the form (take the periodic case and eliminate the x dependence in (6.1)).

(6.5)
$$u_t + \left(\tfrac{1}{2}u^2\right)_\theta + \int_0^1 K(\theta - y)v(y,t)\,dy = 0\,,$$
$$v_t + \left(\tfrac{1}{2}v^2\right)_\theta - \int_0^1 K(-\theta + y)u(y,t)\,dy = 0\,,$$

where u, v and K are periodic of period one, u and v are the amplitudes of the sound waves and K is related to the vorticity/entropy wave, which is not changing at this order (the equation is $(\sigma_3)_t = 0$).

We point out to the (skew-symmetric) convolution coupling of u and v in (6.5), which is typical. The system of equations (6.5) is studied extensively in [MRS]. Numerical calculations show that solutions without wavebreaking are possible. It is also shown that permanent waves with cusp/corner singularities exist (see also [P]) and that a variety of other interesting behaviors can occur.

REMARK 6.2. The theory is clearly very incomplete. Although the restrictions (i) - (iv) on the set of phases are satisfied by many interesting examples (see [AM], [AMR1], [AMR2], [CM], [MA1], [MR3] and [MRA] for some applications), there are many important problems where this is not so. For example, as a wavefront reflects of a caustic, it appears the hypothesis are not satisfied. Furthermore, there is no multiple phase theory for situations involving waves behaving as in §4 so that only the oscillatory case is (partially) understood. The problems that arise in dealing with multiple nonperiodic wavefronts are maybe related to the nonuniformities of this case discussed in §4 - see Remark 4.3.

7. Caustics. In this section we briefly review the situation regarding the failures of W.N.G.O. near the regions where the *Eikonal Equation* rays cross and the phase φ becomes multiple valued, with the wavefronts folding on themselves and the wave amplitudes predicted by W.N.G.O. going to infinity (see Remark 3.8 and equation (3.30)).

The related problems of caustics, singular rays, shadow boundaries, etc., are one of the major open problems in W.N.G.O., although a substantial amount of work exists in the field (see [CS], [H1], [HK2] and [R2] for example). In the context of Linear Geometrical Optics (L.G.O.) the problem has been resolved, and the behavior of the waves at caustics, etc., is well understood (see [BK], [K] and [Lu] for example). This is not so for W.N.G.O. The type of difficulties one must face in nonlinear theory are twofold:

(a) First, in nonlinear situations, different wave modes interact and new modes are generated. These multiple wave situations are very difficult, and presently can be handled only in certain special circumstances (see §6). Clearly, at the places where wavefronts cross and fold, a multiple wave situation arises – even if the original wave was locally monochromatic. In the nonlinear case we can expect new waves to be produced via resonant interactions (see §6); not so in the linear case – where the principle of linear superposition applies.

(b) Second, the simplified, canonical, asymptotic model equations that can sometimes be derived to model the behavior of the waves near caustics, singular rays, etc., in the weakly nonlinear case are not well understood. They are simple, but not sufficiently so as to have known and useful exact solutions (a numerical study seems indicated). It should be pointed out that their linear counterparts can be solved by separation of variables and linear superposition so that this second difficulty is not too different in nature from the first one above.

CAUSTICS (L.G.O.)

Generally the rays associated with the *Eikonal Equation* will intersect, leading to singularities and multiple values in the solution (cusps and folds in the wavefronts). In addition, whenever obstacles are present, limiting ("singular") rays will appear separating regions of space accessible and not reached by the rays.

At all the places indicated in the prior paragraph the expansions in §3 fail, as infinities and other singularities appear. For example at any point where a ray tube collapses, (3.30) predicts an infinite amplitude.

Of particular interest are:
Arêtes: places where a singularity appears for the first time, by focusing of an infinitesimal area of the initial wavefronts. After this first time the wavefronts cross and fold on themselves.
Caustics: these are the locations of the folding places in the wavefronts. Often, but not always, they begin at arêtes.
Foci: same as arêtes, but a finite region of the initial wavefronts is focused.
Singular rays: rays separating "illuminated" zones from "dark" zones due to the presence of obstacles. The expansion fails because it ignores diffraction of the waves from the points of contact of the singular rays and the object.

The resolution of all these difficulties in L.G.O. is well understood by now (see [BK], [K] and [Lu] for example). The multiple values given by the solution of the *Eikonal Equation* simply mean that the wave is no longer monochromatic in those regions. Because of the linear superposition principle this does not represent a problem. Near the singular regions (caustics, etc.) the expansion must be supplemented by local – internal layer – expansions to resolve the infinities. Thus (linear) Geometrical Optics is valid everywhere – including multiple valued regions – provided we take care appropriately of the inner layers appearing near the regions of trouble. For example, near a caustic, the two branches of the wavefront – incident and reflected – are nearly parallel. Thus an expansion very much like the one in §3 , but incorporating deviations from the locally plane mode form via a weak dependence (of $O(1/\sqrt{\epsilon})$ rather than $O(1/\epsilon)$ as in θ) on the transverse direction solves the problem. The wave amplitudes in the caustic layer end up being $O(\epsilon^{-\frac{1}{6}})$ larger than outside, but not ∞.

The main point is that the asymptotic equations valid near the inner layers can be solved by separation of variables and superposition. (For example, near caustics the switch from waves to no waves occurs – in general – via Airy functions). This is an advantage that is lost in the nonlinear case, even though one can still sometimes derive equations that should be valid near (at least some of) the trouble spots.

CAUSTICS (W.N.G.O.)

Hunter and Keller [HK2] have shown that, under certain circumstances, linear caustic theory can be applied to W.N.G.O. However for this they must assume that the incident and reflected wave are nonresonant (rarely true) and that the waves are smooth. Concerning their second restriction: the connection formula they obtain relating the strengths of the incoming and outgoing waves involves a Hilbert transform. Thus an incoming shock leads to the prediction of a logarithmic singularity in the outgoing wave.

One can also use the technique described above of incorporating $O(1/\sqrt{\epsilon})$ scales in the expansions to approach the problem (see [CS], [H1], [R2] and references in those papers). This leads to asymptotic equations which do not seem to develop nonuniformities and infinities. However, these equations are complicated enough that no clear understanding of their behavior exists. We get back to this subject (see equations (7.1) – (7.3)) after discussing some experimental results.

EXPERIMENTAL RESULTS

Sturtevant and Kulkarny undertook a careful experimental investigation of the focusing of weak shocks in inert gases. Their results, reported in [SK], have enormous relevancy to the subject matter here: failure of Weakly Nonlinear Geometrical Optics near arêtes, perfect foci, caustics, etc., particularly in what relates to the expansions in §4.

Sturtevant and Kulkarny produced converging weak shocks of controllable wave front shape by reflecting initially plane front shocks from concave end walls in a large shock tube. We briefly summarize some of their results next. The interested reader should consult [SK] for a full account.

It is clear that as a shock front focuses, its strength increases and, consequently, so does its speed. Thus the parts of the front where more focusing occurs will move at speeds farther (larger) and farther away from the (linear) acoustical speed predicted by the *Eikonal Equation* (3.10). Clearly this (nonlinear) effect works to *prevent* focusing of the front itself – as it occurs in linear theory – not just merely to stop infinities.

It is found in [SK] that beyond a certain critical shock strength (e.g.: of about Mach number M = 1.2 for the incident plane shock in the case of a perfect focus in §3.1 there), the effect in the paragraph above seems to dominate. As the front focuses, the focusing parts speed up and become nearly plane with true focusing entirely avoided. At the ends of this region the shock front develops corners, associated with the appearance of triple points there. The Mach stem and vortex line (the *new waves* being generated by nonlinearity, see (a) at the beginning of this section) associated with the triple points trail the main shock front as the triple points fly apart from each other (see Fig. 18d in [SK]).

As the strength of the shock decreases and approaches criticality, the paths of the triple points – which for large enough strength diverge – become more complicated. They first diverge, then stop, start approaching each other, stop again, and finally diverge again (see Fig. 18c in [SK]). Below the critical strength, after first diverging and stopping, the triple points approach each other and collide (see Fig. 18b in [SK]). After the collision the resulting wavefront takes on an appearance very similar to the "folded on itself" wavefront predicted by the *Eikonal Equation*. However, at the places where the front would merely fold on itself according to the *Eikonal Equation* (caustics) *new waves* (generated by nonlinear interactions, no doubt) are observed and triple point-like structures seem to occur there. For weaker and weaker shocks the "new waves" become fainter and fainter, the interval between formation and collision of the triple points goes to zero and generally the whole observed pattern resembles more and more that predicted by linear theory.

The behavior for strong and moderately strong (well above critical in fact) converging shocks seems to agree reasonable well with that predicted by Whitham's *Geometrical Shock Dynamics* theory [W1], provided one equates the "shock-shocks" of the theory with the triple points. Unfortunately the theoretical foundations of this theory are not well understood. Furthermore, this theory (apparently) cannot predict the transition behaviors described above.

In an effort to explain the observed behavior, up to and possibly including the critical shock strength transition described above, Cramer and Seebass proposed a model in [CS]. Their derivation is based on the idea that (at least for arêtes and provided the original shock front is not too convoluted) even after Weakly Nonlinear Geometrical Optics fails, the wavefronts present are all nearly parallel and thus a nearly one dimensional approximation may apply.

The model proposed in [CS], which is a special case of the models derived in [H1] and [R2], has as its main component the equations

$$(7.1) \qquad \sigma_t + \left\{ \frac{1}{4}\frac{\gamma+1}{\gamma-1}\sigma^2 \right\}_\theta + \frac{\gamma-1}{2}\eta_y = 0, \quad (\gamma-1)\eta_\theta = \sigma_y.$$

Equivalently, in the smooth part of the flow, we have

$$(7.2) \qquad \sigma_{t\theta} + \left\{ \frac{1}{4} \frac{\gamma+1}{\gamma-1} \sigma^2 \right\}_{\theta\theta} + \frac{1}{2} \sigma_{yy} = 0,$$

where we have eliminated η by cross differentiation in (7.1). The y dependence in this model is related to the extra $O(1/\sqrt{\epsilon})$ scale discussed above. We note that this is in fact the same as the Time Dependent Small Disturbance Transonic Flow equation (see [Co] and[KC]).

The problem with the model in [CS], as well as with all similar models proposed, is that very little is known concerning the behavior of equations such as (7.1) in what regards the problems considered here and that motivate their derivation.

While (7.1) has been proposed for arêtes, nonlinear Tricomi equations of the form

$$(7.3) \qquad \sigma_t + (y\sigma + \sigma^2)_\theta + \eta_y = 0, \quad \eta_\theta = \sigma_y$$

have been proposed near caustics and singular rays [H1]. The remarks in the prior paragraph are also valid here.

8. The basic expansion (forced equations, oscillatory case). In this section and the next we consider situations where the system in (2.1) is forced by a small lower order right hand side. Situations like this are of interest in the context of reacting gas dynamics, where the forcing is provided by the combustion (see Example 8.2 at the end of this section). Specifically we consider now the general situation where the governing equations can be written in the form

$$(8.1) \qquad \hat{N}[\mathbf{u}] = \epsilon \mathbf{G}(\mathbf{x}, t, \mathbf{u}, \frac{1}{\epsilon}\mathbf{u}),$$

where $\hat{N}$ is as in (2.1), ϵ is small and $\mathbf{G}$ is smooth.

Again, we start with the ansatz (3.1) and impose the oscillatory condition on the θ dependence. No changes arise till equation (3.18) where the additional term

$$(8.2) \qquad \mathbf{G}(\mathbf{x}, t, 0, \mathbf{v} + \sigma \mathbf{r}^o)$$

appears on the right hand side. The modifications to the leading order asymptotic equations (3.19) and (3.20) are then clear:

Mean Field Equations

$$(8.3) \qquad \frac{\partial}{\partial t}(A_0^o \cdot \mathbf{v})_t + \sum_1^N \frac{\partial}{\partial x_j}(A_j^o \cdot \mathbf{v}) = \overline{\mathbf{G}(\mathbf{x}, t, 0, \mathbf{v} + \sigma \mathbf{r}^o)},$$

where, as usual, the bar signifies mean over θ.

High Frequency Equation

$$(8.4) \qquad \frac{d}{dt}\sigma + \Delta\sigma + \frac{\partial}{\partial\theta}(a\sigma + \frac{1}{2}b\sigma^2) = \boldsymbol{\ell}^\circ \cdot \{\mathbf{G} - \overline{\mathbf{G}}\},$$

where $\mathbf{G}$ is evaluated at $(\mathbf{x}, t, 0, \mathbf{v} + \sigma\mathbf{r}^\circ)$.

The main difference between these equations and those of §3 is that now the *Mean Field Equations* are coupled to the *High Frequency Equation* and cannot be solved independently of it. In particular $\mathbf{v} \equiv 0$ is no longer a solution: generally, *the presence of high frequencies will force mean fields to appear.*

REMARK 8.1 We have here a situation somewhat similar to that in homogenization theory (see [BLG]). There, in order to solve for the macroscopic properties of the solution of an equation with rapidly varying coefficients, one needs certain "effective equations" whose coefficients are not rapidly varying but are computed by special averages over the solutions of the "cell" problems which describe the microstructure. In homogenization theory these "cell" problems are fixed. Here the role of the rapidly varying coefficients is taken over by the high frequencies, the "cell" problem by the "High Frequency Equations" and the "effective" equations by the "Mean Field Equations". Unlike homogenization, however, here the macroscopic and microscopic problems are coupled and the "cell" problem changes dynamically in time. For lack of a better name, we can call (8.3) and (8.4) a set of *"Dynamically Homogenized Equations"*.

EXAMPLE 8.1. General Semilinear System (see [MPT]).

The general semilinear system of strictly hyperbolic equations has the form

$$(8.5) \qquad \sum_0^N A_j^\circ \cdot \frac{\partial}{\partial x_j}\mathbf{U} = \mathbf{F}(\mathbf{x}, t, \mathbf{U}),$$

where the A_j° are as in §2 and we assume $\mathbf{F}$ smooth. Let now $\mathbf{u} = \epsilon\mathbf{U}$, then the equation for $\mathbf{u}$ can be written

$$(8.6) \qquad \sum_0^N \frac{\partial}{\partial x_j}(A_j^\circ \cdot \mathbf{u}) = \epsilon\{\mathbf{F}(\mathbf{x}, t, \frac{1}{\epsilon}\mathbf{u}) + \sum_0^N (\frac{\partial}{\partial x_j}A_j^\circ) \cdot \frac{\mathbf{u}}{\epsilon}\},$$

which has the form (8.1), so that we can use the expansions in this section. Thus, with the notation

$$(8.7) \qquad \mathbf{u} \sim \epsilon(\mathbf{V} + \Sigma\mathbf{r}^\circ), \quad i.e. \ \mathbf{U} \sim \mathbf{V} + \Sigma\mathbf{r}^\circ,$$

(note that $\mathbf{U}$ is *not small* at leading order) we find that (8.3) and (8.4) yield

$$(8.8) \qquad \sum_0^N A_j^\circ \cdot \frac{\partial}{\partial x_j}\mathbf{V} = \overline{\mathbf{F}(\mathbf{x}, t, \mathbf{V} + \Sigma\mathbf{r}^\circ)}$$

and

$$(8.9) \qquad \frac{d}{dt}\Sigma + \tilde{\Delta}\Sigma = \boldsymbol{\ell}^\circ \cdot \{\mathbf{F} - \overline{\mathbf{F}}\}$$

where $\tilde{\Delta} = \sum_0^N \boldsymbol{\ell}^\circ \cdot A_j^\circ \cdot \frac{\partial}{\partial x_j}\mathbf{r}^\circ$. Note that no derivatives of $\Sigma = \Sigma(\mathbf{x}, t, \theta)$ with respect to θ appear; θ is just a parameter in these equations. No shocks occur in this case.

EXAMPLE 8.2. Reacting Gas Dynamics.

Consider a reacting gas, where, for simplicity we will assume that the reaction is controlled by a single concentration Z and a one-step irreversible Arrhenious kinetics is assumed in the limit of high activation energy. Then, after an appropriate nondimensionalization, the equations may be written in the form

(8.10)
$$\rho_t + \text{div}(\rho\mathbf{v}) = 0\,,$$
$$(\rho\mathbf{v})_t + \text{div}(\rho\mathbf{v} \otimes \mathbf{v} + pI) = 0\,,$$
$$(\rho E)_t + \text{div}(\rho E\mathbf{v} + p\mathbf{v}) = \epsilon H_\epsilon\,,$$
$$(\rho Z)_t + \text{div}(\rho\mathbf{v}Z) = -\tfrac{\epsilon}{\beta}H_\epsilon\,,$$

where β is a nondimensional parameter, the reaction term is

(8.11)
$$H_\epsilon = \rho Z e^{\frac{1}{\epsilon}\frac{T-1}{T}}\,,$$

and the notation of Example 2.1 is used. Now

(i) For $\beta = 1$ take $\mathbf{u} = (\rho - 1, \mathbf{v}, T - 1, Z - 1)^T$.

(ii) For $\beta = O(\epsilon)$ take $\mathbf{u} = (\rho - 1, \mathbf{v}, T - 1, \epsilon Z)^T$.

In either case the resulting system has the form (8.1). An extensive study of this system, including multiple wave situations and numerical calculations, can be found in [AMR1], [AMR2], [CM], [MR3] and [MRA]. In this case W.N.G.O. seems to give important information concerning the formation of "Hot Spots", an important topic in combustion theory.

9. The basic expansion (forced equations, single wavefront case).
Here we consider the same set of governing equations as in §8, i.e. (8.1), but use the "definite limits as $\theta \to \pm\infty$" condition on θ of §4, rather than the oscillatory conditions of §8. In Example 9.1 we show how the approximation can be used to obtain a simple model for weak flames in combustion theory. The development here is very similar to that in §4 and §8 and we will only sketch it. Again we have three cases. We use the notation introduced in §4.

CASE 1. $\sigma_+ = \sigma_-$. This case can be reduced to situation in §8.

CASE 2. $\sigma_+ \neq \sigma_-$ but

(9.1)
$$\sum_0^N \frac{\partial}{\partial x_j}(\sigma_\pm A_0^o \cdot \mathbf{r}^o) - \mathbf{G}(\mathbf{x}, t, 0, \mathbf{v} + \sigma_\pm \mathbf{r}^o)$$

are equal. Unlike (4.3), this is harder to satisfy. The asymptotic equations are

Limit Equations

$$(9.2) \qquad \frac{\partial}{\partial t}(A_0^o \cdot \mathbf{v}_\pm) + \sum_1^N \frac{\partial}{\partial x_j}(A_j^o \cdot \mathbf{v}_\pm) = \mathbf{G}(\mathbf{x}, t, 0, \mathbf{v}_\pm).$$

Layer Equation

$$(9.3) \qquad \frac{d}{dt}\sigma + \Delta\sigma + \frac{\partial}{\partial \theta}(a\sigma + \frac{1}{2}b\sigma^2) = \Gamma + S,$$

where Γ is as in (4.5), Δ, a and b are as in (3.21) and

$$(9.4) \qquad S = S(\mathbf{x}, t, \mathbf{v}, \sigma) = \boldsymbol{\ell}^o \cdot \mathbf{G}(\mathbf{x}, t, 0, \mathbf{v} + \sigma \mathbf{r}^o).$$

In this case the expansion is semi-uniform or uniform, depending on the rates of convergence of σ to $\sigma_\pm$.

CASE 3. $\sigma_+ \neq \sigma_-$ and (9.1) does not apply. In this case (9.3) still applies but the expansion is not uniform and (9.2) cannot be enforced. $\mathbf{v}_\pm$ must be determined by matching with an expansion as (4.7). The matching conditions are the same as in §4, except that $(\mathbf{u}_1)_\pm$ satisfy (9.2) rather than (4.4).

EXAMPLE 9.1 If we use the approximations above in the case $\beta = O(\epsilon)$ of Example 8.2, for the particle path characteristics[16], and add diffusive effects – as in §5 – we obtain a simple model for flame propagation. In the simplest version (for plane waves in 1–D) the equations have the form

$$(9.5) \qquad T_t = qZ\Phi(T) + \mu T_{\theta\theta} \ , \quad Z_t = -Z\Phi(T) + \nu Z_{\theta\theta},$$

where T and Z are the two excitation amplitudes and we have replaced the Arrhenious kinetics term in (8.11) by a general term $\Phi(\frac{1}{\epsilon}\frac{T-1}{T})$. In (9.5) T is the leading order in the perturbation to the temperature and Z is the chemical concentration.

Note that, as pointed out in Remark 2.2, most of the terms that would arise in a general multiple-eigenvalues expansion vanish due to the symmetries in the problem.

REMARK 9.1 Finally, we point out that, with special restrictions, it is possible to consider situations where the forcing in (8.1) is not small, $O(\epsilon)$, but in fact $O(1)$. Basically assume the equations have the form

$$(9.6) \qquad \hat{N}[\mathbf{u}] = g(\mathbf{x}, t, \mathbf{u}, \frac{1}{\epsilon}\mathbf{u})\mathbf{R}(\mathbf{x}, t) + \epsilon \mathbf{G}(\mathbf{x}, t, \mathbf{u}, \frac{1}{\epsilon}\mathbf{u}),$$

where g is a scalar function and $\mathbf{R}$ is a common eigenvector to all the the A_j^o

$$(9.7) \qquad A_j^o \mathbf{R} = \eta_j \mathbf{R} \ , \quad \eta_j = \eta_j(\mathbf{x}, t) \ , 0 \leq j \leq n.$$

---

[16]The fact that they are multiple poses no problem and can be taken care with a slight generalization of the expansion. See Remarks 3.7 and 2.2

Then, say, $\lambda_n^o = (\eta_o)^{-1} \sum_1^N \xi_j \eta_j$ and $\mathbf{r}_n^o$ is proportional to $\mathbf{R}$. In this case an expansion where the main excited mode p^{th} is not equal to n, is possible. Waves in the nth mode appear at leading order, forced by the $g\mathbf{R}$ term, but they move at the speed of the pth mode – no new phase appears. The expansion is generally *non-uniform* and applies only for the single wave front case (oscillatory situations do not seem possible, for the " θ structure" of the nth (forced) mode is very rigid).

EXAMPLE 9.2 As special as the circumstances in Remark 9.1 may seem, they apply in reacting gas dynamics. Specifically, consider the situation in Example 8.2 when $\beta = O(\epsilon^2)$. Then the equations can be written in the appropriate form for a proper choice of $\mathbf{u}$. W.N.G.O. provides then a simple model for the propagation of weakly nonlinear detonation waves. The derivations can be found in [RM] and [R2]. In the simplest case the asymptotic equations have the form

(9.8) $$ T_t + (\frac{1}{2}bT^2)_\theta = \mu T_{\theta\theta} + qZ\Phi(T) \ , \quad Z_\theta = -Z\Phi(T) \ , $$

where T measures the amplitude of the main excited mode (an acoustical mode) and Z is the amplitude of the secondary mode (note the equation for Z is an o.d.e., thus the "rigidity" mentioned at the end of Remark 9.1 above). Physically T measures temperature deviation from a base state and Z is the chemical concentration.

REFERENCES

[AM] ARTOLA, M. AND MAJDA, A., *Nonlinear development of instabilities in supersonic vortex sheets. I. The basic kink modes*, Physica D, 28 (1987), pp. 253-281.

[AMR1] ALMGREN, R. F., MAJDA, A. AND ROSALES, R. R., *Rapid initiation in condensed phases through resonant nonlinear acoustics*. In preparation (1989).

[AMR2] ALMGREN, R. F., MAJDA, A. AND ROSALES, R. R., *Dynamic homogenization of reacting materials. II. Numerical results*. In preparation (1989).

[AS] ABLOWITZ, M. AND SEGUR, H., *Solitons and the Inverse Scattering Transform*, SIAM, Philadelphia (1981).

[B] BERGMANN, P.G., *Basic Theories of Physics*, Dover, New York (1962).

[BK] BUCHAL, R. N. AND KELLER, J. B., *Boundary layer problems in diffraction theory*, Comm. Pure Appl. Math., 13 (1960), pp. 85-114.

[BLG] BENSOUSSAN, B., LIONS, J.-L. AND PAPANICOLAU, G., *Asymptotic analysis for periodic structures*, North Holland, Amsterdam (1978).

[Co] COLE, J. D., *Modern developments in transonic flow*, SIAM J. Appl. Math., 29 (1975), pp. 763-787.

[Cr] CRAIK, A. D., Two- and Three-wave Resonance in Nonlinear Waves, in *Nonlinear Waves* (L. Debnath, ed.), Cambridge U.P., New York (1983).

[CB] CHOQUET-BRUHAT, Y., *Ondes asymptotiques et approchées pour des systèmes d'équations aux dérivées partielles non linéaires*, J. Math. Pures et Appl., 48 (1969), pp. 117-158.

[CM] CHOI, Y. S. AND MAJDA, A., *Amplification of small amplitude high frequency waves in a reactive mixture*. To appear in SIAM Review, 1989.

[CR] CEHELSKY, P. AND ROSALES, R. R., *Resonantly interacting weakly nonlinear hyperbolic waves in the presence of shocks: A single space variable in a homogeneous, time independent medium*, Stud. Appl. Math., 73 (1986), pp. 117-138.

[CS] CRAMER, M. S. AND SEEBASS, A. R., *Focusing of weak shock waves at an arête*, J. Fluid Mech., 88 (1978), pp. 209-222.

[DM] DI PERNA, R. AND MAJDA, A., *The validity of Geometrical Optics for weak solutions of conservation laws*, Comm. Math. Physics, 98 (1985), pp. 313-347.

[H1]   HUNTER, J. K., *Transverse diffraction of nonlinear waves and Singular Rays*, SIAM J. Appl. Math., 48 (1988), pp. 1-37.

[H2]   HUNTER, J. K., *A ray method for slowly modulated nonlinear waves*, SIAM J. Appl. Math., 45 (1985), pp. 735-749.

[HK1]   HUNTER, J. K. AND KELLER, J. B., *Weakly nonlinear, high-frequency waves*, Comm. Pure Appl. Math., 36 (1983), pp. 547-569.

[HK2]   HUNTER, J. AND KELLER, J. B., *Caustics of nonlinear waves*, Wave Motion, 9 (1987), pp. 429-443.

[HMR]   HUNTER, J. K., MAJDA, A. AND ROSALES, R. R., *Resonantly interacting, weakly nonlinear, hyperbolic waves. II. Several space variables*, Stud. Appl. Math., 75 (1986), pp. 187-226.

[K]   KELLER, J. B., *Rays, waves and asymptotics*, Bull. Am. Math. Soc., 84 (1978), pp. 727-750.

[KC]   KEVORKIAN, J. AND COLE, J. D., *Perturbation methods in Aplied Mathematics*, Springer-Verlag, New York (1980).

[Lan]   LANDAU, L. D., *On shock waves at large distances from their place of origin*, J. Phys. U.S.S.R., 9 (1945), pp. 495-500.

[Lax]   LAX, P. D., *Hyperbolic systems of conservation laws and the mathematical theory of shock waves*. SIAM Regional Conf. Ser. in Appl. Math., No 11, (1973).

[Li]   LIGHTHILL, M. J., *A technique for rendering approximate solutions to physical problems uniformly valid*, Phil. Mag., 44 (1949), pp. 1179-1201.

[Lu]   LUDWIG, D., *Uniform asymptotic expansions at a caustic*, Comm. Pure Appl. Math., 19 (1966), pp. 215-250.

[M]   MAJDA, A. J., *Criteria for regular spacing of reacting Mach stems*, Proc. Natl. Acad. Sci. USA, 84 (1987), pp. 6011-6014.

[MA]   MAJDA, A. J. AND ARTOLA, M., Nonlinear Geometric Optics for hyperbolic mixed problems, in *Analyse Mathématique et Applications* (volume in honor of J.-L. Lions 60th birthday; C. Agmon, A. V. Ballakrishnan, J. M. Ball and L. Caffarelli, eds.), Gauthier-Villars, Paris (1988), pp. 319-356.

[MP]   MAJDA, A. AND PEGO, R., *Stable viscosity matrices for systems of conservation laws*, Journal of Differential Equations, 56 (1985), pp. 229-262.

[MPT]   MCLAUGHLIN, D., PAPANICOLAU, G. AND TARTAR, L., Weak limits of semi-linear conservation laws with oscillating data, private communication.

[MR1]   MAJDA, A. AND ROSALES, R. R., *A theory for spontaneous Mach stem formation in reacting shock fronts. I. The basic perturbation analysis*, SIAM J. Appl. Math., 43 (1983), pp. 1310-1334.

[MR2]   MAJDA, A. AND ROSALES, R. R., *Resonantly interacting weakly nonlinear hyperbolic waves. I. A single space variable*, Stud. Appl. Math., 71 (1984), pp. 149-179.

[MR3]   MAJDA, A. AND ROSALES, R. R., *Nonlinear Mean Field - High Frequency wave interactions in the Induction Zone*, SIAM J. Appl. Math., 47 (1987), pp. 1017-1039.

[MRA]   MAJDA, A., ROSALES, R. R. AND ALMGREN, R. F., *Dynamic homogenization of reacting materials. I. The asymptotic equations*. In preparation (1989).

[MRS]   MAJDA, A., ROSALES, R. R. AND SCHONBEK, M., *A canonical system of integro-differential equations arising in resonant nonlinear acoustics*, Stud. Appl. Math., 79 (1988), pp. 205-262.

[P]   PEGO, R. L., *Some explicit resonating waves in weakly nonlinear gas dynamics*, Stud. Appl. Math., 79 (1988), pp. 263-270.

[R1]   ROSALES, R. R., Stability theory for shocks in reacting media: Mach stems in detonation waves, in *AMS Lectures in Applied Mathematics*, 24 (1986), G. S. S. Ludford, ed., pp. 431-465.

[R2]   ROSALES, R. R., Diffraction effects in weakly nonlinear detonation waves,to appear in *Proceedings of the "Seminaire International sur les Problèmes Hyperboliques de Bordeaux"*, Bordeaux, June 1988, Springer Verlag "Lecture Notes".

[R3]   ROSALES, R. R., Canonical Equations of Long Wave Weakly Nonlinear Asymptotics, in *Continum Mechanics and its Applications* (G.A.C. Graham and S.K. Malik, ed.), Hemisphere, New York (1989), pp. 365-397.

[RM]   ROSALES, R. R. AND MAJDA, A., *Weakly nonlinear detonation waves*, SIAM J. Appl. Math., 43 (1983), pp. 1086-1118.

[SK]   STURTEVANT, B. AND KULKARNY, V. A., *The focusing of weak shock waves*, J. Fluid Mech., 73 (1976), pp. 651-671.

[W1]   WHITHAM, G. B., *Linear and Nonlinear Waves*, Wiley, New York (1974).

[W2]   WHITHAM, G. B., *The flow pattern of a supersonic projectile*, Comm. Pure Appl. Math., 5 (1952), pp. 301-348.

NUMERICAL STUDY OF INITIATION AND PROPAGATION OF ONE-DIMENSIONAL DETONATIONS

VICTOR ROYTBURD*

In this paper we briefly review some recent work centered on numerical simulation of initiation and propagation of reactive shock waves. This work is a joint project with A. Majda [1-3] with some very recent contributions by Majda's Ph.D. student A. Bourlioux.

The transition to detonation and direct initiation of detonation are extremely complex phenomena. They involve multifaceted interactions between chemistry, strong compressibility and turbulence. Obviously, in physical experiments, all these factors act together so that it is not always possible to determine what particular mechanisms define the course of events. One of the objectives of our research was to isolate the effects that can be satisfactorily described in the framework of one-dimensional inviscid reactive flow with simple chemical kinetics. We focused our attention on the phenomena of generation and evolution of exothermic reaction centers and their role in the development of detonation. The importance of exothermic hot spots as fundamental structures in the secondary stages of transition to detonations has been emphasized by Oppenheim and his coworkers (see, for example, [4]). Such exothermic reaction centers also conveniently approximate initial data for the experiments with direct initiation of detonation from a concentrated energy deposition.

In [2] we demonstrate through careful highly resolved numerical computation that a whole range of experimental phenomena observed in the transient behavior after the formation of a single exothermic reaction center finds transparent qualitative parallels in solutions of the inviscid reactive Euler equations in a single space dimension with one-step irreversible Arrhenius chemical kinetics. An exothermic reaction center (a hot spot) initial data are modeled in our calculations by a rectangular pulse in the temperature profile at constant volume. In addition to initiation from a single hot spot, we consider in [2] a different route to the initiation through resonant interaction of multiple exothermic reaction centers.

During the last twenty years, there has been considerable activity in the numerical investigation of the initiation and propagation of detonations. Two main features distinguish our numerical approach from most of the previous work:

(1) We focus on the details of the transient behavior of the spatial distribution of the state variables;

(2) we pay particular attention to the choice of numerical methods.

In most previous work rather dissipative and/or oscillatory numerical methods were used. As has been documented and analyzed recently [5,6], strong purely numerical artifacts which falsify the actual physical phenomena may be generated by such

*Department of Mathematical Sciences, Rensselaer Polytechnic Institute, Troy, NY 12180

numerical methods. We employ a fractional step numerical scheme consisting of the uniform sampling scheme for the hydrodynamics and a separate fractional step for the chemistry. This choice of numerical scheme is motivated by the careful numerical test cases developed in [5,1,6].

The work on analysis of performance of numerical methods for detonation calculations is continued in [3]. As a test problem we consider a well known phenomenon of unstable galloping detonations (see, for example, [7]). Propagation of these unstable detonations is determined by a rather delicate balance between the shock and the chemical reaction. As a numerical problem, their calculation provides a severe test for numerical methods.

Below we summarize some of the results for each of the topics discussed in preceding paragraphs.

1. Initiation from a Single Exothermic Reaction Center. By varying energetic parameters of the initial hot spot (essentially its height and width), we have been able to simulate several scenarios of initiation that were observed by Lee and coworkers (see [8]) in direct initiation experiments. In particular, we demonstrate an analogue of a supercritical initiation (in Lee's terminology) for which the leading shock and the reaction wave propagate as a single unit: for all times the reaction starts immediately at the shock. Methods of [2] have been applied recently by A. Kapila and the author [9] to a study of strong (supercritical) initiation from a smooth initial hot spot.

By cutting in half the width of the initial hot spot for the strong initiation, one can switch the initiation to a weak regime. For the case of weak initiation, the leading shock and the reaction wave travel completely separated from each other for quite some time. For weak regimes our simulations demonstrate a crucial role of secondary exothermic reaction centers that are formed between the leading shock and the contact discontinuity. It should be noted that the role of the transient spatial structure with exothermic reaction centers in weak and strong initiations has been documented by Oran and coworkers [10] for the full complex hydrogen-oxygen chemistry involving 8 reacting species. Our simulations conclusively demonstrate that purely fluid mechanical effects of pressure and entropy waves play a dominant role in weak regimes rather than the effects of complex chemistry.

2. Resonant Interaction of Multiple Reaction Centers. We illustrate the role of resonant interaction of multiple exothermic reaction centers on two series of simulations [2]. In the first example we consider initial data with three hot spots neither of which individually can initiate detonation. The simulation demonstrates how the resonant wave interaction between the hot spots leads to the initiation of detonation. In another series of simulations we show that high frequency small perturbations over a constant mean flow state enhance tremendously the initiation and formation of secondary reaction centers. The calculations demonstrate a dramatic shortening of the induction time in comparison with the homogeneous induction time. We were also interested in dependence of this effect on the chemical and gas dynamics parameters, in particular on the specific heats ratio γ. For $\gamma = 3$ the induction time is one decimal order of magnitude shorter than for $\gamma = 1.4$. As is well known condense phase explosives are crudely modeled by γ gas laws with large γ. Thus our direct simulations provide supporting evidence for the recent theory of Almgren, Majda, and Rosales [11] which explains the dramatic enhancement of combustion in solid phases through resonant nonlinear acoustics.

3. Propagation of Unstable Detonations.. In a broad range of parameters, steady detonations are dynamically unstable. For the case of detonations overdriven by a piston, a convenient bifurcation parameter is provided by the degree of over-drive. The linearized analysis of Erpenbeck (see, for example, [12] for a summary of his work) shows that as the degree of overdrive decreases, pairs of complex conjugate eigenvalues cross the imaginary axis and the corresponding eigenmodes lose (linearized) stability. Experiments and numerical simulations demonstrate that this loss of stability corresponds to the onset of periodic pulsations.

Numerical study of temporal dynamics of pulsating regimes requires following the detonations for quite long times and distances. In this situation, a full numerical resolution on uniform grids becomes prohibitively expensive (even more so in the case of several space dimensions). For simulations of unstable detonations in a paper with A. Bourlioux and A. Majda [3], we employ a split scheme whose hydrodynamics step utilizes nonreactive shock tracking. The best computational results are obtained with the use of PPM of Colella and Woodward. It is interesting to note that the random choice method which performed extremely well for the weak initiation simulations did not do so well in the unstable case. We suggest that in this case the locally nonconservative nature of the algorithm violates the delicate balance between the chemical reaction and the shock which is necessary for the pulsating regime of propagation.

Simulations in [3] are done in conjunction with the work of Majda and the author [12]. In particular, in [3] we document interesting phenomena when under-resolved calculations may lock solutions into wrong physical regimes. For example, instead of a pulsating detonation, simulations produce a steadily propagating wave. Such phenomena find their qualitative explanation in the framework of the nonlinear theory developed in [12].

Acknowledgements The author's work was supported in part by NSF grant DMS-8603506. This paper was written while the author was on leave at Princeton University. Andy Majda's warm hospitality and support are gratefully appreciated.

REFERENCES

[1] MAJDA, A. AND ROYTBURD, V., *Numerical modelling of the initiation of reacting shock waves*, in *Computational Fluid Mechanics and Reacting Gas Flow*, B. Engquist et al eds.,, I.M.A. Volumes in Mathematics and Its Applications, Vol. 12, 195-217 (1988).

[2] MAJDA, A. AND ROYTBURD, V., *Numerical study of the mechanisms for initiation of reacting shock waves*, Submitted to S.I.A.M. J. Sci. Stat Comput..

[3] BOURLIOUX, A., MAJDA, A. AND ROYTBURD, V., *Numerical modeling of unstable one-dimensional detonations*, (in preparation).

[4] ZAJAC, L.J. AND OPPENHEIM, A.K., *Dynamics of an explosive reaction center*, AIAA Journal, 9, 545-553 (1971).

[5] COLELLA, P., MAJDA, A. AND ROYTBURD, V., *Theoretical and numerical structure for reacting shock waves*, SIAM J. Sci. Stat. Comp. 7, 1059-1080 (1986).

[6] BOURLIOUX, A., *Analysis of numerical method in a simplified detonation model*, (in preparation).

[7] ALPERT, R.L. AND TOONG, T.Y., *Periodicity in hypersonic flows about blunt projectiles*, Astronaut Acta 17, 539-560 (1972).

[8] LEE, J.H.S., *Initiation of gaseous detonation*, Ann. Rev. Phys. Chem. 28, 75-104 (1977).

[9] KAPILA, A. AND ROYTBURD, V., *Transition to detonation: a numerical study*, to appear in Proc. of 3rd International Conference on Numerical Combustion (1989).

[10] ORAN, E.S., YOUNG, T.R., BORIS, J.P. AND COHEN, A., *Weak and strong ignition I. Numerical simulation of shock tube experiments*, Combust. and Flame 48, 135-148 (1982).

[11] ALMGREN, R., MAJDA, A. AND ROSALES, R., *Rapid initiation through high frequency resonant nonlinear acoustics*, in preparation.

[12] MAJDA, A. AND ROYTBURD, V., *A nonlinear theory of low frequency instabilities for reacting shock waves*, in preparation.

RICHNESS AND THE CLASSIFICATION
OF QUASILINEAR HYPERBOLIC SYSTEMS

DENIS SERRE*

Abstract. Rich quasilinear hyperbolic systems are those which possess the largest possible set of entropies. Such systems have a property of global existence of weak solutions, whatever large is the bounded initial data. Although the full gas dynamics is not rich, many physically meaningful systems are. One gives below new examples and properties of the fully linearly degenerate case.

Résumé: Un essai de classification des systèmes quasilinéaires hyperboliques conduit à considérer ceux dont l' ensemble d' entropies est aussi grand que le permettent des considérations immédiates. Ces systèmes, dits riches, ont des solutions faibles globales pour des données initiales bornées. Bien que la dynamique des gaz n' entre pas dans cette catégorie, de nombreux systemes ayant un sens physique sont riches. On donne ci - dessous de nouveaux exemples et on étudie dans cette famille la dégénérescence linéaire des champs caractéristiques.

I. Classification of quasilinear systems. Given an integer $n \geq 1$, one studies systems of the form $u_t + A(u)u_x = 0$, where t is a time-variable, x a space-variable, $u(x,t)$ the unknown belonging pointwise to $\mathbb{R}^n$, and $A(u)$ is an $n \times n$ matrix, depending smoothly on u.

Classically, one is interested with the Cauchy problem for this system. In order to get accurate properties, we shall often restrict to smooth, local in time solutions. In some cases, we shall deal with weak solutions, which means that we choose a conservative form $v_t + (f(v))_x = 0$, even though we do not specify it; it requires that such a form exists.

1. Diagonalization. Hyperbolicity implies that the matrix $A(u)$ is diagonalizable for any value of u. In the linear case (A is constant), the system can thus be reduced to a diagonal form, consisting of uncoupled transport equations. When nonlinearity occurs, the most that one can expect is an equivalent coupled system of transport equations

$$(1.1) \qquad D_i w_i = 0 \ , \ 1 \leq i \leq n \ , \ D_i = \partial_t + \lambda_i(w)\partial_x \ .$$

In that event, the functions $w_i(u)$ are called strict Riemann invariants. The speeds $\lambda_i(w(u))$ are nothing but the eigenvalues of $A(u)$, for which $\mathrm{grad}_u\, w_i$ is a left eigenvector.

It turns out that not all the quasilinear hyperbolic systems (QLH) diagonalize. Assuming for simplicity that Spec $(A(u))$ consists of n distinct real values, a necessary and sufficient condition for the existence of w_i is the well-known Frobenius condition $l_i\{r_j, r_k\} = 0$ for any $j, k \neq i$. Here above, l_i and r_i denotes the left and right eigenvectors related to λ_i, and $\{.,.\}$ is the Poisson bracket of vector fields in u-space.

When $n = 2$, a QLH system is always diagonalizable, but the Frobenius criterion becomes non trivial as $n \geq 3$.

*Ecole Normale Supérieure de Lyon, 46, allée d'Italie, F–69364 Lyon Cedex 07, Serre@frensl61.bitnet

A qualitative way for to introduce diagonalizable system is to consider the interaction of two incident weak shocks for conservative cases. For $n = 2$, two transmitted waves are outgoing. For $n \geq 3$, the interaction produces waves belonging to the other characteristic fields. These waves have strength $O(\alpha\beta)$ where α and β are the strength of incident waves. The necessary and sufficient condition in order that the strength should be weaker is that strict Riemann invariants exist. In that event, the strength turns out to be $O(\alpha\beta(|\alpha| + |\beta|))$. A special case is given by B. Temple's class of systems, for which an i-th and a j-th incident shocks produce only an i-th and a j-th transmitted shock (see [1] and below II.6. for a description of this class).

2. Physical, over-physical systems. Systems which are given by physics, mechanics, chemistry, . . . are often in a conservative form and possess an extra conservation law $E(v)_t + F(v)_x = 0$ where E is a strictly convex function. Following Godunov [2], those systems are hyperbolic and symmetrizable in variables $Q = \text{grad}_u E$.

Some of them may possess many more additional conservation laws. Functions as E are called "entropies" by mathematicians, no matter with their physical sense or their lack of sense; they are not required to be all convex. Trivial entropies are the components v_i and the constants. An example of infinite dimensional set of entropies is provided by Eulerian gas dynamics, with $E = \rho f(S)$, ρ being the mass density, S the physical entropy and f any real-valued function.

It turns out that entropies $E(u)$ are the solutions of a linear partial differential system of the form

$$(1.2) \qquad D_j D_k E = \text{ lower order terms}, \qquad j \neq k \;,$$

where $D_j = r_j \cdot \text{grad}_u$. For $n = 2$, one can choose $D_j = \partial/\partial w_j$ and the system reduces to a single equation for which the Goursat problem is well-posed. Thus the set of entropies is infinite dimensional and parametrized by two arbitrary functions of one variable $F(w_1), G(w_2)$.

3. Rich systems. For $n \geq 3$, the system (1.2) consists of $n(n - 1)$ equations, which can give rise to other equations by combination of their derivatives. Especially, the difference between the equations for the choice (j, k) and (k, j) is a first order equation. Thus generically, a QLH system has not any entropy, except trivial ones in the conservative case.

A simplification occurs in the diagonal case, where there are only $n(n - 1)/2$ equations:

$$(1.3) \qquad \partial(\lambda_i \partial E/\partial w_i)/\partial w_j = \partial(\lambda_j \partial E/\partial w_j)/\partial w_i, \; i \neq j \;.$$

Nevertheless, even this system turns out to be overdetermined and generally prevent the existence of non trivial entropy. Beside the constant coefficient case, where an entropy is uniquely specified by its values on an orthogonal set of reference axis $\mathbf{Re}^i$, one may search for nonlinear systems which endow this property. One calls them "rich hyperbolic systems". An algebraic characterization is given by (see Serre [3] for a description and details about those results which will be given below without proof):

$$(1.4) \qquad D_k((\lambda_j - \lambda_i)^{-1} D_j \lambda_i) = D_j((\lambda_k - \lambda_i)^{-1} D_k \lambda_i) \,,$$

for distinct i, j and k. All along this article, D_i will denote the derivative with respect to w_i.

We again remark that 2×2 systems are trivially rich, so that the theory of rich systems appears to be an attempt of generalization of the theory of 2×2 systems; and actually it is for almost all points of view. An essential one is the method of compensated compactness, developed by L. Tartar [4], R. DiPerna [5], M. Rascle and D. Serre [6], [7],[8] in order to solve the Cauchy problem and to describe the propagation and the interaction of large oscillations. A fundamental consequence holds for genuinely non linear strictly hyperbolic rich systems: a bounded sequence of approximate solutions (namely through artificial viscosity or Godunov's scheme) should converge strongly in L^p for any finite $p \geq 1$ to a weak solution of the system, provided it is given in a conservative form.

The plan of the paper is as follows. The second part is a list of examples and counter-examples arising in physics, mechanics and chemistry. I am particularly grateful to Pr. C. Dafermos for to have driven my attention to a system arising in electrophoresis. It has given me motivation to include the B. Temple's class. The third part deals with the case where all the characteristic fields are linearly degenerate. It contains a short list of formula describing the construction of rich systems and of their entropies, as an abstract of my previous article [3]. A global existence of smooth solutions is given in that case. Its proof is essentially different from the genuine nonlinear case. Finally we discuss the commutability of resolvant operators.

In a forthcoming paper* I shall discuss in more details the linearly degenerate case, including a description of the propagation of large oscillations as in Serre [8] and M. Bonnefille [9]. A group action of resolvant operators on the set of solutions of special ODE systems will be studied. It gives rise to results about (x, t)-almost periodicity or periodicity of solutions of the hyperbolic system, and x-almost periodicity or periodicity for the ODE system.

An alternate approach to rich hyperbolic systems is Tsarev's work [10], where hamiltonian systems are considered. However his class is more restrictive than the rich one. It is essentially due to the definition of symplectic structure: it turns out that naturally hamiltonian 2×2 systems (e.g. isentropic gas dynamics) do not belong to Tsarev's class. Subsequent papers developing the Tsarev's hodograph method or computing higher order densities are due to Y. Kodama [11], Y. Kodama and J.Gibbons [12], P.J. Olver and al. [13].

---

*Systèmes d' EDO invariants sous l'action de systèmes hyperboliques d' EDP. To appear in Annales de l'Institut Fourier.

II. Examples of rich and non rich hyperbolic systems. As discussed in the first section, all the 2×2 systems are rich, so that we shall only consider the case $n \geq 3$. We begin with a well-known non rich example.

1) The gas dynamics. This system consists of balance laws of mass density ρ, momentum ρu and total energy $\rho e + \rho u^2/2$. Thus $n = 3$. A classical result describe all the extra conserved densities by the formula $\rho f(S)$ where f is any real-valued function of one variable, and S is the physical entropy given by the equality

$$T\, dS = de + p\, d(1/\rho).$$

Here above, $p(\rho, e)$ is the hydrodynamic pressure.

On the other hand, the full gas dynamics possesses only one strictly Riemann invariant, namely S. Thus this system appears to be only "one third" rich. In particular, the compensated compactness method is not efficient and the Cauchy problem for large initial data is still open.

Let me give however a consequence of this partial richness related to a linearly degenerate field: because of $\rho_t + (\rho u)_x = 0$ and $S_t + uS_x = 0$, one finds an infinite sequence of quantities satisfying the same transport equation than S. They are inductively defined by $S_{n+1} = \rho^{-1}\partial_x S_n$, $S_0 = S$. Consequently, any function $\rho f(S_0, S_1, \dots)$ is a conserved density. Finally the breakdown of smooth solutions in finite time gives rise to a shock development, without contact discontinuity. In fact, contacts may appear at positive time only as a byproduct of the interaction of two shocks.

2) Nonlinear electromagnetic plane waves. The Maxwell's equations involve four vector fields B, D, E, H depending on time and on a 3-d space variable:

$$(2.1) \qquad B_t + \operatorname{curl} E = 0, \ D_t - \operatorname{curl} H = 0 \ .$$

Following Coleman and Dill [14], the constitutive laws involve an electromagnetic energy density $W(D, B)$:

$$E = \partial W/\partial D \ , \quad H = \partial W/\partial B \ .$$

Hyperbolicity of (2.1) corresponds to the convexity of W. An important conserved density is W.

$$(2.2) \qquad W_t + \operatorname{div}(E \times H) = 0 \ .$$

The system (2.1) is translationally invariant and admit plane waves depending only on t and one space variable, e.g. $x = x_1$. The components B_1 and D_1 are then constant and (2.1) reduces to 4 equations. A natural choice for W is to assume axisymmetry:

$$(2.3) \qquad W(D, B) = W(r) \ , \ r^2 = B^2 + D^2 \ .$$

With that choice, the plane waves obey to a closed 2×2 system, supplemented by two transport equations of the form

(2.4) $\qquad\qquad (pv)_t + (qv)_x = 0 , \quad (v$ is the unknown$)$

where it is known that $p_t + q_x = 0$. Thus this system is endowed with the entropies of the sub-2×2 system and of entropies of the form $pf(v)$ for any f. On the other hand, it has a complete set of strict Riemann invariants: the ones of the subsystem and the unknowns of the transport equations. We conclude that this system is rich.

A complete description may be found in Serre [15], where a global existence theorem is proved for large Cauchy data.

Let us point out that two characteristic fields are linearly degenerate, while the two others may be genuinely nonlinear although the whole system is linear for the most of electromagnetic media as vacuum.

3) Elastic strings. An elastic string lying in a 2 or 3-d physical space can be described in Lagrangian coordinates by $y_{tt} = (T(r)y_x/r)_x$ where the stretching r is the norm of y_x and T is the (scalar) tension. By using $u = y_x$, $v = y_t$, the balance laws become a QLH system of 4 or 6 equations. It is one of the worst of the hyperbolic theory: its entropy set is finite dimensional (6 or 8 independent entropies), while not any strict Riemann invariant exist. There are two linearly degenerate fields, but the interaction of two contact discontinuities produces shocks, yielding to energy dissipation! Even a linear stress-strain relation $T(r) = r - 1$ has been of no help for the existence theory in spite of the linear degeneracy of all the fields.

The Riemann problem for elastic strings has been solved by Keyfitz–Kranzer [16] and Carasso–Rascle–Serre [17]. Analysis and numerical simulations via the Glimm's and Godunov's schemes when the natural type changing occurs have been performed by H. Gilquin, R. Pego and Serre [18], [19], [20].

4) The KdV limit. The behaviour of the Cauchy problem for the Korteweg-de Vries equation $u_t + uu_x = \epsilon u_{xxx}$, as ϵ goes to zero, has been studied independently by P. Lax–D. Levermore [21] by means of the inverse scattering method, and by H.F. Flashka, G. Forest, D. McLaughlin [22] via the formalism of modulated waves. It turns out that the solution sequence u^ϵ does not converge in any norm of Lebesque space in general. But weak convergence holds, which is described by a finite piece-wise constant odd number of functions $w_i(x,t)$, $0 \le i \le 2p$. These functions obey to a diagonal system

$$\partial_t w_i + \lambda_i(w)\partial_x w_i = 0 , \quad 0 \le i \le 2p ,$$

where the speeds are defined in a complicated way by means of hyperelliptic integrals.

Because the KdV equation is known to have infinite sets of independent conserved densities, it is expected that its limit should be rich, and actually it is. The criterion (1.4) follows trivially from a formula of D. Levermore [23] which motivated me in this research. However the relation between the conserved densities of the KdV equation and the ones of its limit had not been yet enlighted.

These systems were proved to have genuinely nonlinear characteristic fields [23] so that the convergence result cited in I.3 would be valid and provide global weak solutions for convenient conservation forms. However such weak solutions have no meaning for the description of KdV phenomena as soon as shocks develop.

An other hierarchy of rich systems, consisting of an even number $N = 2p$ of equations, is the N-reduction of Benney's moment equation (the dispersionless KP hierarchy), see D. Benney [24]. It has been studied in details by Kodama.

5) Electrophoresis. The following system arises in electrophoresis

$$(2.5) \qquad \partial_t u_i + \partial_x (m^{-1} a_i u_i) = 0 \ , \ 1 \le i \le N \ ,$$

where a_i is a positive constant and $m = u_1 + \cdots + u_N$. The unknowns $u_i(x, t)$ should be non-negative functions. Clearly, (2.5) satisfies the corresponding minimum principle. Moreover, it will be asked that $m(x, t) > 0$. This is easily achieved due to the equality

$$\left(\sum_i u_i / a_i \right)_t = 0 \ .$$

We shall assume without generality that $a_1 < a_2 < \cdots < a_N$, otherwise if $a_i = a_{i+1}$, we should replace N by $N - 1$ and (u_i, u_{i+1}) by $u_i + u_{i+1}$. Thus we may define nonlinear functions $d_1(u), \ldots, d_{N-1}(u)$ by the formulas

$$(2.6) \qquad \sum_i u_i / (a_i - d_k(u)) = 0 \ , \quad a_k < d_k < a_{k+1} \ .$$

Finally, $d_0(u)$ will denote $\left(\left(\prod_i a_i \right) \sum_i u_i / a_i \right)^{-1}$. The d_k's are strict Riemann invariants of the system, which is transformed to

$$(2.7) \qquad \begin{cases} \partial_t d_0 = 0 \\ \partial_t d_k + m^{-1} d_k \partial_x d_k = 0 \ , & 1 \le k \le N - 1 \ . \end{cases}$$

In order to close the system, it remains to relate m to the d_k's. Let us define the rational fraction of one variable

$$F(X) = X \sum_i \frac{u_i}{a_i - X} = \sum_i \frac{a_i u_i}{a_i - X} - m \ .$$

One may rewrite F as a ratio P/Q of two polynomials of degrees not greater than N. But we know the poles a_i and the zeros 0 and $d_k, k \ge 1$, of F. Thus there exists a real constant α such that

$$F(X) = \alpha X \prod_{k=1}^{N-1} (d_k - X) \cdot \prod_{k=1}^{N} (a_i - X)^{-1} \ .$$

Because of $F(\infty) = -m,\quad \alpha = m$. Finally

$$F'(0) = \sum_i \frac{u_i}{a_i} = m \prod_{k=1}^{N-1} d_k \left(\prod_{i=2}^{N} a_i \right)^{-1} ,$$

so that $m^{-1} = d_0 \prod_{k=1}^{N-1} d_k$, and $\lambda_i = d_0 d_i \prod_{k=1}^{N-1} d_k$. It is now easy to check by hand the criterion (1.4).

6) The B. Temple's class. The electrophoresis system, as the one of chromatography, actually belongs to the B. Temple's class. Let me give a rather rigid definition. A conservative system will be said to belong to the Temple's class if any point of the u-space is the intersection of $(N-1)$-dimensional linear characteristic manifolds. The existence of such (linear or nonlinear) manifolds is nothing but the existence of strict Riemann invariants. Their linearity is related to the fact that they will be invariant for weak solutions.

An alternate criterion of algebraic type is as follow. Consider the equation $l.u = c$ of a linear manifold. Then this manifold is characteristic for the conservative system $u_t + f(u)_x = 0$ if and only if the function $l.f(u)$ assumes a constant value b on it. In electrophoresis, $l_i = (a_i - d)^{-1}$ is convenient for any d (take $c = 0$), and this gives N distinct families of linear manifolds, depending on the position of d with respect to the a_i's.

For another such system, let w_i the Riemann invariants and $l^i(w_i)u = c_i(w_i)$ corresponding equations of the characteristic linear manifolds. Then it is easily seen that $(l^i.u - c_i)^+$ is a non trivial (i.e. non affine) entropy for any choice of i and w_i. Thus we can construct N infinite families of entropies, parametrized by $1 - d$ bounded measures

$$(2.8) \qquad E_{i,\sigma}(u) = \int (l^i(w).u - c_i(w))^+ d\sigma(w) ,$$

with corresponding fluxes

$$F_{i,\sigma}(u) = \int (l^i(w).f(u) - b_i(w)) \operatorname{sgn}(l^i.u - c_i) d\sigma(w) .$$

We have thus proved the

THEOREM. *Any conservative systems, such that all its left eigenvectors $l_i(u)$ are normal vector fields to linear real manifold, is rich.*

$\square$

For such systems, we shall not pay attention to the nonlinearity of characteristic fields because the compensated compactness method is superfluous. A global existence theorem has been proved in the BV class for large initial data by Serre [25], Leveque and Temple [26]. See Temple [1] for a systematic construction in case $N = 2$.

The following subsection is devoted to a special 2×2 example, for which we use the formula (2.8) for to exhaust the list of entropies.

7) An example in Temple's class.

The system $u_t + (uv)_x = 0$, $v_t + (v^2 + u)_x = 0$ belongs to the Temple's class, because the trivial entropy $E_a = u - av - a^2$ is an algebraic divisor of its flux $F_a = (v - a)E_a$, for any choice of the real parameter a. Actually we have to restrict the above assertion, because the system is hyperbolic only on the zone $v^2 + 4u > 0$. It becomes elliptic as the dependent variables enter inside the parabola $v^2 + 4u < 0$. Thus the aforementioned linear manifolds (straight lines in this 2×2 case) cover only the domain $v^2 + 4u \geq 0$. An important consequence of this remark is that the formula (2.8) will not be usable for constructing entropies inside the elliptic zone. We thus shall restrict to the hyperbolic one.

The Riemann invariants are clearly the two (real) roots w and z of the quadratic equation

$$X^2 + vX - u = 0 .$$

Then $E_a = (w - a)(a - z)$, where we assume $w \geq z$. A similar idea to (2.8) is that $(w - a)^+(a - z)$ is again an entropy, so that the following formulae define an entropy and its flux for any choice of a bounded measure m:

$$E = \int_{-\infty}^{w} (w - a)(a - z) dm(a)$$

$$F = \int_{-\infty}^{w} (v - a)(w - a)(a - z) dm(a) .$$

We shall keep in mind that $v = -w - z$. Defining functions $f, g, h, k(w)$ as the antiderivatives of $a^p dm(a)$, $0 \leq p \leq 3$, we rewrite E and F as

$$E(w, z) = -wz f(w) + (w + z)g(z) - h(w) ,$$
$$F(w, z) = (w + z)wz f(w) - (w^2 + wz + z^2)g(w) + k(w) .$$

The definition of f, g, h and k allows us to introduce a function $T(w)$ satisfying the following four equalities:

$$f = T''' \qquad\qquad k = w^3 T''' - 3w^2 T'' + 6wT' - 6T$$
$$g = wT''' - T'' \qquad\qquad h = w^2 T''' - 2wT'' + 2T'$$

So that we get an infinite family of pairs entropy-flux, parametrized by a real function T of one variable:

(2.9) $$\begin{cases} E = (w - z)T''(w) - 2T'(w) \\ F = (z - w)(z + 2w)T''(w) + 6wT'(w) - 6T(w) \end{cases}$$

The above formula actually does not give all the entropies of our system because w and z did not play the smale role in our calculations. We need to supplement it by the symmetric formula depending on a real function $S(z)$, so that the general entropy would be

(2.10) $$(w - z)(T''(w) - S''(z)) - 2T'(w) - 2S'(z) .$$

It turns out that this formula gives all the entropies of the system, as it can be checked by hand, using the entropy equation

$$((2w + z)E_w)_z = ((2z + w)E_z)_w .$$

Conversely, (2.10) does not give any information about entropies in the elliptic zone, except the case where $S = T$ is a polynomial. Then the formula makes sense and defines a smooth function on the whole plane. For instance, the choice $S = T = X^5/10$ gives $E = v^4 + 6v^2 u + 6u^2$.

REMARK. In the formula (2.8), the special entropy $(l_i.u - c_i)^+$ appears to be an extremal one in the cone of convex entropies, so that $E_{i,\sigma}$ will be convex if and only if σ is a non-negative measure. This fact relies to the scalar example where the entropies $|u - k|$ or $(u - k)^+$, used by Kruzkhov [27] and Tartar [4] generate all the convex functions of one variable by means of the integrals

$$\int_{-\infty}^{u} (u - k)d\sigma(k) .$$

Coming back to the more explicit formula (2.10), we get a convex entropy if and only if $T = -S$ and T^{IV} is non-negative. The condition $T = -S$ comes from the fact that $(w - a)^+(a - z)$ is not convex, so that we have to apply formula (2.8).

The next subsection is devoted to the compensated compactness theory, applied to this system, and we shall pay attention to the elliptic zone.

8) Compensated compactness with an elliptic zone. The compensated-compactness theory is a tool which has been powerful in the study of the convergence of the artificial viscosity method for the Cauchy problem:

(2.11) $$\begin{cases} u_t^\epsilon + f(u^\epsilon)_x = \epsilon\, u_{xx}^\epsilon \\ u^\epsilon(x,0) = u_0(x) \end{cases}$$

There are essentially two *a priori* requirements in order to be allowed to apply this theory:

i) to have an L^∞ uniform estimate for u^ϵ. In all the known examples, it has been proved by means of the positively invariant domains of Chuey–Conley–Smoller [28].

ii) to have an entropy-flux pair (η, q) where η is uniformly strictly convex on every compact set of the u-space.

It is important to notice that none of these requirements occurs for the system introduced in the former section, so that I shall not prove the convergence of (u^ϵ, v^ϵ) to a solution, but only the following alternative:

THEOREM. *Let (u^ϵ, v^ϵ) a solution of the Cauchy problem*

$$u_t^\epsilon + (u^\epsilon v^\epsilon)_x = \epsilon u_{xx}^\epsilon \ , \ v_t^\epsilon + ((v^\epsilon)^2 + u^\epsilon)_x = \epsilon v_{xx}^\epsilon$$
$$u^\epsilon(x,0) = u_0(x) \ , \ v^\epsilon(x,0) = v_0(x) \ .$$

Then, as $\epsilon \to 0_+$, for a suitable subsequence,

i) *either the maximal time-interval $(0, T(\epsilon))$ of existence shrinks to zero,*

ii) *either the solution does not remain bounded in L^∞,*

iii) *either $\epsilon^{1/2}(u_x^\epsilon, v_x^\epsilon)$ does not remain bounded in L^2 ,*

iv) *either strong convergence in L_{loc}^p , $1 \le p < \infty$, holds, so that the limit (u, v) is a solution of $u_t + (uv)_x = 0$, $v_t + (v^2 + u)_x = 0$*

$\square$

Proof. If i), ii) and iii) do not hold, then we are allowed to apply the compensated compactness theory. So that the Young measure $\nu_{x,t}$ which describes the oscillations of the sequence of approximated solutions has a compact support and satisfies almost everywhere (x, t) the Tartar's equality

(2.12) $$\nu(e_1 f_2 - e_2 f_1) = \nu(e_1)\nu(f_2) - \nu(e_2)\nu(f_1) \ ,$$

for all entropy-flux pairs (e_i, f_i), $i = 1, 2$. We shall only deal with the entropies $E_a = E_a^+ - E_a^-$ and $|E_a| = E_a^+ + E_a^-$.

We begin by assuming that for any real number a, Supp ν is not contained in the straight line $u - av - a^2 = 0$. It implies that either $\nu(E_a^+)$ or $\nu(E_a^-)$ is positive, so that $\nu(|E_a|) > 0$. Due to the equality $E_a^+ F_a^- = E_a^- F_a^+$, we get ($F_a^\pm$ are not equal to positive and negative parts of F_a)

$$\nu(E_a^+)\nu(F_a^-) = \nu(E_a^-)\nu(F_a^+) \ ,$$

so that we may define uniquely a real number $c(a)$ by the formulae

$$\nu(F_a^\pm) = c(a)\nu(E_a^\pm) \ .$$

By linearity, $\nu(F_a) = \nu(E_a)c(a)$. Furthermore, $c(\cdot)$ is clearly a continuous function defined on the real line, and $\nu(F_a^+ + F_a^-) = c(a)\nu(|E_a|)$.

We then apply (2.12) to $|E_a|$ and $|E_b|$, with b different from a:

$$(a - b)\nu(|E_a| \ |E_b|) = (c(b) - c(a))\nu(|E_a|)\nu(|E_b|)$$

Letting now $b \to a$, we conclude that $c(\cdot)$ is continuously differentiable, with

(2.13) $$c'(a) = -\nu(E_a^2)\nu(|E_a|)^{-2}$$

By Cauchy-Schwarz inequality, $c' \leq -1$. The second step is to remark that $c(\cdot)$ is a rational fraction, expanding at $\pm\infty$ as

$$c(a) = -a + \text{ constant } + O\left(\frac{1}{a}\right) ,$$

so that the non-increasing function $c(a) + a$ assumes equal values at $\pm\infty$. Thus it is a constant, and $c'(a) \equiv -1$. Returning now to (2.13), we conclude that for any real value a, the support of the Young measure is contained in the union of three parallel lines $u - av = $ constant. Using two different values of a, we find a set consisting of 9 points, which contains Supp ν. Then almost all choice of a third value of a will reduce this set to three points. Now remarking that the aforementioned parallel lines are $u - av - a^2 = 0$ or $\pm\sigma$, a fourth choice of a reduces the support to one point. Thus the preclude assumption is false.

We thus now deal with the case where Supp ν is contained in one line $u - av - a^2 = 0$. Then the analysis becomes one-dimensional and is nothing more than that Tartar did in [4]. Thus Supp ν reduces to one point and $\nu_{x,t}$ is a Dirac mass for almost all x, t.

$\square$

REMARK. The first part of this proof can be mimic for any system belonging to the Temple's class, until a formula similar to (2.13). It would be interesting to consider the general case. The essential question is to understand the role of the genuine nonlinearity and to decide how many inflections are allowed in order to get the result that the Young measure is a Dirac mass.

III. Systems with linearly degenerate characteristic fields.

1) Facts about rich systems. This subsection gathers some facts about the construction of rich systems and their entropies. All of them are proved in [3].

a) Given a rich system, the condition (1.4) gives us a set of positive functions $N_i(w)$, $1 \leq i \leq N$, such that

$$(3.1) \qquad D_j\lambda_i = (\lambda_j - \lambda_i)D_j(\text{Log } N_i) , \qquad i \neq j .$$

Let us define quantities $c_{ij} = N_j^{-1}D_jN_i$ and $y_i = \lambda_iN_i$. Then one gets the relations

$$(3.2) \qquad D_kc_{ij} = c_{ik}c_{kj} \qquad , \; i \neq j \neq k, \; i \neq k ,$$
$$(3.3) \qquad D_jy_i = c_{ij}y_j \qquad , \; i \neq j .$$

Reciprocally, let us choose functions $\gamma_{ij}(w_i, w_j)$, $i \neq j$, then the system (3.2) has a unique local solution $\{c_{ij} , \; i \neq j\}$ assuming the Goursat data

$$(3.4) \qquad c_{ij}(w) = \gamma_{ij}(w_i, w_j) , \text{ as } w_k = w_k^* , \; k \neq i, j ,$$

where w_k^* are given constants. This solution is global provided the following norms are small enough:

$$\|\gamma\| = \underset{i,j}{\text{Max}} \int_{\mathbf{R}} \text{Sup}\{|\gamma_{ij}|; w_j \in \mathbf{R}\} \, dw_k$$

The next step is to solve (3.3) and the similar system $D_j N_i = c_{ij} N_j$, $i \neq j$. It turns out that (3.2) is the set of compatibility conditions for this overdetermined system, so that a global existence and uniqueness result holds for Goursat data:

(3.5) $\qquad\qquad N_i(w) = \quad$ given $n_i(w_i)$ as $w_k = w_k^*$, $k \neq i$,

and similarly for y. A rich system is then completely defined by the formula $\lambda_i = y_i/N_i$.

b) The derivation of entropies is as follow. Clearly the transposed $\mathbf{c}'$ satisfies (3.2) too, so that the following system has a global existence and uniqueness property:

(3.6) $\qquad\qquad\qquad\qquad D_j p_i = c_{ji} p_j$

(3.7) $\qquad\qquad p_i(w) = \quad$ given $q_i(w_i)$ as $w_k = w_k^*$, $k \neq i$.

Such p's are used to construct irrotational fields $g_i = p_i N_i$. The potential E of g is then a generic entropy of the rich system. Its flux F is a potential for the irrotational field $h_i = p_i u_i$.

c) The last feature I want to recall here is the proof of blow up of smooth solutions in the genuinely nonlinear case. It mimics the Lax's proof for 2×2 systems [29]. It appeared in details for the KdV limit in Levermore [23]. Let me first define the operator $L_i = \partial_t + \lambda_i \partial_x$, so that $L_i w_i = 0$, $1 \leq i \leq N$. Taking the x-derivative, and denoting $\partial_x w_i$ by z_i, it comes

(3.8) $\qquad\qquad\qquad\qquad L_i z_i + z_i \partial_x \lambda_i = 0$.

Let us compute the last term:

$$\partial_x \lambda_i = \sum_j z_j D_j \lambda_i = z_i D_i \lambda_i + \sum_j (\lambda_j - \lambda_i) z_j D_j (\text{Log } N_i) .$$

Now, one uses the equality $(\lambda_j - \lambda_i) z_j = -L_i w_j$, which gives

$$\partial_x \lambda_i = z_i D_i \lambda_i - \sum_j (L_i w_j) D_j (\text{Log } N_i) = z_i D_i \lambda_i - L_i (\text{Log } N_i)$$

and finally (3.8) becomes

(3.9) $\qquad\qquad\qquad L_i(z_i/N_i) + N_i^{-1} z_i^2 \partial_i \lambda_i = 0$

This fundamental equality will be used intensively in the linearly degenerate case below. With genuinely nonlinearity, it is a Ricatti-like equation along the i-th characteristic curves of the (x,t) plane, for the unknown $z_i N_i^{-1}$. Because of the transport equations $L_i w_i = 0$, a smooth solution takes its values in a fixed parallelipiped of $\mathbf{R}^N$, so that the coefficients $N_i \partial_i \lambda_i$ are bounded and bounded away from zero. Thus $z_i N_i^{-1}$ becomes infinite in a finite time provided its initial value has the opposite sign to $\partial_i \lambda_i$. It is clearly the case for some x-point if we assume the initial data $w(x)$ to be compactly supported.

2) Construction of fully degenerate examples.

a) Let us consider a rich system satisfying the degeneracy $\partial_i \lambda_i \equiv 0$ for all $1 \leq i \leq n$. We shall assume that none of the λ_i's has a critical point, and also the strict hyperbolicity.

We introduce the diagonal entries c_{ii} of the $\mathbf{C}$ as usual

$$\partial_i N_i = c_{ii} N_i \ .$$

Then deriving the equality $N_i \lambda_i = u_i$ with respect to w_i gives $\partial_i u_i = c_{ii} u_i$. Note that $\lambda_i \neq 0$ and $u_i \neq 0$ almost everywhere.

Let us choose two distinct integers $1 \leq i, j \leq n$ and derive the equality $u_j \partial_j N_i = N_j \partial_j u_i$ with respect to w_j. Using the former, one gets

$$(\lambda_j - \lambda_i)(\partial_j c_{ii} - c_{ij} c_{ji}) = 0 \ ,$$

so that

(3.10)
$$\partial_j c_{ii} - c_{ij} c_{ji} = 0 \ , \qquad \forall \, j \neq i \ .$$

Finally, we derive $\partial_i u_i = c_{ii} u_i$ with respect to w_i. It comes

$$u_i(\partial_j c_{ii} - c_{ij} c_{ji}) = u_j(\partial_i c_{ij} - c_{ii} c_{ij}) \ ,$$

so that

(3.11)
$$\partial_i c_{ij} - c_{ii} c_{ij} = 0 \ , \qquad \forall \, j \neq i$$

Finally, we gather all the known equalities in:

(3.12)
$$\partial_k c_{ij} = c_{ik} c_{kj} \quad \text{for any } i, j, k \text{ such that } k \neq j \ .$$

b) Conversely, we suppose that (3.12) holds, and then we construct a family of fully degenerate systems related to the matrix $\mathbf{C}$. We first construct a convenient filed $(N_1, \ldots, N_n)$. It is uniquely defined by the system $\partial_j N_i = c_{ij} N_j$ $(j \neq i)$ and the prescribed values of each N_i on the i-th axis. In order to satisfy $\partial_i N_i = c_{ii} N_i$ everywhere, it is sufficient to impose this differential equation to the prescribed value of N_i along the i-th axis, thanks to the equality

$$\partial_j(\partial_i N_i - c_{ii} N_i) = 0 \ , \qquad j \neq i \ .$$

We now construct the u_i's. We choose a constant vector $(\alpha_1, \ldots, \alpha_n)$ and prescribe the value $\alpha_i N_i$ to u_i on the i-th axis. Then the equation $\partial_i u_i = c_{ii} u_i$ is again true on the i-th axis and thus everywhere, so that $\partial_i \lambda_i \equiv 0$ for $\lambda_i = u_i/N_i$.

To summarize these results, let us say that each solution of (3.12) gives rise to an n-dimensional real vector space of fields $(N_1, \ldots, N_n)$ satisfying $\partial_j N_i = c_{ij} N_j$

for any i and j. Furthermore, any such N_i gives rise to an n-dimensional vector space of λ's, satisfying

$$(3.13) \qquad \partial_j \lambda_i = (\lambda_j - \lambda_i)\partial_j \operatorname{Log} N_i \quad , \quad \forall\, i,j \; .$$

These λ's defined rich hyperbolic systems with linearly degenerate fields. The above construction suggest to parametrize the speeds as λ^α and the systems as (φ^α). Note that the special choice $\alpha = (a,\ldots,a)$ gives the simple decoupled transport equations

$$\partial_t w_i + a\partial_x w_i = 0 \;, \quad 1 \le i \le n \;.$$

c) As was shown in [3], the entropies are given explicitly in the linearly degenerate case by the formula

$$E = N_1 f_1(w_1) + \cdots + N_n f_n(w_n) \;,$$
$$F = u_1 f_1(w_1) + \cdots + u_n f_n(w_n) \;,$$

where $f_1,\ldots,f_n$ are arbitrary functions of one variable.

3) An explicit example.

a) A particular solution of (3.12) is

$$(3.14) \qquad \begin{cases} c_{ij} = \displaystyle\prod_{k\neq i,j}(w_i - w_k) \cdot \left[\prod_{k\neq j}(w_j - w_k)\right]^{-1} \;, \quad j\neq i \\[2ex] c_{ii} = \displaystyle\sum_{k\neq i}(w_k - w_i)^{-1} \;. \end{cases}$$

A related field $N(w)$ is

$$(3.15) \qquad N_i = \prod_{k\neq i}(w_i - w_k)^{-1} \;.$$

This field and $\mathbf{C}$ are not globally defined, so that we shall restrict ourselves to a parallelipiped which does not meet any hyperplanes $w_i = w_j$, $j \neq i$. Then we may construct solutions of (3.13) by choosing a symmetric polynomial of $n-1$ variables P, being of partial degree 1 with respect to each variable, next defining $\lambda_i(w) = P(\hat{w}_i)$. Here, $\hat{w}_i$ means that we have removed the i-th component w_i from w.

b) Knowing that, we see that another choice for N is $N_i' = N_i Q(\hat{w}_i)$ where Q has the same properties than P. This yields to new speeds $\lambda_i' = (PQ^{-1})(\hat{w}_i)$. Finally the following system is rich which all its fields being linearly degenerate:

$$(3.16) \qquad \begin{cases} Q(\hat{w}_i)\partial_t w_i + P(\hat{w}_i)\partial_x w_i = 0 \quad, \quad 1\le i \le n \\ \text{with symmetric } P \text{ and } Q \text{ s.t. partial degrees} \\ \text{of } P \text{ and } Q \text{ are } \le 1. \end{cases}$$

Let us remark that for any Q, this gives an n-dimensional vector space of speeds, so that all the fully linearly degenerate systems related to $N_i' = N_i Q(\hat{w}_i)$ belong to the family described by (3.16).

c) Because we know some of the speeds $\mu_i = R(\hat{w}_i)/Q(\hat{w}_i)$ which are related to the same N' than λ', one would try to use Tsarev's hodograph formula [10] to find explicit solutions of (3.16). Here we have to solve

$$(3.17) \qquad P(\hat{w}_i)t = R(\hat{w}_i) + Q(\hat{w}_i)x , \quad 1 \le i \le n .$$

In order to keep strict hyperbolicity, we ask for solutions of (3.17) such that $w_i \ne w_j (j \ne i)$. Then the difference of two equations of (3.17) gives

$$P'(\hat{w}_{ij})t = R'(\hat{w}_{ij}) + Q'(\hat{w}_{ij})x , \quad i \ne j ,$$

where P' denotes the derivative of P with respect to one of its variables, and $\hat{w}_{ij}$ has only $n-2$ components. By induction, we get a non-trivial equation with constant coefficients, which is absurd. Thus Tsarev's hodograph formula does not give the expected solution when applied to speeds μ_i such that $\partial_i \mu_i = 0$. However, other speeds are available among the ones which satisfy $\partial_j \mu_i = (\mu_j - \mu_i)\partial_j \operatorname{Log} N_i'$, and they generally give nontrivial results.

4. Global existence of smooth solutions. The analysis carried out in subsection 1.c gives an opposite result for degenerate systems, namely that smooth initial data yields to globally defined smooth solution, with the same regularity.

The proof consists essentially in four steps: local existence, L^∞ estimate, Lipschitz-type estimates, C^m estimates. We begin with a C^m initial data $w^\circ(x), m \ge 2$, being bounded on the whole line. Boundedness of the derivatives is not required. An essential assumption is that the field N is well-defined on the parallelipiped

$$K = \prod_i [\operatorname{Min} w_i^0, \ \operatorname{Max} w_i^0] ,$$

and that each of the N_i's assumes positive values on K. We require also that the speeds λ_i are smooth functions of w.

The local existence of a smooth solution of the Cauchy problem follows from Kato [30], where a result is stated in the Sobolev class H_{loc}^m. Thus we get a $C_{\text{loc}}^{m-1/2}$ solution. The transport equation $L_i w_i = 0$ shows then that $w(x,t)$ belongs to K, so that each $N_i(w(x,t))$ is bounded by above and bounded away from zero.

Now we use relations (3.9), which reduce here to

$$(3.18) \qquad L_i(z_i/N_i) = 0 , \quad z_i = \partial_x w_i$$

By repeating the argument in (1.c), we find a hierarchy of transport equations involving higher derivatives of w_i, namely:

$$(3.19) \qquad \begin{cases} L_i W_i^k = 0, \\ W_i^{k+1} = N_i^{-1}\partial_x W_i^k , \quad W_i^0 = w_i \end{cases}$$

This transport equation makes sense for $0 \leq k \leq m$, because it involves no more that products of an $H^{m-\frac{1}{2}}$ function and a $H^{m-k-3/2}$ distribution, and

$$m - \frac{1}{2} + m - k - \frac{3}{2} \geq m - 2 \geq 0 .$$

Thus (3.19) propagate the local boundedness of the k-th derivatives of w, up to the m-th. This is the Lipschitz-type estimate.

It remains to prove the continuity of $\partial_x^m w_i$ (estimates of mixed derivatives $\partial_t^k \partial_x^l w_i$ follow trivially from (3.19) and the knowledge about the x-derivatives). We have only to remark that the speeds λ_i are locally Lipschitz continuous, so that the i-th transport equation defines an homeomorphism of the space line at each time $0 \leq t < T$, provided that the solution exists on $(0, T)$. Then $W_i^m(T)$, being the composition of its initial value by the aforementioned homeomorphism, is continuous. Since $W_i^m = N_i^{-m} \partial_x^m w_i + l.o.t$, the continuity of $\partial_x^m w_i$ is proved up to T. Using again the local existence result, we conclude that the solution can be globally defined.

In addition, we remark that uniqueness holds thanks to [30], and that the time may be reversed because we deal with classical solutions. Thus we have proved:

THEOREM. *Consider a rich system* $\partial_t w_i + \lambda_i(w) \partial_x w_i = 0, \quad 1 \leq i \leq n,$ *with smooth speeds* λ_i, *satisfying* $\partial \lambda_i / \partial w_i = 0$. *Assume that the* N_i's *are smooth positive functions on* $K = \prod_{i=1}^{n} [w_i^-, w_i^+]$. *Then there exists a unique one-parameter group of operators* $S(t)$, *acting on* $C^m(\mathbf{R}; K)$, *such that for any Cauchy data* $w^0 \in C^m(\mathbf{R}; K)$, *the unique smooth solution of the Cauchy problem is given by* $w = S(t)w^0$.

□

REMARK. Let us notice that the strict hyperbolicity is not required, essentially because weak hyperbolicity always holds. However, it is essential that the N_i's do not vanish. For instance, $N_1 = N_2 = 0$ when $w_1 = w_2$ for the system

$$\partial_t w_1 + w_2 \partial_x w_1 = 0 , \quad \partial_t w_2 + w_1 \partial_x w_2 = 0 ,$$

and the Cauchy data $w_1(x, 0) = w_2(x, 0) = z_0(x)$ gives rise to a solution $z(x, t) = w_1 = w_2$ of the Burgers equation $z_t + z z_x = 0$, for which global existence of smooth solutions does not hold.

5) Commutability of resolvant operators. In this subsection, we shall fix the parallelipiped K and the field N as above. Thanks to 2.b, we know about a system (φ^a), given by speeds λ_i^α, for any $\alpha \in \mathbf{R}^n$, such that the above theorem holds. We denote by $S^\alpha(t)$ its group of resolvant operators and we retain $S^\alpha = S^\alpha(1)$. Any $S^\alpha(t)$ can be reduced to $S^{t\alpha}$, so that it is sufficient to study the S^α's.

THEOREM. *For* K *and* N *as above, the map* $\alpha \longmapsto S^\alpha$ *is a group homomorphism:*

$$S^\alpha \circ S^\beta = S^\beta \circ S^\alpha = S^{\alpha+\beta} .$$

☐

Proof. We begin with the commutation relation. Let us fix a C^m Cauchy data w^0, $m \geq 3$. We define $w^1 = S^\alpha w^0$, $w^2 = S^\beta w^1$, $w^3 = S^\beta w^0$, $w^4 = S^\alpha w^3$. Our goal is to compare w^2 and w^4 up to the second order as α and β go to zero. Using the fact that $S^\alpha = S^\gamma(t)$ where $t = |\alpha|$ and $\gamma = t^{-1}\alpha$, one derives from $\partial_t w_i = -\lambda_i(w)\partial_x w_i$ the expansion:

$$w_i^1 = w_i^0 - \lambda_i^\alpha(w^0)\partial_x w_i^0 + O(\alpha^2) .$$

Combining similar formulae, one gets

$$(3.20) \quad w_i^2 - w_i^4 = \sum_j [(\lambda_j^\alpha - \lambda_i^\alpha)\partial_j \lambda_i^\beta - (\lambda_j^\beta - \lambda_i^\beta)\partial_j \lambda_i^\alpha]\partial_x w_j^0 \partial_x w_i^0 + O(|\alpha|^3 + |\beta|^3)$$

The functions appearing inside the above brackets apply to w^0. We now use the hypothesis that the speeds λ^α and λ^β are related to the same field N, so that the brackets vanish identically. Finally, for $|\alpha|, |\beta| \leq 1$, one finds

$$(3.21) \quad \mathrm{Sup}_x |S^\alpha \circ S^\beta(w^0) - S^\beta \circ S^\alpha(w^0)| \leq C(|\alpha|^3 + |\beta|^3) .$$

Note that the constant in the left hand side depends only on the C^3 norm of w^0, because of the estimates of the former subsection.

The next step is to choose a positive integer r and to write the commutator $[S^\alpha, S^\beta]$ as the sum of r^2 terms of the form $P\Delta Q$, where Δ is the commutator $[S^{\alpha/r}, S^{\beta/r}]$ and P, Q are monomials in $S^{\alpha/r}, S^{\beta/r}$ with $d^0 P + d^0 Q = 2r - 2$. Using again the estimates of 2), one gets a bound valid for each term, with a constant depending on w^0 but not on r:

$$\mathrm{Sup}_x |P\Delta Q(w^0)| \leq C \left(|\frac{\alpha}{r}|^3 + |\frac{\beta}{r}|^3 \right) .$$

Summing all these inequalities, we then deduce

$$\mathrm{Sup}_x |[S^\alpha, S^\beta](w^0)| \leq Cr^{-1}(|\alpha|^3 + |\beta|^3) ,$$

which gives $[S^\alpha, S^\beta] = 0$ by letting $r \to \infty$.

The last trick is the proof of the additivity formula. To this end, we first note an estimate similar to (3.21):

$$(3.22) \quad \mathrm{Sup}_x |(S^\alpha \circ S^\beta - S^{\alpha+\beta})(w^0)| \leq C(|\alpha|^2 + |\beta|^2) ,$$

where C depends only on the C^2 norm of w^0. We next remark that, due to the commutativity property, $S^\alpha \circ S^\beta - S^{\alpha+\beta}$ can be rewritten as the sum of $r + 1$ terms of the form $P\theta$, where $\theta = S^{\alpha/r}S^{\beta/r} - S^{\frac{\alpha+\beta}{r}}$ and P is a monomial in $S^{\alpha/r}, S^{\beta/r}$ and $S^{\frac{\alpha+\beta}{r}}$ with suitable degree. As above, we conclude that

$$\mathrm{Sup}_x |(S^\alpha \circ S^\beta - S^{\alpha+\alpha})(w^0)| \leq C(r+1) \left(|\frac{\alpha}{r}|^2 + |\frac{\beta}{r}|^2 \right) ,$$

so that $r \to \infty$ gives $S^\alpha \circ S^\beta = S^{\alpha+\beta}$.

☐

Acknowledgements. I am grateful to the IMA for its invitation. This work has been significantly improved during the 3 weeks I spend there. I had stimulated conversations on this topic with D. Levermore, D. McLaughlin, P.J. Olver and C. Dafermos.

REFERENCES

[1] B. TEMPLE, *Systems of conservation laws with invariant submanifolds*, Trans. of AMS, 280 (1983), pp. 781–795.

[2] S.K. GODUNOV, *Lois de conservation et intégrales d' énergie des equations hyperboliques*, Saint-Etienne (1986). Carasso, Raviart, Serre eds. Springer Lecture Notes in Maths # 1270. Heidelberg (1987).

[3] D. SERRE, *Systèmes hyperboliques riches de lois de conservation*, Collège de France Seminar, XI or XII. J-L. Lions, H. Brézis eds. (to appear).

[4] L. TARTAR, *Compensated compactness and applications to PDE; in Nonlinear analysis*, Heriot-Watt symposium IV (1979), R.J. Knops ed. Pitman Research notes in Maths, London (1980).

[5] R. DI PERNA, *Convergence of approximate solutions to conservation laws*, Arch. Rat. Mech. Anal. 82 (1983), pp. 27–70.

[6] M. RASCLE, *Thèse d' Etat*, Saint-Etienne (1983).

[7] M. RASCLE, D. SERRE, *Compacité par compensation et systèmes hyperboliques de lois de conservation Applications*, C.R.A.S., A 299 (1984), pp. 673–676.

[8] D. SERRE, *La compacité par compensation pour les systèmes hyperboliques non linéaires de deux équations à une dimension d'espace*, J. de Maths. Pures et Appl. 65 (1986), pp. 423–468
Propagation des oscillations dans les systèmes hyperboliques non linéaires. Saint. Etienne (1986). Carasso, Raviart, Serre Eds. Springer Lecture Notes in Maths # 1270. Heidelberg (1987).

[9] M. BONNEFILLE, *Propagation des oscillations dans deux classes de systèmes hyperboliques (2 × 2 et 3 × 3)*, Comm. Partial Diff. Equ., 13 (1988), pp. 905–925.

[10] S.P. TSAREV, *On Poisson brackets and one-dimensional hamiltonian systems of hydrodynamic type*, Soviet Math. Dokl, 31 (1985), pp. 488–491.

[11] Y. KODAMA, *Exact solutions of hydrodynamics type equations having infinitely many conserved densities*, IMA Preprint # 478.

[12] J. GIBBONS, Y. KODAMA, *A method for solving the dispersionless KP hierarchy and its exact solutions II*, IMA Preprint # 477.

[13] M. ARIK, F. NEYZI, Y. NUTKU, P.J. OLVER, J.M. VEROSKY, *Multi Hamiltonian structure of the Born-Infeld equation*, IMA Preprint # 497.

[14] B.D. COLEMAN, E.H. DILL, Z. Angew. Math. Phys., 22 (1971), pp. 691–702.

[15] D. SERRE, *Les ondes planes en électromagnétisme non linéaire*, Physica D, 31 (1988), pp. 227–251.

[16] B. KEYFITZ, H. KRANZER, *A system of non strictly hyperbolic conservation laws arising in elasticity theory*, Arch. Rat. Mech. Anal., 72 (1980), pp. 219–241.

[17] C. CARASSO, M. RASCLE, D. SERRE, *Etude d' un modèle hyperbolique en dynamique des cables*, Math. Mod. and Num. Anal. 19 (1986), pp. 573–599.

[18] R.L. PEGO, D. SERRE, *Instabilities in Glimm's scheme for two systems of mixed type*, SIAM J. of Numer. Anal., 25 (1988), pp. 965–988.

[19] H. GILQUIN, *Glimm's scheme and conservation laws of mixed type*, SIAM J. Scc. Stat. Comput., 10 (1989), pp. 133–153.

[20] H. GILQUIN, D. SERRE, *Well-posedness of the Riemann problem; consistency of the Godunov's scheme*, To appear in AMS Series. Proceeding of Brunswick (1988).

[21] P.D. LAX, C.D. LEVERMORE, *The small dispersion limit for the Korteweg-de Vries equation, I, II, III*, Comm. Pure and Applied Maths, 36 (1983), pp. 253–290, 571–593, 809–830.

[22] H. FLASHKA, G. FOREST, D.W. MCLAUGHLIN, Comm. Pure and Applied Math. 33 (1979), pp 379–.

[23] C.D. LEVERMORE, *The hyperbolic nature of the zero dispersion KdV limit. Comm. Partial Diff. Equ., 13 (1988), pp. 495–514.*

[24] D.J. BENNEY, Studies in Appl. Maths., 52 (1973), pp. 45–.

[25] D. SERRE, *Solutions à variations bornées pour certains systèmes hyperboliques de lois de conservation,* J. Diff. Equ., 68 (1987) pp. 137–168.

[26] R.J. LEVEQUE, B. TEMPLE, *Stability of Godunov's scheme for a class of 2 × 2 systems of conservation laws,* Trans. of AMS, 288 (1985), pp. 115–123.

[27] S.N. KRUZKHOV, *Generalized solutions for the Cauchy problem in the large for nonlinear equation of first order,* Soviet Math. Dokl., 10 (1969), pp. 785–788.

[28] K. CHUEH, C. CONLEY, J. SMOLLER, *Positively invariant regions for systems of nonlinear diffusion equations,* Indiana U. Math. J., 26 (1977), pp. 373–392.

[29] P.D. LAX, *Development of singularities of solutions of nonlinear hyperbolic partial differential equations,* J. Math. Physics, 5 (1964), pp. 611–613.

[30] T. KATO, *The Cauchy problem for quasi-linear symmetric hyperbolic systems,* Arch. Rat. Mech. Anal., 58 (1975), pp. 181–205.

A CASE OF SINGULARITY FORMATION IN VORTEX SHEET MOTION STUDIED BY A SPECTRALLY ACCURATE METHOD

M. J. SHELLEY*

Abstract. Moore's asymptotic analysis of vortex sheet motion predicts that the Kelvin-Helmholtz instability leads to the formation of a weak singularity in the sheet profile at a finite time. The numerical studies of Meiron, Baker & Orszag, and of Krasny, provide only a partial validation of his analysis. In this work, the motion of periodic vortex sheets is computed using a new, spectrally accurate approximation to the Birkhoff-Rott integral. As advocated by Krasny, the catastrophic effect of round-off error is suppressed by application of a Fourier filter, which itself operates near the level of the round-off. It is found that to capture the correct asymptotic behavior of the spectrum, the calculations must be performed in very high precision. The numerical calculations proceed from the initial conditions first considered by Meiron, Baker & Orszag. For the evolution of a large amplitude initial condition, the results indicate that Moore's asymptotic analysis is valid only at times well before the singularity time. Near the singularity time the form of the singularity departs significantly away from that predicted by Moore. Convergence of the numerical solution beyond the singularity time is not observed.

1. Introduction. A vortex sheet is a surface in an inviscid and incompressible fluid across which the velocity is discontinuous, and is a simple model of a high Reynolds number shear layer. Ideally, the vortex sheet could be used as the first term of the outer solution in a matched asymptotic expansions approach to studying high Reynolds number shear flows. Since the global existence and smoothness of a two-dimensional shear flow evolving from smooth initial conditions is guaranteed (McGrath (1967)), a necessary condition for the success of such a program would be the global existence of this outer solution. However, there is now strong evidence that the two-dimensional vortex sheet acquires a singularity at a finite time well before the appearance of the large-scale roll-up commonly associated with shear layers. The possible nature and existence of vortex sheet motion beyond the singularity time is an open question.

The first analytical evidence of singularity formation was provided by Moore (1979, 1985). For a sinusoidally perturbed sheet with uniform strength, Moore performed a small amplitude perturbation analysis of the Birkhoff equation, which governs vortex sheet motion. His analysis indicated that at a finite time the curvature of the sheet profile diverged, with the sheet strength remaining finite, but acquiring a cusp. More specifically, his approximate analysis indicated the presence of a branch singularity of order 3/2, in the extended Lagrangian variable plane, which moved to the real axis in a finite time. For specially chosen entire initial data, Meiron, Baker & Orszag (1982) (hereafter referred to as MBO) found consistent results by studying the Taylor series in time constructed numerically from the Birkhoff equation. Using the point vortex approximation and a Fourier filter to control the errors due to round-off error, Krasny (1986a) (hereafter referred to as Kr) studied direct simulations of vortex sheet motion from an unstable, linear eigenfunction initial condition, and also found results consistent with Moore (1979).

---

*Department of Mathematics, University of Chicago, Chicago, IL 60637

Moreover, before the singularity time, the numerical solution converged, but convergence was lost beyond the singularity time. Neither of these studies was able to convincingly identify the order of the branch. Caflisch & Orellana (1988) have found a continuum of explicit solutions to the Birkhoff equation which display finite time singularities. However, these solutions have initial data which are not entire, but which begin with branch type singularities in the extended lagrangian variable plane. The order of the branch singularity is a free parameter, and the mechanism by which a singularity is chosen by the general initial value problem is not known. None of these studies indicate whether the sheet still exists beyond its singularity time. Other results concerning existence and well-posedness of vortex sheet motion can be found in Sulem, Sulem, Bardos & Frisch (1981), Caflisch & Orellana (1986), and Caflisch & Orellana (1988).

A precise understanding of singularity formation in vortex sheet motions is important for several reasons. Firstly, vortex sheets are often used in models of fluid motions, and it is important to understand the the limits of their applicability. For example, vortex sheets have been used successfully in the study of large amplitude surface waves (see, for example, Longuett-Higgins & Cokelet, 1976, or Baker, Meiron, & Orszag, 1982), and in the Rayleigh-Taylor instability for a fluid falling into vacuum (Baker, Meiron & Orszag, 1980, Baker, McCrory, Verdon & Orszag, 1987). These all correspond to interfacial flows with an Atwood ratio of 1 or -1. Conversely, researchers have had considerable difficulty in simulating the Rayleigh-Taylor instability in fluids with non-unit Atwood ratio (Baker, et al, 1980, for example). It is believed that these difficulties are related to the formation of singularities in the vortex sheets modelling the interface between the two fluids of different densities. Secondly, Krasny (1986b) and Baker & Shelley (1989) have raised the intriguing possibility that the vortex sheet may exist after its singularity time as a doubly-branched spiral. Clearly then, a detailed knowledge of the formation of the singularity is crucial in deducing the form of the spiral, or if it is even an allowable possibility. Lastly, the mathematical analysis of the vortex sheet singularity is greatly simplified by knowing that it actually does have the form suggested by Moore (Caflisch, private communication).

It is possible that vortex sheet motion exists beyond its singularity time, perhaps as some yet weaker solution to the Euler equations. However, consistent discretizations to the Birkhoff equation have failed to yield reliable results beyond the singularity time, so alternative methods have been employed. Krasny (1986b) modified the Birkhoff equation by smoothing the singular Biot-Savart kernel through convolution with an approximate delta-function of width δ. Past the singularity time for the vortex sheet, and for a fixed value of δ, solutions to the smoothed equation reveal a doubly branched spiral structure, with the number of turns of the spiral increasing, apparently without limit, as δ is reduced. While convergence in the outer arms was observed, the behavior in the center was nonuniform, and the limit may not exist.

In Baker & Shelley (1989), the vortex sheet is replaced by constant and finite vorticity contained within a thin layer of mean width H. The limiting behavior of such vortex layers is then studied numerically, as H is reduced, to determine the possible nature of the vortex sheet past its singularity time. Beyond the singularity time, for each value of H, the vortex layer forms an elliptical core with attached, trailing arms. In the region of the boundary where the arms attach to the cores, the curvature shows very rapid growth. A natural conjecture then is that the curvature on the boundary of the layer diverges, and that these curvature singularities collapse to the vortex sheet singularity in the limit. However, a careful examination of the computed Fourier spectrum suggests that this is not the case, and that the boundaries of the layer remain smooth. Instead, it seems the singularity forms only in the limit of zero H, as a consequence of accumulations of vorticity in the center region. It is this accumulation, for a finite value of H, which engenders the appearance of the elliptical core, with attached, trailing arms. As H is reduced at a fixed time past the singularity time, the approximation to the vortex sheet strength associated with the core width appears to increase without bound. However, the total vortex sheet strength associated with the core, or equivalently, the core circulation, appears to go to zero as H is reduced. Such behavior does not preclude the existence of a classical weak solution in the zero H limit (Diperna & Majda 1987). Concomitant with this, the arms attached to the core converge away from the core, but become increasingly tightly wound in the core region, and convergence was nonuniform.

Using a new, spectrally accurate approximation to the Birkhoff-Rott integral, singularity formation in vortex sheet motion is re-examined, with the intent of acquiring more precise information on the singularity structure. The infinite order of the quadrature follows from the fact that the asymptotic error expansion of the point-vortex approximation, as used by Kr, has only one term of algebraic order, namely $O(h)$, with the remainder being of infinite order. A simple Richardson extrapolation removes this first-order error term, yielding a spectrally accurate quadrature rule. Unlike other such quadrature rules (Conte (1979)), no singularity subtraction is performed, nor do derivative approximations need to be computed.

Due to the ill-posedness of the linearized motion (the disturbance growth rate scales linearly with its wave number), care must be taken to control the effect of round-off error. To this end, we employ the Fourier filter advocated in Kr, which, at each time-step, zeroes any Fourier modes with amplitude less than some given tolerance. The filter operates near the level of the machine round-off. The calculations presented here were performed in 30 digits of precision, with the filter tolerance set to 10^{-25}. It will be seen that this level of precision and filter level are necessary to discern the asymptotic behavior of the Fourier spectrum, which in turn reveals the structure of the nascent singularity.

The evolution of the vortex sheet from an initial condition also considered by MBO is studied in detail. This initial condition is a large perturbation from the equilibrium, and as such, Moore's analysis is not strictly valid in predicting its subsequent singularity formation. However, we are interested in the genericity of the Moore's results, i.e., does Moore's analysis give the general form of singularities arising in vortex sheet evolution. We find that it does not. For the initial condition considered here, Moore's analysis appears to be valid at times well before the singularity time. Near the singularity time, a transition in behavior takes place, and the real and imaginary parts of the solution become differentiated in their behavior, with the imaginary part becoming smoother than the real part. This is not predicted by Moore's analysis. A more complete study of the range of validity of Moore's analysis will be given elsewhere (Shelley, 1989).

Section 2 provides further background of the problem. Section 3 discusses the modified point vortex approximation, its properties, and its relationship with other approximations. Section 4 gives the results of a numerical study of vortex sheet evolution using the modified point vortex approximation. Section 5 gives concluding remarks.

2. Background of the problem. The geometry of a planar vortex sheet is illustrated in Figure 2.1.

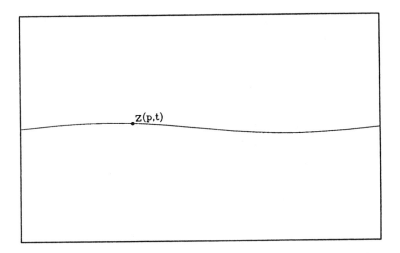

Figure 2.1: A schematic of two-dimensional vortex sheet geometry.

The sheet position is parametrized by its Lagrangian variable p as $z(p,t) = x(p,t) + iy(p,t)$, $-\infty < p < +\infty$, and its motion governed by the Birkhoff-Rott equation,

$$(2.1) \qquad \frac{\partial z^*}{\partial t}(p,t) = \frac{1}{2\pi i} \, P \int_{-\infty}^{+\infty} \frac{\gamma(q)}{z(p,t) - z(q,t)} \, .$$

Here γ is called the vortex sheet strength, z^* denotes the complex conjugate of z, and the Birkhoff-Rott integral is of principal value type. That γ does not depend on time expresses the fact that the circulation is conserved along particle paths. The jump in tangential velocity across the interface is called the true vortex sheet strength, and is given by

$$(2.2) \qquad \omega(p,t) = \frac{\gamma(p)}{|z_p(p,t)|} \, ,$$

where the subscript p refers to partial differentiation. If the sheet initial data is 2π periodic in the x direction, that is $z(p + 2\pi, t = 0) = 2\pi + z(p, t = 0)$, and $\gamma(p + 2\pi) = \gamma(p)$, then the subsequent motion will be also, and is governed by the periodic form of (2.1),

$$(2.3) \qquad \frac{\partial z^*}{\partial t}(p, t) = \frac{1}{4\pi i} P \int_0^{2\pi} \gamma(q) \cot \frac{1}{2}(z(p, t) - z(q, t)) dq \ .$$

It is the periodic form of the Birkhoff-Rott integral that shall be considered in this study. As for smooth vorticity distributions, there are constants of the motion associated with vortex sheet motion. These include the total circulation and the first moments of the vorticity in a periodic strip, and the energy,

$$(2.4) \qquad E = \int_0^{2\pi} dp\gamma(p) \int_0^{2\pi} dq\gamma(q) \ln(\cosh[y(p) - g(q)] - \cos[x(p) - x(q)]).$$

There will sometimes be analagous constants of the motion for the discrete systems arising as approximations to vortex sheet motion.

The chief difficulty, which makes both the analytical and numerical treatment of the Birkhoff-Rott equation very delicate, is the ill-posedness of its linearized motion, as a consequence of the Kelvin-Helmholtz instability. A simple equilibrium is given by a flat sheet with uniform sheet strength. Letting $\gamma(p) = 1 + \varepsilon\tilde{\gamma}(p)$, and $z(p, t) = p + \varepsilon\mu(p, t)$, with $\varepsilon \ll 1$, the linearized Birkhoff-Rott equation is given by

$$\frac{\partial \mu^*}{\partial t}(p, t) = \frac{1}{4\pi i} P \int_0^{2\pi} [\tilde{\gamma}(q) - \mu_q(q, t)] \cot \frac{1}{2}(p - q) dq$$

$$(2.5) \qquad\qquad = -\frac{i}{2} H (\mu_q - \tilde{\gamma})(p, t) \ ,$$

where H denotes the Hilbert transform (Carrier, Krook and Pearson, p. 417). Letting $\tilde{\gamma}(p) = \beta_k^c \cos kp + \beta_k^s \sin kp$ yields

$$\mu(p, t) = [\alpha_k^c \left(1 - i\frac{\sigma}{|\sigma|}\right) e^{\sigma t} - \frac{1}{|\sigma|}\beta_k^s] \cos kp$$

$$(2.6) \qquad\qquad + [\alpha_k^s \left(1 - i\frac{\sigma}{|\sigma|}\right) e^{\sigma t} + \frac{1}{|\sigma|}\beta_k^c] \sin kp \ ,$$

where

$$(2.7) \qquad\qquad \sigma^2(k) = \frac{1}{4} k^2 \ .$$

Thus, there is a positive growth rate $\sigma(k) = k/2$, with the linearized modes having arbitrarily large growth rates, implying an ill-posed linear motion. This is known as the Kelvin-Helmholtz instability. Birkhoff & Fisher (1959) conjectured that for the full nonlinear motion, the linear ill-posedness would cause initial data analytic in its Lagrangian variable to lose its analyticity in a finite time. Birkhoff (1962) also conjectured that such analytic initial data for (2.1) would remain analytic for at least short times. Sulem, Sulem, Bardos, & Frisch (1981) have proved Birkhoff's conjecture (Birkhoff, 1962), and Caflisch & Orellana (1986) have proved long-time existence, within the class of analytic functions, for analytic initial data that is slightly perturbed from a flat sheet. Caflisch & Orellana (1988) and Ebin (1988) have proved the ill-posedness for nonlinear vortex sheet motion in some non-analytic function spaces.

Moore (1979,1984) has given the first analytical evidence that a vortex sheet can lose its analyticity after a finite time, through the formation of a weak singularity in the sheet profile. Moore examined the evolution from the initial condition

$$(2.8) \qquad z(p, t = 0) = p + i\varepsilon \sin p \ , \ \gamma(p) = 1 \ ,$$

for $\varepsilon \ll 1$. This initial condition is composed of both a growing and a decaying linear mode, given by (2.6). The interface may be represented as the Fourier series

$$(2.9) \qquad z(p,t) = p + \sum_{k=-\infty}^{k=+\infty} A_k(t)e^{ikp} \ ,$$

with $A_k = A_{-k}$. Substitution of (2.9) into (2.1) yields an infinite system of ordinary differential equations for the evolution of $A_k(t)$, with each equation itself containing an infinite number of terms. For small ε, the amplitudes $A_k(t)$ can be expanded as

$$(2.10) \qquad A_k(t) = \varepsilon^k A_{k,0}(t) + \varepsilon^{k+2} A_{k,2} + \varepsilon^{k+4} A_{k,4} + \cdots \ ,$$

and by equating powers of ε, it is found that evolution of $A_{k,0}(t)$ depends only upon $A_{n,0}(t)$ for $n \leq k$. This system, for $A_{k,0}(t)$, is then studied for large k. In an analysis presumeably valid for $1 \ll k \ll t$, Moore found the asymptotic behavior is given by

$$(2.11) \qquad \varepsilon^k A_{k,0}(t) \approx \frac{(1+i)}{t\sqrt{2\pi}} \ k^{-5/2} \exp\left(k\left(1 + \frac{1}{2}t + \ln\frac{\varepsilon t}{4}\right)\right).$$

If $t_c(\varepsilon)$ is defined by

$$(2.12) \qquad 1 + \frac{1}{2}t_c + \ln t_c = \ln\frac{4}{\varepsilon} \ ,$$

then for $t < t_c(\varepsilon)$, the leading order Fourier coefficients, $\varepsilon^k A_{k,0}$, decay exponentially fast. However, for $t = t_c(\varepsilon)$, this exponential decay is lost, and $\varepsilon^k A_{k,0}$ decays only algebraically as $k^{-5/2}$. This is the crucial result of Moore's analysis; if the full evolution of the vortex sheet is well described by this leading order behavior, then at $t = t_c(\varepsilon)$, the analyticity of the solution is lost.

The behavior of the Fourier coefficients can be reinterpreted in terms of the spatial behavior of the sheet profile. The approximate summation of the Fourier series given by $\varepsilon^k A_{k,0}(t)$ yields

$$(2.13) \qquad z(p,t) = p + \frac{2\sqrt{3}}{3t}(1+i)[(1 - e^{ip}\varepsilon\theta(t))^{3/2} - (1 - e^{-ip}\varepsilon\theta(t))^{3/2}] + \psi,$$

where ψ contains less singular terms. As $t \to t_c(\varepsilon)^-$, we have that $\varepsilon\theta(t) \to 1^-$, and the profile acquires singularities at $p = 0, \pm 2\pi, \pm 4\pi$, etc. Thus, the singularity formation can be interpreted as the approach to the real p-axis, from above and below, of a pair of branch singularities of order $3/2$. The analyticity strip width of the solution, or the distance of the singularity pair from the real axis, is given by

$$\alpha(t) = -\left(1 + \frac{1}{2}t + \ln\frac{\varepsilon t}{4}\right).$$

MBO also examined vortex sheet evolution, but from the initial condition

$$(2.14) \qquad\qquad z(p, t = 0) = p, \gamma(p) = 1 + a\cos p \ .$$

Again, for small a this initial condition is a combination of a growing and a decaying linear mode, given in (2.6). MBO also found evidence of an isolated singularity forming in the sheet profile at a finite time. In addition to repeating Moore's asymptotic analysis for this initial condition, they also studied the singularity formation using extended time series methods. For Moore's analysis, the expansion of the Fourier amplitudes is in a, rather than ε. They found that to leading order the analysis gave results for initial condition (2.14) identical to those for (2.8); The MBO results are gotten by merely substituting a for ε in (2.10), (2.11), and (2.12). Their study using extended time series gave results different from those predicted by Moore's analysis. The series methods indicated that at the singularity time, $|A_k(t_c)1 \approx k^{-2.7\pm 0.2}$, rather than k^{25}, as given by Moore's analysis. They were unable to assign a definitive cause to the discrepency, attributing it perhaps to higher order corrections dropped in Moore's analysis. They also found that $t_c(a)$, as calculated from (2.12), was typically an underestimate of the singularity time.

Krasny (1986a) also studied vortex sheet singularity formation, now proceeding from initial conditions that are unstable eigenfunctions for the Kelvin-Helmholtz instability. In agreement with Moore and MBO, he found evidence for the appearance of a weak singularity in the sheet profile at a finite time. In particular, he studied the evolution of vortex sheets from initial conditions of the form

$$(2.15) \qquad\qquad z(p, t = 0) = p + (1 - i)\varepsilon\sin p, \gamma(p) = 1 \ .$$

In addition to performing Moore's analysis for (2.15), Kr also performed direct simulations of vortex sheet motion using the point vortex approximation to the Birkhoff-Rott equation, and a Fourier filtering technique to control the growth of errors introduced by the round-off of the calculation. The Fourier amplitudes, $A_k(t)$, were approximated from the numerical data using the discrete Fourier transform, and the behavior of the approximate amplitudes was studied. Kr assumed that in some range of sufficiently large k, the approximate amplitudes, call them $\widehat{A}_k(t)$, had the form suggested by Moore's analysis, or

$$(2.16) \qquad |\widehat{A}_k(t)| \approx C(t)k^{-\beta(t)}e^{-\alpha(t)k} \ .$$

Thus, if $\alpha(t)$ goes to zero at some finite time, as Moore's analysis indicates is the case, the decay of the Fourier spectrum is purely algebraic, indicating some derivative singularity whose order is determined by the value of β at this critical time. The parameters C, β, and α were estimated using a least squares fit over some range in k, and the behavior in time of these approximated parameters is examined. This anaysis suggested that $\alpha(t)$ did become zero at a finite time. However, Kr was unable to obtain reliable estimates for the value of β, which Moore's analysis indicated should be $\beta(t) = 5/2$, independently of time. We believe that this may be due to the low-order of the point vortex approximation and insufficient precision (15 digits).

In section 4, singularity formation in vortex sheets is re-examined, focusing on evolution of a large amplitude initial condition from the set of initial conditions examined by MBO. As in Kr, direct simulations of vortex sheet motion will be studied, and a Fourier filter used to control the effect of round-off error. However, a spectrally accurate discretization of the Birkhoff-Rott integral will be used, rather than the point vortex approximation, which is only first-order in space. Combining this high-order approximation with a different analysis method of the spectrum, we will provide evidence that at times before the singularity time β is indeed $5/2$. However, this behavior changes markedly near the singularity time, and the motion diverges from the behavior predicted by Moore's analysis.

3. The modified point vortex approximation. In this section the numerical approximations used for calculating vortex sheet motion are discussed. This discussion will focus mostly on quadrature methods for the Birkhoff-Rott integral, together with the properties of these approximations. Here will be given a new, spectrally accurate quadrature of the Birkhoff-Rott integral. This quadrature is closely related to the point-vortex approximation, and preserves many of the favorable properties of that approximation. We begin with a discussion of the point vortex approximation.

Following Rosenhead (1931), Krasny (1986a) used the point vortex approximation (subsequently referred to as the PVA) to study vortex sheet motion. Discretizing $z(p, t = 0)$ and $\gamma(p)$ uniformly in the Lagrangian parameter p as $z_j(t = 0) = z(jh, t = 0)$ and $\gamma_j = \gamma(jh)$, with $h = \dfrac{2\pi}{N}$ and $j = 0(1)N$, (2.1) is approximated by the set of ordinary differential equations

$$(3.1) \qquad \frac{d}{dt}z_j^*(t) = \frac{1}{4\pi i} \ h \sum_{\substack{k=0 \\ k \neq j}}^{N-1} \gamma_k \cot\frac{1}{2}(z_j(t) - z_k(t)) \ .$$

The PVA is the trapezoidal rule approximation to the Birkhoff-Rott integral, omitting the singular contribution at $k = j$. This set of discrete equations is then numerically integrated, and its behavior studied, as far as is practically possible, in the limit of large N. The spatial consistency error, or the quadrature error, of the PVA is $O(h)$. Caflisch & Lowengrub (1988) have proved that for initial data close to the flat equilibrium, the PVA converges to the motion of vortex sheet.

Presupposing the strict positivity (or negativity) of $\gamma(p)$, the discrete set of equations, (3.1), forms a Hamiltonian system with Hamiltonian

$$(3.2) \qquad H_N = h \sum_{j=0}^{N-1} \gamma_j \, h \sum_{\substack{k=0 \\ k \neq j}}^{N-1} \gamma_k \ln[\cosh(y_j - y_k) - \cos(x_j - x_k)] \,,$$

where the conjugate variables p_m and q_m are related to the spatial variables x_m and y_m by

$$p_m = (8\pi h \gamma_m)^{1/2} y_m, \text{ and } q_m = (8\pi h \gamma_m)^{1/2} x_m \,.$$

Consequently, H_N is a constant of the motion. H_N can also be related to the energy E of the vortex sheet by noting that H_N is an $O(h)$ accurate quadrature of the energy given in (2.4), with the quadrature being of a form consistent with the PVA discretization of the Birkhoff-Rott equation. In addition to the Hamiltonian, the PVA has as constants of the motion the circulation and the first moments of vorticity in a periodic strip.

As with the vortex sheet, a simple equilibrium of (3.1) is given by $z_k(t) = kh$, $\gamma_k = 1$. Associated with this equilibrium is an discrete dispersion relation, analogous to (2.7), given by

$$(3.3) \qquad \sigma_N^2(k) = \frac{1}{4} k^2 \left(1 - \frac{k}{N}\right)^2 \,,$$

for $k = 0(1)\dfrac{N}{2}$ (Lamb, 1932). Note that the linearized discrete system (3.1) can evince a high wave number growth very similar to that evinced by the vortex sheet, and that (2.7) is recovered as $N \to \infty$. The largest discrete growth rate is at the highest wave number allowed on the mesh, $k = \frac{1}{2}N$, with growth rate $\sigma_N(k) = \dfrac{N}{8}$, or half the growth rate for the same wave number in the sheet motion. Practically speaking, this means that perturbations from the round-off error of the calculation can (and will) lead to the rapid and spurious growth of the high wave-number amplitudes, causing a rapid departure of the discrete system from approximating the continuous system. For this reason, Kr employed a Fourier filter that, at each time-step, zeroed any Fourier amplitude whose modulus was less than some preassigned tolerance. By choosing this tolerance to be close to the round-off error of the calculation, it should, in principle, effect only those modes whose amplitude is being determined by the action of round-off error, rather than the nonlinear dynamics of the motion. The utility of such a device for calculating vortex sheet motion with finite precision is well-documented by Kr, and we direct the interested reader there for further details. This device is also used in this work, but in contrast to Kr, it is not found that the use of the filter is equivalent to calculating in a higher precision, and find that care must be taken though to determine the effect of the filter level, and of the precision which determines it, on the calculated solution.

There is another approximation to the Birkhoff-Rott equation, based upon the trapezoidal rule at alternate points, which is closely related to the PVA, does not require the approximation of any derivatives, and is yet spectrally accurate. Specifically, this quadrature rule gives

$$(3.4) \qquad \frac{d}{dt} z_j^*(t) = \frac{(2h)}{4\pi i} \sum_{\substack{k=0 \\ j+k \text{ odd}}}^{N-1} \gamma_k \cot \frac{1}{2}(z_j(t) - z_k(t)) \, ,$$

as an approximation to (2.1). Here N is assumed to be even. We refer to (3.6) as the modified PVA, or MPVA.

To see the connection between the PVA and the MPVA, their respective errors as approximations to the Birkhoff-Rott integral must be examined. Not only will it be seen that the PVA is of first order accuracy, while the MPVA is of infinite order, but that the MPVA arises through one Richardson extrapolation of the PVA error expansion.

We begin with a simple statement of the quadrature error of the trapezoidal rule to the integral of a periodic, analytic function over its period. Let $g(p)$ be periodic on $[0, 2\pi]$, and analytic in the strip $[-i\rho, +i\rho]$ about the real p axis, with $0 < \rho < \infty$. Then

$$|\Delta_h| = |\int_0^{2\pi} g(p)dp - h \sum_{k=0}^{N-1} g(kh)| \leq C(\rho) \frac{e^{-\rho N}}{1 - e^{-\rho N}} \, ,$$

and consequently,

$$(3.5) \qquad\qquad |\Delta_h| \leq 2C(\rho)e^{\rho N}$$

for sufficiently large N. This follows from expressing Δ_h as the aliasing error of the zeroeth mode of the discrete Fourier transform of g_k from the zeroeth mode of the Fourier transform of $g(p)$. This may then be easily bounded through an application of Cauchy's theorem. The approximation error expressed in (3.5) is a typical realization of spectral accuracy; The error decreases faster than any algebraic power of $\frac{1}{N}$. The Birkhoff-Rott integral will now be rewritten so that this result becomes applicable.

Without being specific, assume that $\gamma(q)$, and $z(q)$ are analytic in some strip about the real q axis. Assume also that $z(q)$ is a single-valued function of q, and that $z_q(q) \neq 0$ for any q. All of these assumptions are justified, for at least short times in the small amplitude regime, by the various regularity results previously given for vortex sheet motion (Caflisch & Orellana 1986, and Sulem et al 1981). Without loss of generality, set $p = 0$, and assume $z(0) = 0$. Letting $f(q) = \gamma(q)\cot\frac{1}{2}z(q)$, and using the periodicity of the integrand, now consider the integral centered about the origin, or

$$I = P \int_{-\pi}^{\pi} f(q)dq$$

and its two approximations,

$$I_1^h = h \sum_{\substack{k=-N/2 \\ k \neq 0}}^{N/2-1} f_k \ , \ \text{and} \ I_2^h = 2h \sum_{\substack{k=-N/2 \\ k \ \text{odd}}}^{N/2-1} f_k \ .$$

Clearly, I_1^h corresponds to the PVA, and I_2^h to the MPVA. Now, let $f(q) = \text{Ev}(q) + \text{Od}(q)$, where $\text{Ev}(q) = \frac{1}{2}(f(q) + f(-q))$ and $\text{Od}(q) = \frac{1}{2}(f(q) - f(-q))$. Note that $\text{Od}(q)$ is an odd function about the origin, and has a simple pole there. On the other hand, $\text{Ev}(q)$ is analytic in a strip about the real q axis, with a removeable singularity at $q = 0$. Now using the fact that the principal value integral over $[-\pi, \pi]$ of an odd function is zero yields

$$I = \int_{-\pi}^{\pi} \text{Ev}(q)dq$$

Both the MPVA and the PVA preserve this feature, yielding

$$I_1^h = h \sum_{\substack{k=-N/2 \\ k \neq 0}}^{N/2-1} \text{Ev}_k \ , \ \text{and} \ I_2^h = 2h \sum_{\substack{k=-N/2 \\ k \ \text{odd}}}^{N/2-1} \text{Ev}_k \ .$$

Thus the MPVA is the trapezoidal rule approximation, over alternate points, to the integral over the period of a periodic, analytic function. The spectral accuracy follows from the previous remarks. The PVA is also a trapezoidal rule approximation, now over all points except $kh = 0$. It is the omission of this term that yields the $O(h)$ error. The inclusion of this term leads to the approximation of Van de Vooren (1980),

$$(3.6) \qquad \frac{d}{dt}z_j^*(t) = h\frac{1}{4\pi i} \sum_{\substack{k=0 \\ k \neq j}}^{N-1} \gamma_k \cot\frac{1}{2}(z_j(t) - z_k(t)) + h\frac{\gamma_j z_{ppj}(t) - 2\gamma_{pj} z_{pj}(t)}{z_{pj}^2(t)} \ .$$

Here, the order of the quadrature error is determined by how accurately z_{pj} and z_{ppj} are approximated. For example, discrete Fourier transforms could again be used to yield spectral accuracy. Another high-order quadrature of the Birkhoff-Rott integral uses the identity

$$\frac{1}{4\pi i} P \int_0^{2\pi} z_q(q) \cot\frac{1}{2}(z(p) - z(q))dq = 0 \ .$$

Equation (2.1) is then rewritten as

$$(3.7) \qquad \frac{\partial z^*}{\partial t}(p,t) = \frac{1}{4\pi i} \int_0^{2\pi} \frac{\gamma(q)z_p(p,t) - \gamma(p)z_q(q,t)}{z_p(p,t)} \cot\frac{1}{2}(z(p,t) - z(q,t))eq \ .$$

The integrand in (3.7) is now smooth and periodic, having made the singularity at $q = p$ removeable. If $z(p)$ and $\gamma(p)$ are again discretized uniformly in their parametric variable p, the trapezoidal rule may be applied to yield spectral spatial accuracy. Here $z_q(q)$ at the mesh points would be found using the discrete Fourier transform (again, spectrally accurate). The trapezoidal rule can be applied in at least two ways. One may choose to perform the quadrature over all the points, taking the appropriate limiting value at the point where $p = q$ (and for which $z_{qq}(q)$ must be approximated), or the quadrature may be performed over alternate points (see Baker & Shelley 1989), thus avoiding the evaluation of the integrand at the indeterminate point. An alternate interpretation would be to note that the only algebraic error term in the quadrature error for the PVA comes from the $k = 0$ term. A single Richardson extrapolation removes this order h term, and yields the MPVA and spectral accuracy. To the author's knowledge, the first application of the alternate point trapezoidal rule to such integrals is due to Sidi & Israeli (1988), in their study of the solution of integral equations for conformal mapping. It is from this work that the author realized that such quadrature methods could be applied to vortex sheet motion.

Finally, note that these arguments are only concerned with the spatial consistency of the MPVA, not with its convergence to the time-dependent motion of the sheet. Hou, Lowengrub & Krasny (1989) have now proved that, for initial data close (in the appropriate sense) to the flat equilibrium, the MPVA converges to the motion of vortex sheet.

As the MPVA arises from the PVA through a Richardson extrapolation, it is hardly surprising that the MPVA is also a Hamiltonian system. A Hamiltonian for this system can again be found by a consistent discretization of the energy E, or

$$(3.8) \qquad \widehat{H}_N = h \sum_{j=0}^{N-1} \gamma_j 2h \sum_{\substack{k=0 \\ j+k \ \text{odd}}}^{N-1} \gamma_k \ln[\cosh(y_j - y_k) - \cos(x_j - x_k)] \ ,$$

with the same conjugate variables as the PVA. Note that (3.8) is only an $O(h)$ approximation to E, and it would seem contradictory that a spectrally accurate approximation to the Birkhoff-Rott integral could arise from it. Fortunately, $\widehat{H}_N$ differs from being a spectrally accurate approximation to E by only a time independent term, which can be given explicitly. A Hamiltonian which is a spectrally accurate approximation to E is given by

$$(3.9)$$

$$H_N = \widehat{H}_N + h \sum_{j=0}^{N-1} \left(\int_0^{2\pi} \gamma(q)\ln[1 - \cos(jh - q)]dq - 2h \sum_{\substack{k=0 \\ j+k \ \text{odd}}}^{N-1} \gamma_k \ln[1 - \cos(jh - kh)] \right)$$

The correction term is independent of the solution $x_j(t)$ and $y_j(t)$, and does not modify the equations of motion. In principle, it can be calculated to as high an accuracy as desired, as $\gamma(q)$ is assumed known. For the initial data for x and y considered in this work, the correction term can be calculated in closed form. Thus, H_N is a constant of the motion for the discrete system, and a spectrally accurate discretization of the energy E. Additionally, as with the vortex sheet and the PVA, the circulation and first moments of the vorticity in a period are constants of the motion for the MPVA.

Again, a simple equilibrium is given by $z_k(t) = kh$ and $\gamma_k = 1$. As pointed out by G. Baker, its discrete dispersion relation matches exactly that of the continuous system, or

$$(3.10) \qquad\qquad \sigma_N^2(k) = \frac{1}{4} k^2 \ .$$

for $k = 0(1)\dfrac{N}{2}$. This is a reflection of the spectral accuracy of the approximation, but is also an argument for the necessity of the Fourier filter in controlling the behavior of the high modes in the calculation. Again, the fastest growing discrete mode is at the largest wave-number on the mesh, $k = \dfrac{N}{2}$.

4. Numerical Results. The results of a numerical study of singularity formation in vortex sheet motion are given in this section. In particular, vortex sheet evolution from initial conditions (2.13) is studied for a particular value of a. As discussed in detail in section 2, MBO found strong evidence for singularity formation from these initial conditions, by using both Moore's analysis, and by studying Taylor series in time that were constructed numerically. Here, using the MPVA, the vortex sheet is numerically evolved forward in time, and its behavior is studied as the numerical parameters of the approximation are varied. A fourth-order, Adams-Moulton, predictor-corrector method is used for time integration. The numerical parameters are not only N, the number of ODEs, and the time-step Δt, but also the tolerance level δ of the Fourier filter. The minimum value of δ is dictated by the available precision of the calculation. It is both δ and N that determine the number of Fourier modes that participate in the calculation of the Birkhoff-Rott integral. If the decay of the spectrum is assumed to be strictly monotonic, that is, $|A_{k+1}(t)| < |A_k(t)|$ (and of course that $A_k(t) \to 0$ as $k \to \infty$), then given a $\delta > 0$ there exists a K_δ, such that $|A_k| < \delta$ for all $k > K_\delta$. All such modes are set to zero by the Fourier filter. Thus, the number of modes participating in the integral evaluation is $\min(K_\delta, N/2)$. The value of δ can have a large effect on the behavior of the spectrum. Indeed, δ must be treated as a convergence parameter, with its effect upon the calculation studied as its value is reduced as far as is practically possible. Thus, the calculations are performed in 29 digit arithmetic (Cray double precision), and the range of filter levels ($\delta = 10^{-14}, 10^{-20}$, and 10^{-25}) varied. It was found that this level of precision was necessary to uncover the asymptotic behavior of the spectrum.

Additionally, a much different technique is used for analyzing the numerically computed spectrum than previous studies (Sulem et al, 1983, Kr, and Baker &

Shelley, 1989). Rather than estimating the decay of the spectrum through a least squares fit over a range of wave numbers, to the form given by (2.16), the spectrum is fit point-wise with a form which includes (2.16), but also attempts to capture higher order effects.

As a result, at times well before the singularity time, we provide strong evidence that the Fourier series of the sheet decays asymptotically as indicated by (2.11). Secondly, we present numerical evidence that at the singularity time, Moore's analysis is no longer sufficient in explaining the form of the singularity. At this time, the singularity has a form significantly different from that predicted by Moore's analysis. To make this point more clearly, we present some illustrative numerical results, before proceeding with the detailed analysis of the data.

As has been remarked previously (Moore 1979, MBO, Kr), the singularity appears in the sheet profile well before the occurence of any large scale roll-up typically associated with shear layers. Using the modified point vortex approximation, the vortex sheet was evolved forward from the initial conditions

$$(4.1) \qquad z(p, t = 0) = p, \ \gamma(p) = -1 + a \cos p \ ,$$

with $a = 1/2$. This is really just the initial condition (2.13), after a π translation, and a change in sign of the vortex sheet strength, and is used simply for historical reasons. The only difference in behavior is that the singularities now occur at $p = \ldots, -\pi, \pi, \ldots$.

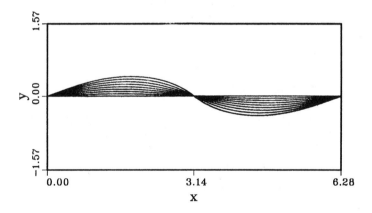

Figure 4.1: The evolution of the vortex sheet, with $a = 1/2$, for the MBO initial conditions, from $t = 0$ to 1.6 at intervals of 0.2.

Figure 4.1 shows the sheet profile from $t = 0$ to $t = 1.6$, at intervals of 0.2. The calculation used $N = 256$, $\Delta t = 0.0025$, and $\delta = 10^{-25}$. This time-step is sufficiently small so that the Hamiltonian was conserved to at least 11 digits throughout the calculation. At the last time shown, $t = 1.6$, the sheet is not much perturbed from its initial condition, and still appears quite smooth. Nonetheless, the sheet is very close to its singularity time. Moore's analysis predicts that the spatial form of the singularity forming from (4.1), behaves as

(4.2a)
$$x_{pp}(p, t_c) \approx c(p - \pi)^{-1/2} ,$$

and

(4.2b)
$$y_{pp}(p, t_c) \approx c(p - \pi)^{-1/2} ,$$

for $0 < p - \pi \ll 1$. This follows from examining (2.13), appropriately translated for this initial condition, in the neighborhood of $p = \pi$ at the critical time. The interest here is not that the singularity time is captured precisely, but rather that the spatial form of the singularity is well described.

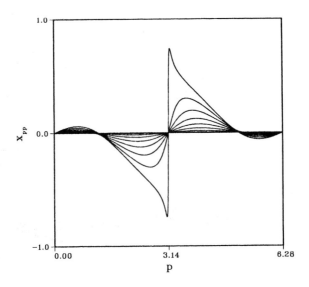

Figure 4.2a: The evolution of $x_{pp}(p, t)$ at the same times as Figure 4.1.

Figures 4.2a and b show the evolution of $x_{pp}(p, t)$ and $y_{pp}(p, t)$, respectively, at the same times as in Figure (4.1). These derivatives are calculated using the discrete Fourier transform. In apparent agreement with (4.2a), $x_{pp}(p, t)$ seems to be diverging at $p = \pi$. And indeed, evidence will be presented that $x_{pp}(p, t)$ diverges as the inverse square root. Quite a different behavior is observed for $y_{pp}(p, t)$. Rather than diverging as in (4.2b), it appears that $y_{pp}(p, t)$ is remaining bounded. Evidence will be presented that for $y(p, t)$ there is a transition in behavior from that predicted by Moore's analysis, and that $y_{pp}(p, t)$ acquires only a step discontinuity at the singularity time. By extrapolating $(x_{ppp}(\pi,t))^{-1}$ and $(yppp(\pi,t))^{-1}$ to zero, the critical time is estimated to be $t_c \approx 1.615 \pm 0.01$. This is quite close to the estimate made by MBO of $t_c \approx 1.6$, and greater than the estimate made by (2.12) of $t_c \approx 1.44$.

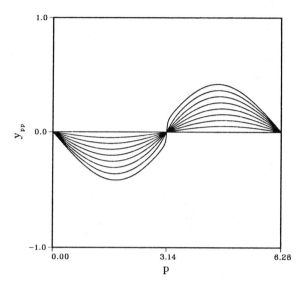

Figure 4.2b: The evolution of $y_pp(p,t)$ at the same times as Figure 4.1.

4.1 The validity of Moore's analysis before the singularity time. It is apparent from the preceding results that, at least for large amplitude initial data, Moore's analysis is not sufficient in predicting the actual form of the singularity which forms in the sheet profile. Nonetheless, the analysis does enjoy some validity as it is predictive of the occurence of the singularity, captures the correct scaling of the singularity time with disturbance amplitude (see Caflisch & Orellana, 1986), and as we will attempt to show here, describes well the behavior of the sheet at times before the singularity time.

Moore's analysis for the initial condition (4.1) gives that the Fourier amplitudes, $A_k(t)$, should behave as

$$(4.3) \qquad A_k(t) \approx (-1)^k \frac{(1-i)}{t\sqrt{2\pi}} k^{-5/2} \exp\left(k \left(1 + \frac{1}{2}t + \ln\frac{at}{4} \right) \right) ,$$

for $t < t_c$ and $k \gg 1$. As in Kr, this form is used as an Ansatz for the behavior of the numerically computed spectrum, though now the behavior of the the real and imaginary parts of the solution must be considered separately. Figures 4.3a and b show the evolution of the the discrete Fourier transforms, $X_k(t)$ and $Y_k(t)$, of $x_j(t) - jh$ and $y_j(t)$, respectively, from $t = 0$ to $t = 1.6$ at intervals of 0.1. As the solutions x and y from the MBO initial conditions can each be represented as sine series, their Fourier transforms have a symmetry about $k = 0$, and thus only positive k need be considered.

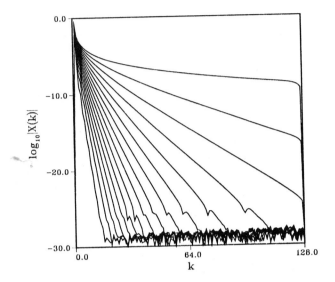

Figure 4.3a: The evolution of $X(k)$, the Fourier transform of $x(p, t) -$ p at the same times as Figure 4.1.

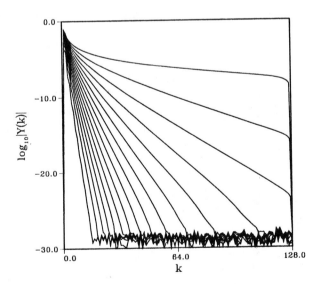

Figure 4.3b: The evolution of $Y(k)$, the Fourier transform of $y(p, t)$ at the same times as Figure 4.1.

Now assume that

(4.4a)
$$|X_k(t)| = C_X k^{-\beta_X} e^{-\alpha_X k} \; ,$$

and

(4.4b)
$$|Y_k(t)| = C_Y k^{-\beta_Y} e^{-\alpha_Y k} \; ,$$

Rather than estimate, for example, the values of C_X, α_X, and β_X by a least squares fit to X_k over some range of k (as in Sulem et al 1983, Kr, and Shelley & Baker 1989), these values are instead estimated by requiring that the form (4.4) hold not only at k, but also at $k-1$ and $k+1$. This technique has also been used by Pugh in his studies of similar problems (private communication). By taking a logarithm of both sides of (4.4), this yields three linear equations for the three unknowns $\alpha_X(k)$, $\beta_X(k)$, and $C_X(k)$. As (4.4) is only asymptotic in k, the behaviors of these three unknowns should be examined as $k \to \infty$. Of course, in a numerical simulation there are the added constraints of approximation errors to be dealt with, which limits the range of k over which one expects the fit to be accurate. However, this technique allows for a more precise study of the behavior of the spectrum by removing some of the subjective biases present in a least squares fit (such as choosing the range in k). It will be seen presently that (4.4) is not fully adequate in deducing the behavior of the spectrum, and will have to be modified.

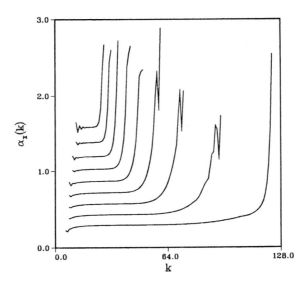

Figure 4.4a: The evolution of $\alpha_X(k)$, from $t = 0.6$ to $t = 1.4$ at intervals of 0.1, fit using formula (4.4). Decreasing values in the k independent region corresponds to increasing time.

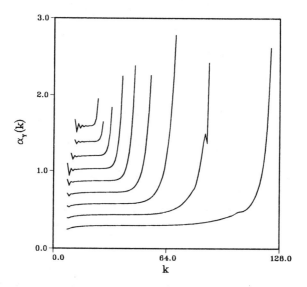

Figure 4.4b: The evolution of $\alpha_Y(k)$, fit using formula (4.4) at the same times as Figure 4.4a.

Note that Moore's analysis yields the values of both β_X and β_Y to be 5/2, independently of k and t, and that α_X and α_Y are equal, independent of k, and monotonically decreasing functions of t. Figures 4.4a and b show the fits to α_X and α_Y as a function of k, respectively, at times from $t = 0.6$ to 1.4 at intervals of 0.1. Note that for a large range of k, both α_X and α_Y show little dependence on k, and that as time increases, the k independent values of α_X and α_Y tend towards zero. Also, the values of α_X and α_Y in the k independent range coincide in value. All of these features are consistent with Moore's analysis.

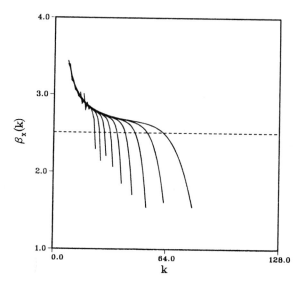

Figure 4.5a: The evolution of $\beta_X(k)$, fit using formula (4.4) at the same times as Figure 4.4a. The range of upward concavity of the fit increases as time increases. The dashed line is at $5/2$.

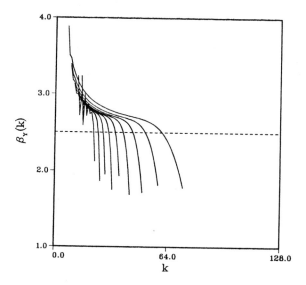

Figure 4.5b: The evolution of $\beta_Y(k)$, fit using formula (4.4) at the same times as Figure 4.4a.

The fits to β_X and β_Y are shown in Figures 4.5a and b, respectively, at the same times as were α_X and α_Y. At first glance, the behavior seems quite different from a simple 5/2 predicted by Moore's analysis. However, it appears that over some range of k, the fit to β_X and β_Y is decreasing and concave upwards. If this were the true behavior of the fit in the absence of approximation errors, it might suggest that β_X and β_Y are asymptotic to some value. The value of 5/2 has been included suggestively as a dashed line in the graphs. Recall that Moore's analysis is only asymptotic in large k, and that higher order effects excluded in the analysis may influence the fit.

The partial inclusion of these higher order effects is effected by modifying the Ansatz (4.4). Consider x as an example. Assume that $x(p) - p$ has a branch singularity in its analytic continuation of the form (2.13), but now of an unspecified order $\beta_X - 1$, rather than 3/2. The asymptotic decay of X_k is governed by this singularity. The first two terms of the large k expansion for X_k have the form

$$|X_k(t)| = C_X k^{-\beta_X} e^{-\alpha_X k} + D_X k^{-(\beta_X + 1)} e^{-\alpha_X k}$$
$$= C_X k^{-\beta_X} e^{-\alpha_X k} \left(1 + \frac{D_X}{C_X} \frac{1}{k}\right) ,$$

or

$$\ln|X_k(t)| = \ln C_x - B_X \ln k - \alpha_X k + \ln\left(1 + \frac{D_X}{C_X} \frac{1}{k}\right)$$

(4.5)
$$= \ln C_X - \beta_X \ln k - \alpha_X k + \frac{D_X}{C_X} \frac{1}{k} + O\left(\frac{1}{k^2}\right) .$$

Here α_X is the distance of the branch singularity from the real p-axis. Dropping the $O\left(\dfrac{1}{k^2}\right)$ term, (4.5) is the form used in the fit to the spectrum, and requires fitting to four consecutive points to yield four linear equations for the four unknowns.

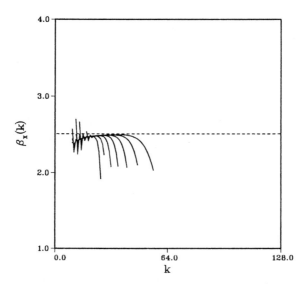

Figure 4.6a: The evolution of $\beta_X(k)$, of 0.1, fit using formula (4.5)

at the same times as Figure 4.4a. The range over which the fit is approximately 5/2 (dashed line) increases as time increases.

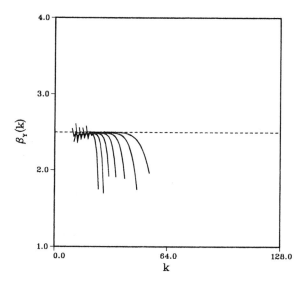

Figure 4.6b: The evolution of $\beta_Y(k)$, fit using formula (4.5) at the same times as Figure 4.4a.

Figures 4.6a and b show β_X and β_Y, respectively, as calculated from (4.5). It is now clear that in the range for which β_X and β_Y were concave upwards in the old fit, they were asymptotic to 5/2. The values for α_X and α_Y from (4.5) are virtually identical to those from the old fit and are not shown.

It is appropriate here to examine the influence of N and δ on the results. As discussed at the beginning of this section, it is both these quantities that determine the number of modes that participate in the calculation at some time. Again, assume that the spectrum is given and is monotonically decreasing as a function of k. If $\delta > 0$ is fixed, then for sufficiently large N it will be that $K_\delta = \min(K_\delta, N/2)$, that is, the number of nonzero modes (and the accuracy of the approximation) is determined solely by the value of δ, not by the value of N. This result is very much realized in these calculations, and has a marked effect on the ability to deduce delicate information about the singularity formation, such as the asymptotic decay of the Fourier spectrum.

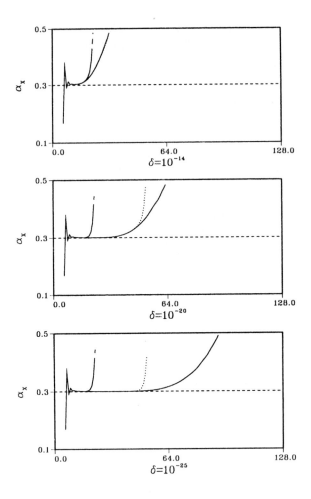

Figure 4.7a: The fit to $\alpha_X(k)$ at $t = 1.4$ for different values of N and δ. Each box shows the fit for a fixed value of δ ($10^{-14}, 10^{-20}$, and 10^{-25}), as N is varied over 64 (solid-dashed), 128 (dot-dot), and 256 (solid).

Figure 4.7a shows the influence of both N and δ upon the fit to $\alpha_X(k)$ at $t = 1.4$. In each box, δ is kept fixed, and N is varied ($N =$64 (solid-dashed), 128 (dot-dot), and 256 (solid)). The dashed line is at what seems to be the the k independent value of α_X from the fit (0.30). Consider the top box, for which $\delta = 10^-14$ and $K_\delta \approx 65$. It appears that as N is increased, the fits to $\alpha_X(k)$ collapse onto a limiting curve, which if the above argument is valid, should be obtained for those fits in which $K_\delta \leq N/2$. This is the case for both $N = 128$ and 256, for which the fits lie upon one another. Now, examining the other two boxes for which $\delta = 10^{-20}$ and

$\delta = 10^{-25}$, it is clear that the limiting behavior is strongly dependent upon δ, and that as δ is decreased the limiting curve has a broader and broader range of k independent values. This same behavior is even more pronounced in the fit to β_X, shown in Figure 4.7b.

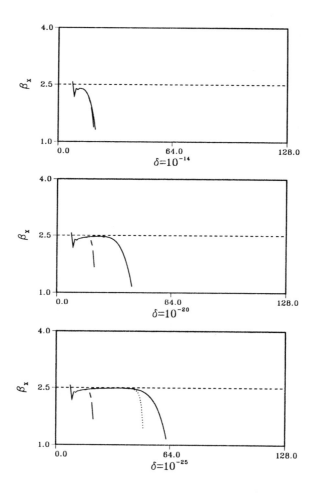

Figure 4.7b: The fit to $\beta_X(k)$ at $th = 1.4$ for different values of N and δ, as in Figure 4.7a.

Here the dashed line is at $5/2$, the k independent value of $\beta_X(k)$ suggested by Moore's analysis. As in Figure 4.7a, as N is increased, and δ decreased, $\beta_X(k)$ evinces a broader and broader range of k independent behavior, and apparently asymptotes to $5/2$.

If the behavior observed in Figure s 4.7a and b persists as N becomes larger and δ smaller, then it provides a very strong validation of Moore's analysis for the intermediate time behavior of the vortex sheet. It is also clear that the high precision was necessary to convincingly demonstrate that the spectrum behaves asymptotically as in (2.11).

4.2 Vortex sheet behavior near the singularity time. We now turn to the behavior of the vortex sheet near its singularity time. The results given here have $N = 513$, $\Delta t = 0.0025$, and $\delta = 10^{-25}$. The initial condition for the MPVA was generated by using the discrete Fourier transform to double the $N = 256$ data at $t = 1.3$. This was chosen as the doubling time because $K_\delta(t = 1.3) < N/2$, and errors should be determined by the value of δ rather N. It was deemed too expensive to calculate the entire evolution from $t = 0$ at the higher resolution. Doubling in this way did lead to some improvement in the results near the critical time, when $K_\delta(t = 1.3) \gg N/2$. Again, the Hamiltonian is conserved to at least eleven digits throughout the calculation.

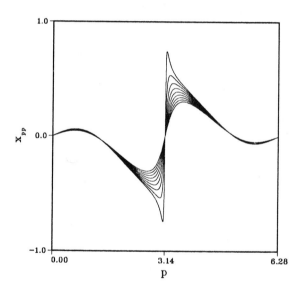

Figure 4.8a: The evolution of $x_{pp}(p, t)$ from $th = 1.4$ to 1.6 at intervals of 0.025.

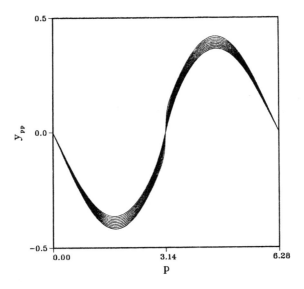

Figure 4.8b: The evolution of $y_{pp}(p,t)$ at the same times as in Figure 4.8a.

Figures 4.8a and 4.8b show the behavior of $x_{pp}(p,t)$ and $y_{pp}(p,t)$, respectively, from $t = 1.4$ to $t = 1.6$ at intervals of 0.025 . Again, the difference in behavior between x_{pp} and y_{pp} is observed as $t \to t_c^-$. In particular, $x_{pp}(p,t)$ appears to be diverging at $p = \pi$, in the form of an infinite jump discontinuity, while $y_{pp}(p,t)$ is plainly not diverging, though its derivative is seemingly becoming infinite. To make more clear the difference in behavior, Figures 4.9a and b show $x_{ppp}(p,t)$ and $y_{ppp}(p,t)$ at the same times. Both $(x_{ppp}(\pi,t))^{-1}$ and $(y_{ppp}(\pi,t))^{-1}$ appear to be approaching zero simultaneously at some finite time, which through extrapolation, is estimated to be $t_c \approx 1.615 \pm .01$. Further, the behavior of $y_{ppp}(p,t)$ in the neighborhood of $p = \pi$ suggests that $y_{pp}(p,t)$ is approaching a finite jump discontinuity as $t \to t_c^-$.

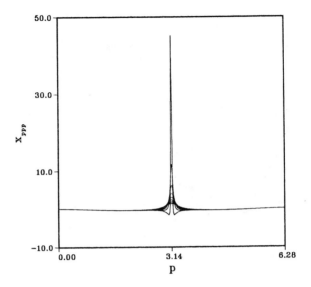

Figure 4.9a: The evolution of $x_{ppp}(p,t)$ at the same times as in Figure 4.8a.

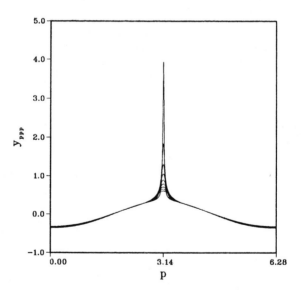

Figure 4.9b: The evolution of $y_{ppp}(p,t)$ at the same times as in Figure 4.8a.

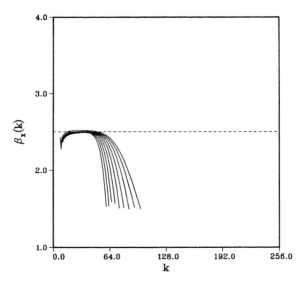

Figure 4.10a: The evolution of $\beta_X(k)$, fit using formula (4.5) from $t = 1.4$ to 1.6 at intervals of 0.05. The range over which the fit is approximately 5/2 (dashed line) increases as time increases.

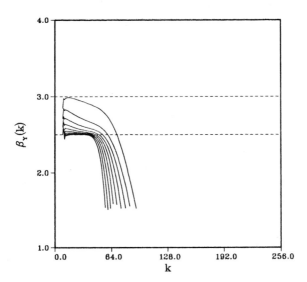

Figure 4.10b: The evolution of $\beta_Y(k)$, fit using formula (4.5) at the same times as Figure 4.10a. Dashed lines are at 5/3 and 3. The approach to 3 corresponds to increasing times.

This behavior is consistent with that of the spectrum. Figures 4.10a and 4.10b show β_X and β_Y, respectively, from $t = 1.4$ to $t = 1.6$ at intervals of 0.025, using the fit in (4.5). The difference in the behavior of these fits is apparent. β_X is still well fit by a value of 2.5, which would yield a divergent second derivative, while the fit to β_Y shows a transition away from 2.5. Indeed, the last time shown, $t = 1.6$, suggests that there is a transition from an algebraic decay of $k^{-5/2}$ for $Y_k(t)$, to a k^{-3} decay. Such an algebraic decay at the singularity time would be be consistent with $y_{pp}(p, t_c)$ containing a step discontinuity.

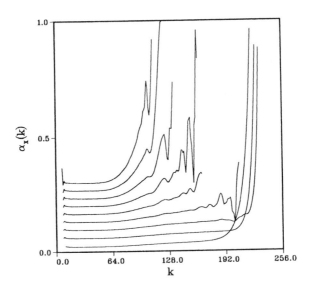

Figure 4.11: The evolution of $\alpha_X(k)$, fit using formula (4.5) at the same times as Figure 4.10a.

Figure 4.11 shows α_X at the same times. It is clear that it is becoming more difficult to resolve the sheet as evidenced by the noisiness in the fit, but α_X still shows the tendency towards zero as the singularity time is approached. While there are differences in the singular behavior of $x(p, t)$ and $y(p, t)$, resp., no such difference in their singularity times is evident; α_Y is not shown as its values are practically identical to those of α_X. Choosing $k = 32$ as a representative of the k independent portion of the α_X fit, Figure 4.12 shows its behavior as a function of time. The approach to zero is obvious. The dashed line is the extrapolation to zero at $t = 1.615$.

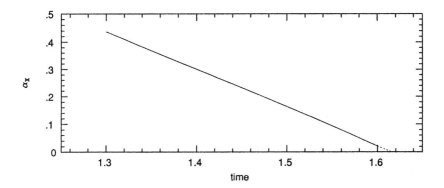

Figure 4.12: $\alpha_X(kh = 32)$ as a function of time. The dashed portion is an extrapolation to an approximated singularity time of $t_c h = 1.615$.

While there are differences between analytic prediction and numerical results, the manner in which vortex sheets become singular does not change. At a point along the sheet, there is a very rapid compression of the vorticity through the Lagrangian motion of the marker particles. It is because the vorticity must remain confined to the sheet, without the additional degrees of freedom of smoother vorticity distributions, that this compression leads to the appearance of singularities. This is illustrated by the behavior of the true vortex sheet strength ω, shown in Figure 4.13 from $t = 1.3$ to $t = 1.6$ at intervals of 0.05, as a function of the signed arclength of the sheet from $p = \pi$. ω gives the jump in tangential velocity across the sheet, and is given initially by $\omega(p, t = 0) = -p + \frac{1}{2}\cos(p)$ (dashed). Around the local extrema at $p = \pi$ ($s = 0$), $\omega(p, t)$ becomes concentrated, and increasing in amplitude. Moore's analysis predicts that at $t = t_c$, $\omega(p, t)$ is finite, with a square-root cusp at $p = \pi$. As the singularity in $y(p, t)$ remains of the form predicted by Moore, and that in $x(p, t)$ only weakens, this conclusion is unchanged.

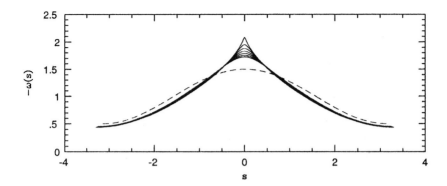

Figure 4.13: The evolution of $\omega(s,t)$, the true vortex sheet strength, from $th = 1.3$ to 1.6 at intervals of 0.05. The dashed graph is the initial true vortex sheet strength. Increasing amplitude corresponds to increasing time.

Section 5. Concluding remarks. The intent of this work was to examine the generality of Moore's analysis, presumeably valid only for small amplitude data, in predicting the form of vortex sheet singularities. By studying in detail the evolution of a large amplitude, entire initial condition, it was found that Moore's analysis was valid in predicting the behavior of the sheet at times well away from the singularity time, but near the singularity time the sheet behavior underwent a transition leading to a change of form in the nascent singularity. A more complete study (Shelley 1989), is to be given elsewhere.

This work does not address the possible existence and nature of the vortex sheet after the singularity time, but has instead focused on gaining precise information on the form of the singularity. As has also been observed by Kr, no convergence of the numerical solution was observed after the singularity time. Of course, the spectral accuracy of the MPVA is lost in the presence of singularities. At these later times, the motion becomes dominated by grid scale interactions, and is apparently chaotic. It appears that mollification of some sort is necessary to numerically study behavior past the singularity time (Krasny 1986b, Baker & Shelley 1989). Such studies indicate that the solution, if it exists, may have the form of doubly branched spiral. It is known that measured-valued solutions exist globally for vortex sheet initial data, but the notion of such a solution is so general that it gives little information about its specific nature. The scaling of vorticity concentrations in the study of thin vortex layers by Baker & Shelley (1989) suggests that the vortex sheet may actually exist as a classical weak solution after the singularity time (Diperna & Majda, 1987).

Explicit singular solutions have been constructed by Caflisch & Orellana (1988), with $\gamma(p) = 1$, which have the form

$$z(p, t) = p + s_0 + r \ ,$$

where

$$s_0 = \varepsilon(1 - i) \left(\left(1 - e^{-\frac{t}{2} - ip} \right)^{1+\nu} - \left(1 - e^{-\frac{t}{2} + ip} \right)^{1+\nu} \right) \ ,$$

ε is small, r is a correction term, and $\nu > 0$. $\nu = 1/2$ would give the spatial structure of Moore's singularity at $t = 0$. The singularity found in this work is not of this form, though it is quite possible that such a singularity could be constructed analytically.

Acknowledgements. The author would like to thank G. R. Baker, R. E. Caflisch, A. Majda, D. I. Meiron and S. A. Orszag for useful discussions. This work was partially supported under contracts ONR/DARPA N00014-86-K-0759 and ONR N00014-82-C-0451. Some of the computations were carried out on the Cray-XMP at Argonne National Laboratory.

REFERENCES

BAKER, G. R., MCCRORY, R. L., VERDON, C. P. & ORSZAG, S. A., *Rayleigh-Taylor instability of fluid layers*, J. Fluid Mech. 178 (1987) 161.

BAKER, G. R., MEIRON, D. I., & ORSZAG, S. A., *Vortex simulations of the Rayleigh-Taylor instability*, Phys. Fluids 23 (1980), 1485.

BAKER, G. R., MEIRON, D. I., & ORSZAG, S. A., *Generalized vortex methods for free-surface flow problems*, J. Fluid Mech. 123 (1982), 477.

BAKER, G. R. & SHELLEY, M. J., *Boundary integral techniques for multi-connected domains*, J. Comp. Phys. 64 (1986), 112.

BAKER, G. R. & SHELLEY, M. J., *On the connection between thin vortex layers and vortex sheets*, To appear in J. Fluid Mech. (1989).

CAFLISCH, R. & LOWENGRUB J., *Convergence of the Vortex Method for Vortex Sheets*, to appear where (1988)?.

CAFLISCH, R. & ORELLANA, O., *Long time existence for a slightly perturbed vortex sheet*, Comm. Pure Appl. Math. XXXIX (1986), 807.

CAFLISCH, R. & ORELLANA, O., *Singular solutions and ill-posedness of the evolution of vortex sheets*, (1988).

DUCHON, R. & ROBERT, R., *Global vortex sheet solutions to Euler equations in the plane*, to appear in Comm. PDE. (1989).

DIPERNA, R & MAJDA, A., *Concentrations in regularizations for 2-d incompressible flow*, Comm. Pure Appl. Math. 40 (1987), 301.

EBIN, D., *Ill-posedness of the Rayleigh-Taylor and Helmholtz problems for incompressible fluids*, CPAM (1988).

KRASNY, R., *A study of singularity formation in a vortex sheet by the point-vortex approximation*, J. Fluid Mech. 167 (1986a), 65.

KRASNY, R., *Desingularization of periodic vortex sheet roll-up*, J. Comp. Phys. 65 (1986b), 292.

LONGUETT-HIGGINS, M. S. & COKELET, E. D., Proc. R. Soc. Lond. A 350 (1976), 1.

MAJDA, A., *Vortex dynamics: Numerical analysis, scientific computing, and mathematical theory*, In Proc. of the First Intern. Conf. on Industrial and Applied Math., Paris (1987).

McGRATH, F. J., *Nonstationary plane flow of viscous and ideal fluids*, Arch. Rat. Mech. Anal., 27 (1967), 329.

MEIRON, D. I., BAKER, G. R., & ORSZAG, S. A., *Analytic structure of vortex sheet dynamics. 1. Kelvin-Helmholtz instability*, J. Fluid Mech. 114 (1982), 283..

MOORE, D. W., *The spontaneous appearance of a singularity in the shape of an evolving vortex sheet*, Proc. R. Soc. Lond. A 365, (1979) 105.

MOORE, D. W., *Numerical and analytical aspects of Helmholtz instability*, In Theoretical and Applied Mechanics, Proc. XVI IUTAM, eds. Niodson and Olhoff (1985), 263.

PULLIN, D. I. AND PHILLIPS, W. R. C., *On a generalization of Kaden's problem. J. Fluid Mech. 104 (1981), 45.*

ROSENHEAD, L., *The formation of vortices from a surface of discontinuity*, Proc. R. Soc. Lond. A 134 (1931), 170.

SHELLEY, M. J., *A study of singularity formation in vortex sheet motion by a spectral accurate vortex method*, To appear in J. Fluid Mech. (1989).

SIDI, A. & ISRAELI, M., *Quadrature methods for periodic singular and weakly singular Fredholm integral equations*, J. Sci. Comp. 3 (1988), 201.

SULEM, C., SULEM, P. L., BARDOS, C. & FRISCH, U, *Finite time analyticity for the two and three dimensional Kelvin-Helmholtz instability*, Comm. Math. Phys. 80 (1981), 485.

SULEM, C., SULEM, P. L. & FRISCH, H., *Tracing complex singularities with spectral methods*, J. Comp. Phys. 50 (1983), 138.

VAN DER VOOREN, A. I., *A numerical investigation of the rolling up of vortex sheets*, Proc. Roy. Soc. A 373 (1980), 67.

VAN DYKE, M., *Perturbation Methods in Fluid Mechanics*, The Parabolic Press (1975).

THE GOURSAT-RIEMANN PROBLEM FOR PLANE WAVES IN ISOTROPIC ELASTIC SOLIDS WITH VELOCITY BOUNDARY CONDITIONS*

T. C. T. TING†AND TANKIN WANG†

Abstract. The differential equations for plane waves in isotropic elastic solids are a 6×6 system of hyperbolic conservation laws. For the Goursat-Riemann problem in which the initial conditions are constant and the constant boundary conditions are prescribed in terms of stress, the wave curves in the stress space are uncoupled from the wave curves in the velocity space and the equations are equivalent to a 3×3 system. This is not possible when the boundary conditions are prescribed in terms of velocity. An additional complication is that, even though the system is linearly degenerate with respect to the c_2 wave speed, the c_2 wave curves cannot be decoupled from the c_1 and c_3 wave curves. Nevertheless, we show that many features and methodology of obtaining the solution remain essentially the same for the velocity boundary conditions. The c_1 and c_3 wave curves are again plane polarized in the velocity space although the plane may not contain a coordinate axis of the velocity space. Likewise, the c_2 wave curves are circularly polarized but the center of the circle may not lie on a coordinate axis of the velocity space. Finally, we show that the c_2 wave curves can be treated separately of the c_1 and c_3 wave curves in constructing the solution to the Goursat-Riemann problem when the boundary conditions are prescribed in terms of velocity.

Key words. Goursat-Riemann problem, wave curves, elastic waves

AMS(MOS) subject classifications. 35L65, 73D99

1. Introduction. In a fixed rectangular coordinate system x_1, x_2, x_3, consider a plane wave propagating in the x_1 - direction. Let σ, τ_2, τ_3 be, respectively, the normal stress and two shear stresses on the $x_1 = $ constant plane. Also, let u, v_2, v_3 be the particle velocity in the x_1, x_2, x_3 direction, respectively. The equations of motion and the continuity of displacement can be written as a 6×6 system of hyperbolic conservation laws[1,2,3]

$$\mathbf{U}_x - \mathbf{F}(\mathbf{U})_t = \mathbf{0},$$
(1.1)
$$\mathbf{U} = (\sigma, \tau_2, \tau_3, u, v_2, v_3),$$
$$\mathbf{F}(\mathbf{U}) = (\rho u, \rho v_2, \rho v_3, \varepsilon, \gamma_2, \gamma_3).$$

In the above, $x = x_1$, t is the time, ρ is the mass density in the undeformed state, $\varepsilon, \gamma_2, \gamma_3$ are, respectively, the longitudinal strain and the two shear strains. For isotropic elastic solids, the stress-strain laws have the form [1]

$$\sigma = f(\varepsilon, \gamma^2),$$
(1.2)
$$\tau_2 = \gamma_2 \, g(\varepsilon, \gamma^2),$$
$$\tau_3 = \gamma_3 \, g(\varepsilon, \gamma^2),$$
$$\gamma^2 = \gamma_2^2 + \gamma_3^2,$$

*This work has been supported by the U.S. Air Force Office of Scientific Research under contract AFOSR-89-0013.

†Department of Civil Engineering, Mechanics and Metallurgy, University of Illinois at Chicago, Box 4348, Chicago, IL 60680.

where f and g are functions of ε and γ^2. We see that γ is the total shear strain. If τ is the total shear stress on the x = constant plane, we obtain from $(1.2)_{2,3}$,

$$
\tau = \gamma \, g(\varepsilon, \gamma^2),
$$
$$
\tau^2 = \tau_2^2 + \tau_3^2.
$$
(1.3)

In the region of ε and γ^2 where equations $(1.2)_1$ and $(1.3)_1$ have an inversion, we have

$$
\varepsilon = h(\sigma, \tau^2),
$$
$$
\gamma = \tau \, q(\sigma, \tau^2),
$$
(1.4)

where h and q are functions of σ and τ^2. Equations $(1.2)_{2,3}$ can then be written as

$$
\gamma_2 = \tau_2 \, q(\sigma, \tau^2),
$$
$$
\gamma_3 = \tau_3 \, q(\sigma, \tau^2).
$$
(1.5)

We study the Goursat-Riemann problem of (1.1) in which the strains $\varepsilon, \gamma_2, \gamma_3$ are assumed to be known functions of σ, τ_2, τ_3 as given in $(1.4)_1$ and (1.5). In Section 2 the characteristic wave speeds and the right eigenvectors of (1.1) are presented. The simple wave curves and shock wave curves in the stress space are presented in Section 3 and that in the velocity space are examined in Section 4. Up to this point, the material is general isotropic Cauchy elastic solids, i.e., the existence of strain energy function is not assumed. In Section 5 we consider hyperelastic solids. In particular, the simple wave curves for second order hyperelastic materials are presented which are used in Section 6 as an illustration to solve the Goursat-Riemann problem with stress boundary conditions. In Section 7 the Goursat-Riemann problem is solved in which the boundary conditions are prescribed in terms of velocity.

2. Characteristic wave speeds and right eigenvectors. For the Riemann problem and the Goursat-Riemann problem, the solution $\mathbf{U}$ depends on one parameter x/t only [4-9]. If $\mathbf{U}$ is continuous in x/t, we have a simple wave (or rarefaction wave) solution in which x/t = c, the characteristic wave speed. In this case, $(1.1)_1$ is reduced to

$$
\{\mathbf{I} + c(\nabla \mathbf{F})^T\}\mathbf{U}' = \mathbf{0},
$$
(2.1)

where $\mathbf{I}$ is a unit matrix, ∇ is the gradient with respect to the components of $\mathbf{U}$, the superscript T stands for the transpose, and the prime denotes differentiation with c. If we introduce the vectors

$$
\mathbf{s} = (\sigma, \tau_2, \tau_3),
$$
$$
\mathbf{p} = (\varepsilon, \gamma_2, \gamma_3),
$$
$$
\mathbf{u} = (u, v_2, v_3),
$$
(2.2)

equation (2.1) can be written as two equations

$$
\mathbf{s}' + \rho c \mathbf{u}' = \mathbf{0},
$$
$$
\mathbf{u}' + c \mathbf{G} \mathbf{s}' = \mathbf{0},
$$
(2.3)

where the components of the 3×3 matrix $\mathbf{G}$ are

$$(2.4) \qquad\qquad G_{ij} = \partial p_i / \partial s_j.$$

Elimination of $\mathbf{u}'$ in (2.3) yields

$$(2.5) \qquad\qquad (\mathbf{G} - \eta \mathbf{I})\mathbf{s}' = \mathbf{0},$$
$$(2.6) \qquad\qquad \eta = (\rho c^2)^{-1}.$$

Thus η and $\mathbf{s}'$ are, respectively, the eigenvalue and eigenvector of $\mathbf{G}$. Assuming that η_i, $i = 1, 2, 3$, are positive, the wave speed c_i come in three pairs of positive and negative values. We let

$$(2.7) \qquad\qquad c_1^2 \ge c_2^2 \ge c_3^2 > 0.$$

Hence,

$$(2.8) \qquad\qquad 0 < \eta_1 \le \eta_2 \le \eta_3.$$

From equations $(1.4)_1$, (1.5), (2.2) and (2.4), it is readily shown that the second and third columns of the matrix $(\mathbf{G} - \eta \mathbf{I})$ are linearly dependent when $\eta = q$. Hence $\eta = q$ is an eigenvalue of $\mathbf{G}$. The other two eigenvalues can be shown to satisfy the quadratic equation

$$(2.9) \qquad \eta^2 - (\varepsilon_\sigma + \gamma_\tau)\eta + (\varepsilon_\sigma \gamma_\tau - \varepsilon_\tau \gamma_\sigma) = 0,$$

in which the subscript σ and τ denote differentiation with these variables. We therefore have, using (2.8),

$$(2.10) \qquad \begin{aligned} \eta_1 &= \frac{1}{2}\{(\varepsilon_\sigma + \gamma_\tau) - Y\}, \\ \eta_2 &= q(\sigma, \tau^2) = \gamma/\tau, \\ \eta_3 &= \frac{1}{2}\{(\varepsilon_\sigma + \gamma_\tau) + Y\}, \\ Y &= \{(\varepsilon_\sigma - \gamma_\tau)^2 + 4\gamma_\sigma \varepsilon_\tau\}^{1/2}. \end{aligned}$$

The second equality for η_2 follows from $(1.4)_2$. By substituting

$$(2.11) \qquad \begin{aligned} \mathbf{s}' &= (0, \tau_3, -\tau_2), & for \ \eta = \eta_2 \\ \mathbf{s}' &= \{\tau(\eta - \gamma_\tau), \ \tau_2\gamma_\sigma, \ \tau_3\gamma_\sigma\}, & for \ \eta = \eta_1, \eta_3, \end{aligned}$$

in (2.5) and making use of $(2.10)_2$ for $\eta = \eta_2$ and (2.9) for $\eta = \eta_1, \eta_3$, it can be verified that equation (2.5) is satisfied. Equations (2.11) therefore provide the eigenvectors $\mathbf{s}'$.

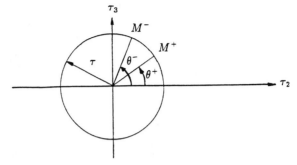

(a) The stress space

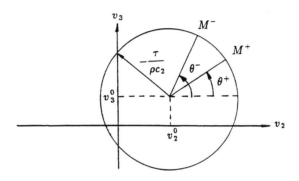

(b) The velocity space $(c_2 < 0)$

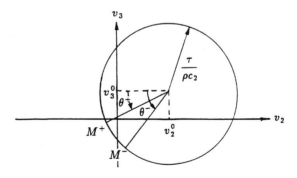

(c) The velocity space $(c_2 > 0)$

Fig.1 The c_2 simple wave(or V_2 shock wave) curve
for which σ, τ and u are constants and $c_2 = V_2$.

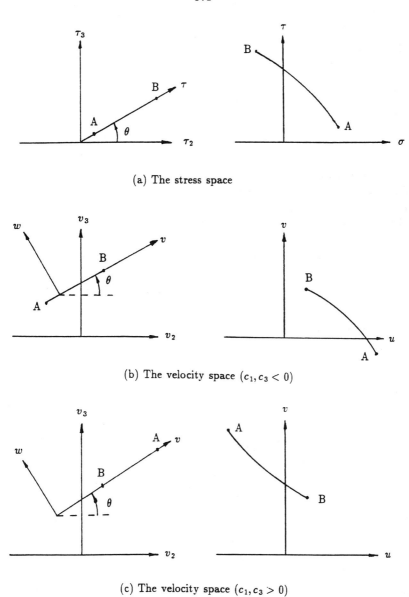

(a) The stress space

(b) The velocity space $(c_1, c_3 < 0)$

(c) The velocity space $(c_1, c_3 > 0)$

Fig.2 The c_1 or c_3 simple wave curves on $\theta = $ constant plane.

3. Simple wave curves and shock wave curves in the stress space. The differential equation for simple wave curves associated with c_2 is given in $(2.11)_1$ which can be written as

$$\frac{d\sigma}{0} = \frac{d\tau_2}{\tau_3} = \frac{d\tau_3}{-\tau_2}.$$

Hence,

(3.1) $$\sigma = constant, \qquad \tau^2 = \tau_2^2 + \tau_3^2 = constant.$$

In the stress space (σ, τ_2, τ_3), (3.1) represents a circle with its center on the σ-axis. The c_2 simple wave curve is therefore "circularly polarized", Fig.1(a). Moreover, from (3.1) and $(2.10)_2$, η_2 and hence c_2 is a constant along the c_2 simple wave curve. Thus the system is linearly degenerate with respect to c_2 [8] and the c_2 simple wave curve is in fact a shock wave curve.

The simple wave curve for c_1 or c_3 is given in $(2.11)_2$ which is rewritten as

(3.2) $$\frac{d\sigma}{\tau(\eta - \gamma_\tau)} = \frac{d\tau_2}{\tau_2 \gamma_\sigma} = \frac{d\tau_3}{\tau_3 \gamma_\sigma}.$$

The last equality yields

(3.3) $$\tau_2/\tau_3 = constant.$$

If we let

(3.4) $$\tau_2 = \tau\cos\theta, \qquad \tau_3 = \tau\sin\theta,$$

we have

(3.5) $$\theta = constant.$$

Equations (3.2) now reduce to

$$\frac{d\sigma}{\eta - \gamma_\tau} = \frac{d\tau}{\gamma_\sigma},$$

or

(3.6) $$\frac{d\tau}{d\sigma} = \frac{\gamma_\sigma}{\eta - \gamma_\tau} = \frac{\eta - \varepsilon_\sigma}{\varepsilon_\tau},$$

the second equality follows from (2.9). Equation (3.6), when integrated, provides simple wave curves for c_1 and c_3 on the (σ, τ) plane. If (σ, τ_2, τ_3) is regarded as a rectangular coordinate system, (σ, τ, θ) is a cylindrical coordinate system. Since simple wave curves for c_1 and c_3 are on a $\theta = constant$ plane, they are "plane polarized", Fig.2(a).

If **U** as a function of x/t is discontinuous at x/t = V, we have a shock wave with shock velocity V. The Rankine-Hugoniot jump conditions for $(1.1)_1$ are

(3.7) $$[\mathbf{U}] + V[\mathbf{F}(\mathbf{U})] = \mathbf{0},$$

where

$$[\mathbf{U}] = \mathbf{U}^- - \mathbf{U}^+$$

denotes the difference in the value of $\mathbf{U}^+$ in front of and $\mathbf{U}^-$ behind the shock wave. Using the notations of (2.2), (3.7) is equivalent to

(3.8)
$$[\mathbf{s}] + \rho V[\mathbf{u}] = \mathbf{0},$$
$$[\mathbf{u}] + V[\mathbf{p}] = \mathbf{0}.$$

Equations (3.8) are the Rankie-Hugonint jump conditions for (2.3). Elimination of [**u**] leads to

(3.9) $$[\mathbf{s}] = \rho V^2[\mathbf{p}].$$

From $(1.4)_1$, (1.5) and (2.2), we may write (3.9) in full as

(3.10)
$$[\sigma] = \rho V^2[h(\sigma, \tau^2)],$$
$$[\tau_2] = \rho V^2[\tau_2\, q(\sigma, \tau^2)],$$
$$[\tau_3] = \rho V^2[\tau_3\, q(\sigma, \tau^2)].$$

If we eliminate ρV^2 between $(3.10)_{2,3}$, it can be shown that [1]

(3.11) $$[q][\tau_2/\tau_3] = 0.$$

There are two possibilities for this equation to hold. We discuss them separately below.

One possibility is [q] = 0. If the shock wave speed for this case is V_2, we see that (3.10) are satisfied if

(3.12)
$$(\rho V_2^2)^{-1} = q^+ = q^-,$$
$$[\sigma] = 0 = [\tau],$$
$$\tau_2^\pm = \tau\cos\theta^\pm, \qquad \tau_3^\pm = \tau\sin\theta^\pm.$$

This is identical to the circularly polarized simple wave curve discussed earlier. Hence $V_2 = c_2$ and the V_2 shock wave curve is identical to the c_2 simple wave curve, Fig.1(a).

The other possibility for (3.11) to hold is

$$[\tau_2/\tau_3] = 0.$$

This is identical to (3.3). It follows from (3.4) that

(3.13) $$\theta^+ = \theta^-,$$

and $(3.10)_{2,3}$ are reduced to the same equation

$$[\tau] = \rho V^2 [\tau\, q(\sigma, \tau^2)] = \rho V^2 [\gamma\, (\sigma, \tau^2)].$$

This and $(3.10)_1$ can be written as

(3.14) $$\rho V^2 = \frac{[\tau]}{[\gamma]} = \frac{[\sigma]}{[\varepsilon]}, \qquad for\ V = V_1\ or\ V_3.$$

For a fixed (σ^+, τ^+), the second equality provides a shock wave curve for (σ^-, τ^-) on the (σ, τ) plane which is a θ = constant plane. The shock wave curves are therefore plane polarized as shown by the double solid lines in Fig.3(a). Since there are two shock wave curves emanating from the point (σ^+, τ^+) (only one is shown in Fig.3), the associated shock wave speeds are denoted by V_1 and V_3 as indicated in (3.14).

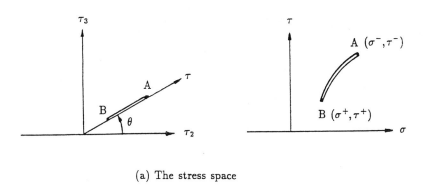

(a) The stress space

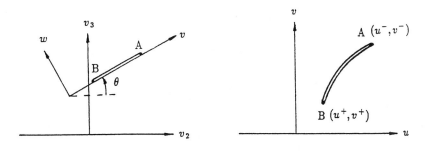

(b) The velocity space ($V_1, V_3 < 0$)

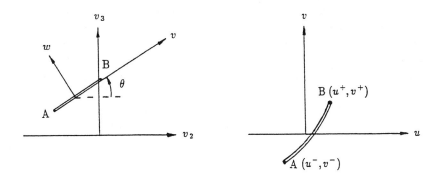

(c) The velocity space ($V_1, V_3 > 0$)

Fig.3 The V_1 (or V_3) shock wave curve on $\theta = $ constant plane.

4. Simple wave curves and shock wave curves in the velocity space. The differential equation (2.5) contains the stress **s** only which enables us to determine simple wave curves in the three-dimensional stress space without considering the full six-dimensional stress-velocity space. This is not possible for simple wave curves in the velocity space. We write $(2.3)_1$ in full as

(4.1)
$$du = -\frac{1}{\rho c}d\sigma,$$
$$dv_2 = -\frac{1}{\rho c}d\tau_2,$$
$$dv_3 = -\frac{1}{\rho c}d\tau_3.$$

For the velocity space we have to distinguish positive wave speed c_i from negative wave speed c_i.

When $c = \pm c_2$, equations (3.1) apply and hence σ, τ and c_2 are all constant. Using (3.4), (4.1) lead to

(4.2)
$$u = constant,$$
$$v_2 - v_2^0 = -\frac{\tau}{\rho c_2}cos\theta,$$
$$v_3 - v_3^0 = -\frac{\tau}{\rho c_2}sin\theta,$$

where v_2^0 and v_3^0 are the integration constants. The determination of the integration constants will be illustrated in Section 7. We see that the c_2 simple wave curves in the velocity space are also circularily polarized, Fig.1(b). The radius of the circle is $\tau/\rho|c_2|$. However, unlike in the stress space, the center of the circle is not necessarily on the u-axis. Moreover, the point on the simple wave curve assumes a different position depending on whether c_2 is a negative (Fig.1(b)) or a positive wave speed (Fig.1(c)).

When $c = \pm c_1$ or $\pm c_3$, substitution of (3.4) into $(4.1)_{2,3}$ and noticing that θ is a constant in this case, we obtain

$$dv_2 = -\frac{1}{\rho c}cos\theta\, d\tau,$$
$$dv_3 = -\frac{1}{\rho c}sin\theta\, d\tau,$$

or

$$d(v_2\, cos\theta + v_3\, sin\theta) = -\frac{1}{\rho c}d\tau,$$
$$d(-v_2\, sin\theta + v_3\, cos\theta) = 0.$$

If we let

(4.3)
$$v = v^0 + v_2\, cos\theta + v_3\, sin\theta,$$
$$w = w^0 - v_2\, sin\theta + v_3\, cos\theta,$$

where v^0 and w^0 are the integration constants, (4.1) are equivalent to

(4.4)
$$du = -\frac{1}{\rho c} d\sigma,$$
$$dv = -\frac{1}{\rho c} d\tau,$$
$$w = 0.$$

The determination of the integration constants v^0 and w^0 will also be illustrated in Section 7. From (4.4), the c_1 and c_3 simple wave curves in the velocity space are also plane polarized on the w = 0 plane, (Fig.2(b), 2(c)). Unlike in the stress space, the plane may not contain the u-axis. Equation $(4.4)_{1,2}$ can be combined to give

$$\frac{dv}{du} = \frac{d\tau}{d\sigma},$$

and hence the slope of the simple wave curve in the velocity space and in the stress space are identical. This does not mean that the simple wave curves in the two spaces are identical. From $(4.4)_{1,2}$, the infinitesimal arclength of the simple wave curve in the velocity space is equal to the corresponding arclength in the stress space divided by the factor ρc. For $c < 0$, therefore, the curves in the velocity space can be obtained from that in the stress space by dividing every infinitesimal line segment of the curve by the factor ρc without changing the orientation of the line segment, Fig.2(b). For $c > 0$, the same procedure applies except that the direction of the wave curve is reversed, Fig.2(c).

We next present the shock wave curves in the velocity space. Using (3.4), $(3.8)_1$ written in full are

(4.5)
$$[\sigma] + \rho V[u] = 0,$$
$$[\tau\cos\theta] + \rho V[v_2] = 0,$$
$$[\tau\sin\theta] + \rho V[v_3] = 0.$$

For $V = V_2 = c_2$, σ and τ are constant. Hence

(4.6)
$$[u] = 0,$$
$$[v_2] = -\frac{\tau}{\rho c_2}[\cos\theta],$$
$$[v_3] = -\frac{\tau}{\rho c_2}[\sin\theta].$$

Equations (4.2) which represent the c_2 simple wave curves in the velocity space satisfy (4.6). Therefore, the V_2 shock wave curves and c_2 simple wave curves are also identical in the velocity space.

For $V = V_1$ or V_3, θ is a constant and $(4.5)_{2,3}$ can be written as

$$[\tau]\cos\theta + \rho V[v_2] = 0,$$
$$[\tau]\sin\theta + \rho V[v_3] = 0.$$

By linearly combining the two equations and using (4.3), we obtain

(4.7)
$$[\tau] + \rho V[v] = 0,$$
$$w = 0.$$

Therefore, the shock wave curves in the velocity space for $V = V_1$ or V_3 are also plane polarized on $w = 0$ plane, Fig.3(b,c). From $(4.5)_1$ and $(4.7)_1$, we have

(4.8)
$$\frac{[v]}{[u]} = \frac{[\tau]}{[\sigma]} = \frac{[\gamma]}{[\varepsilon]},$$

the last equality follows from $(3.14)_2$. The first equality implies that the slope of the line connecting (σ^+, τ^+) to (σ^-, τ^-) on the shock wave curve in the stress space is identical to the slope of the line connecting (u^+, v^+) to (u^-, v^-) on the shock wave curve in the velocity space, Fig.3(b,c).

5. Hyperelastic solids. For hyperelastic solids, there exists a complementary strain energy [10] $W(\sigma, \tau^2)$ whose gradients with respect to σ and τ provide the strains ε and γ, i.e.,

(5.1)
$$\varepsilon = W_\sigma, \qquad \gamma = W_\tau.$$

The characteristic wave speeds are, from (2.6) and (2.10),

(5.2)
$$(\rho c_1^2)^{-1} = \frac{1}{2}\{(W_{\sigma\sigma} + W_{\tau\tau}) - Y\},$$
$$(\rho c_2^2)^{-1} = W_\tau/\tau,$$
$$(\rho c_3^2)^{-1} = \frac{1}{2}\{(W_{\sigma\sigma} + W_{\tau\tau}) + Y\},$$
$$Y = \{(W_{\sigma\sigma} - W_{\tau\tau})^2 + 4W_{\sigma\tau}^2\}^{1/2}.$$

The differential equation (3.6) for the c_1 and c_3 simple wave curves in the stress space is

(5.3)
$$\frac{d\tau}{d\sigma} = \frac{2W_{\sigma\tau}}{(W_{\sigma\sigma} - W_{\tau\tau}) \mp Y} = \frac{-(W_{\sigma\sigma} - W_{\tau\tau}) \mp Y}{2W_{\sigma\tau}},$$

where the upper (or lower) sign is for c_1 (or c_3) simple wave curves. The simple wave curves for c_1 and c_3 are now orthogonal to each other[1,2].

The simplest nonlinear hyperelastic solids are the second order materials for which ε and γ are functions of σ and τ of the order up to two. This means that W must be a function of σ and τ of order up to three. Noticing that W is a function of σ and τ^2 and that the constant terms produce no strains while the linear terms would have yielded non-zero strains when the stresses vanish, we write

(5.4)
$$W = \frac{a}{2}\sigma^2 + \frac{d}{2}\tau^2 + \frac{b}{6}\sigma^3 + \frac{e}{2}\sigma\tau^2,$$

where a, d, b and e are constants. d and a are positive and have the property [1,2]

(5.5) $$1 < \delta = d/(d-a) < 4$$

Equations (5.2) become

(5.6)
$$(\rho c_1^2)^{-1} = \frac{1}{2}\{[(d+a) + (b+e)\sigma] - Y\},$$
$$(\rho c_2^2)^{-1} = d + e\sigma,$$
$$(\rho c_3^2)^{-1} = \frac{1}{2}\{[(d+a) + (b+e)\sigma] + Y\},$$
$$Y = \{[(d-a) - (b-e)\sigma]^2 + 4e^2\tau^2\}^{1/2},$$

and (5.3) gives

(5.7)
$$\frac{d\tau}{d\bar{\sigma}} = \frac{1}{-\phi \mp (\phi^2 + 1)^{1/2}} = \phi \mp (\phi^2 + 1)^{1/2},$$
$$\phi = \frac{k\bar{\sigma}}{2\tau}, \qquad k = 1 - \frac{b}{e},$$
$$\bar{\sigma} = \sigma - \sigma_*, \qquad \sigma_* = \frac{d-a}{b-e}.$$

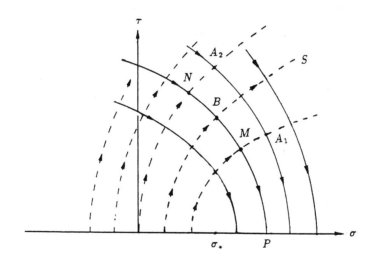

Fig.4 The simple wave curves for c_1 (solid line) and c_3 (dashed lines) for $k \le -1$. $(\sigma_*, 0)$ is the umbilic point at which $c_1 = c_3$.

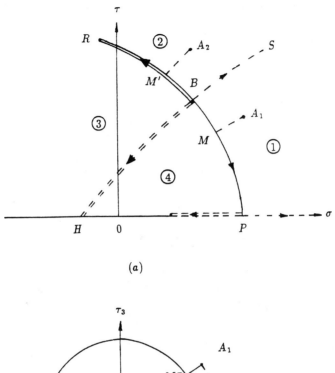

(a)

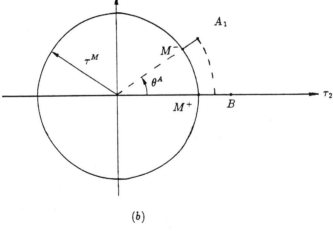

(b)

Fig.5 Wave curves in the stress space.

In the above, we have assumed $e > 0$. If $e < 0$, we change the sign of e and σ, and (5.4) remains the same. Therefore, the solution for $e < 0$ can be obtained from the solution for $e > 0$ by reversing the sign of σ. We see that the simple wave curves given by $(5.7)_1$ depend on one non-dimensional parameter k. At the point $(\sigma, \tau) = (\sigma_*, 0)$, we have $c_1 = c_3$ and hence it is an umbilic point at which the system loses its strict hyperbolicity [11-15]. The simple wave curves are not necessarily orthogonal to each other at the umbilic point.

Equation $(5.7)_1$ can be integrated explicitly [16]. There are four different geometries of the simple wave curves depending on the value of k [1,2]. In Fig.4 we show the simple wave curves for $k \leq -1$. The solid line (or dashed lines) are for the c_1 (or c_3) simple wave curves. The arrows indicate the direction along which the absolute value of the associated wave speed decreases. According to $(5.7)_1$, the simple wave curves are similar. This means that the family of simple wave curves can be obtained by enlarging or diminishing one simple wave curve. In the next two sections, we use Fig.4 to illustrate how one constructs solutions to the Goursat-Riemann problem.

6. The Goursat-Riemann problem for stress boundary conditions. Consider a half space $x \leq 0$. The Goursat-Riemann problem prescribes constant initial conditions s and u at t = 0 and constant boundary conditions s or u at x = 0. In this section we consider the constant stress boundary conditions.

For the initial conditions, we assume without loss in generality,

(6.1)
$$\sigma(x, 0) = \sigma^B, \qquad \tau(x, 0) = \tau^B, \qquad \theta(x, 0) = \theta^B = 0,$$
$$\mathbf{u}(x, 0) = \mathbf{u}^B = 0, \qquad x < 0,$$

where σ^B, τ^B are constants. The superscript B stands for "before". If the initial velocity $\mathbf{u}^B$ are not zero, we simply superimpose $\mathbf{u}^B$ to the final solution $\mathbf{u}(x, t)$. If $\theta^B \neq 0$, we re-orient the coordinate axes x_2 and x_3 so that $\tau_3^B = 0$ and hence $\tau_2^B = \tau^B$. As to the stress boundary conditions, we let

(6.2) $\sigma(0, t) = \sigma^A, \qquad \tau(0, t) = \tau^A, \qquad \theta(0, t) = \theta^A, \qquad t > 0,$

in which $\sigma^A, \tau^A, \theta^A$ are constants. The superscript A stands for "after".

We use Fig.4 to illustrate how one constructs the solution to satisfy the initial conditions (6.1) and the boundary conditions (6.2). Suppose that the initial conditions are represented by the point B in Fig.4 and the boundary condition (6.2) are represented by the point A_1 (see also Fig.5(a)). The problem is to connect from point B to point A_1 an admissible wave curve which consists of a series of simple wave curves and/or shock wave curves. Since we are starting from point B (the initial conditions) to point A_1 (the boundary conditions), an admissible wave curve is the one along which the associated wave speed decreases. In Fig. 4, an admissible simple wave curve follows the arrows on the curves. Also, since $c_1^2 \geq c_3^2$, we can switch from a solid line to a dashed line. Thus the admissible wave curve from B to A_1 is the c_1 simple wave curve BM and the c_3 simple wave curve MA_1. If $\theta^A \neq 0$, we have to consider the (τ_2, τ_3) plane, Fig. 5(b). The admissible wave curve consists of the c_1 simple wave curve BM^+, the V_2 shock wave curve M^+M^-, and the c_3 simple wave curve M^-A_1.

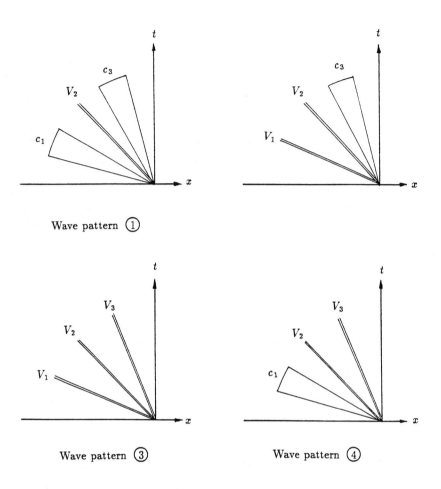

Fig.6 Wave patterns for the wave curves in Fig.5 .

383

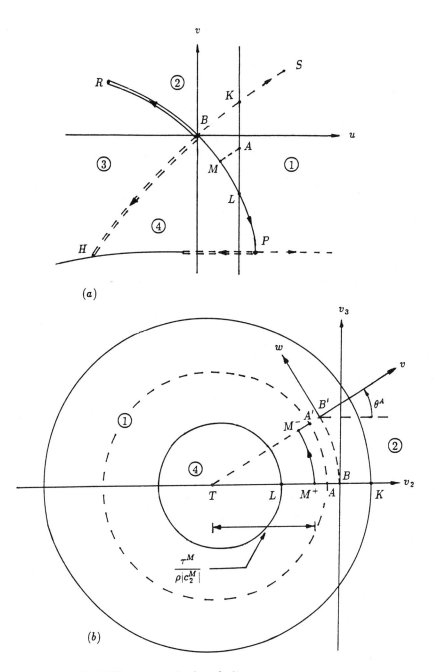

(a)

(b)

Fig.7 Wave curves in the velocity space

The wave curve in Fig. 5(a) is still given by BMA_1, regardless of whether θ^A vanishes or not.

We see that the above solution applies as long as point A_1 is in the region right of the simple wave curves BS and BP in Fig. 4. This region is identified as region 1 in Fig. 5(a) and the corresponding wave pattern in the (x,t) plane is referred to as wave pattern 1 in Fig. 6. In Fig. 6, a double line stands for a shock wave while a circular sector denotes a simple wave.

If the boundary conditions (6.2) are represented by the point A_2 in Fig. 4, the curve BNA_2 is not an admissible wave curve because along the c_1 simple wave curve BN the wave speed c_1 increases. We therefore should have a V_1 shock wave curve given by (3.14)$_2$. With the stress-strain laws of (5.1) and (5.4), (3.14)$_2$ provides the shock wave curve BR in Fig. 5(a). The arrow on BR stands for the direction along which the shock wave satisfies Lax stability condition [4]. When $\theta^A = 0$, the admissible wave curve consists of the V_1 shock wave curve BM' and the c_3 simple wave curve $M'A_2$. If $\theta^A \neq 0$, there will be an additional V_2 shock wave curve. For point A_2 in the region 2 bounded by BS and BR in Fig. 5(a), the associated wave pattern in the (x,t) plane is the wave pattern 2 shown in Fig. 6.

If we use (5.1), (5.4) and (3.14)$_2$ to construct the V_3 shock wave curve BH in Fig. 5(a), we separate the remaining region into region 3 and region 4. Depending on whether the boundary conditions (6.2) are represented by a point in region 3 or region 4, the admissible wave curves can be obtained in the same way. The corresponding wave patterns in the (x,t) plane are shown in Fig. 6.

7. The Goursat-Riemann problem for velocity boundary conditions. We now consider the Goursat-Riemann problem in which the boundary conditions are prescribed in terms of the velocity. As before, we assume without loss in generality the initial conditions given in (6.1). The boundary conditions are

$$(7.1) \qquad u(0,t) = u^A, \qquad v_2(0,t) = v_2^A, \qquad v_3(0,t) = v_3^A, \qquad t > 0,$$

where u^A, v_2^A, v_3^A are constants.

We first study the special case $v_3^A = 0$. With the initial conditions given in (6.1), it is not difficult to see that v_3, τ_3 and θ vanish for $x < 0$ and $t > 0$. From (4.3)$_1$, $v = v_2$ for $x < 0$ and $t > 0$. Using (4.4), the simple wave curves in the velocity space can be determined from the simple curves in the stress space. The simple wave curves in the velocity space associated with the simple wave curves BS and BP in Fig. 5(a) are shown in Fig. 7(a). Likewise, the shock wave curves associated with BR and BH in Fig. 5(a) can be obtained from (4.5)$_1$ and (4.7)$_1$ and are shown in Fig. 7(a). We have again divided the velocity plane (u,v) into four regions. The method of finding the solution is identical to that for the stress boundary conditions with $\theta^A = 0$. Thus, depending on whether the boundary conditions (7.1) with $v_3^A = 0$ and $v_2^A = v^A$ are represented by a point in region 1, 2, 3 or 4, we have the wave pattern 1, 2, 3 or 4 shown in Fig. 6 in which the V_2 shock wave is absent.

When $v_3^A \neq 0$, the solution cannot be obtained by a simple superposition of a V_2 shock wave. As shown in [17], the V_2 shock wave does not commute with c_1

and c_3 simple waves when the wave curves in the velocity space are considered. Nevertheless, we will show that one can make use of Fig. 7(a) to construct the solution when $v_3^A \neq 0$.

First of all, we show that the wave curve BMA in Fig. 7(a) corresponds to a family of solutions with $v_3^A \neq 0$. When $v_3^A = 0$, the c_1 simple wave curve BM and the c_3 simple wave curve MA are, in the (v_2, v_3) plane, the segments BM^+ and M^+A in Fig. 7(b) which are on the v_2-axis. When $v_3^A \neq 0$, we may introduce a V_2 shock wave curve M^+M^- which is a circle with its center at T and radius $\tau^M/\rho|c_2^M|$. This is the circularly polarized shock wave given in $(4.2)_{2,3}$ in which v_2^0 and v_3^0 are determined by the location of T. If we draw a circle AA' concentric with M^+M^-, any point A' on the circle can be the location of the new boundary conditions (v_2^A, v_3^A). The wave curves for this case will be the c_1 simple wave curve BM^+, the V_2 shock wave curve M^+M^- and the c_3 simple wave curve M^-A'. The angle the line TA' makes with the v_2-axis is θ^A. The new coordinates v,w defined in (4.3) are obtained by rotating the v_2, v_3 axes about T an angle θ^A. The location B' of the origin of (v, w) coordinates determines the constants (v^0, w^0) in (4.3). In the (u, v) plane, Fig. 7(a), the wave curve is still BMA. The corresponding wave pattern is the wave pattern 1 in Fig. 6. Thus the wave curve BMA in Fig. 7(a) corresponds to a family of solutions for which the boundary conditions (v_2^A, v_3^A) are on the circle AA' shown by the dotted line in Fig. 7(b).

With the V_2 shock wave considered separately, one can determine the admissible wave curve for the velocity boundary conditions when $v_3^A \neq 0$ by an iteration scheme. However, one should be able to determine whether the wave pattern belongs to wave pattern 1, 2, 3 or 4 before employing the iteration scheme. This is presented next.

When (u^A, v_2^A, v_3^A) are given, we draw the vertical line KL in the (u, v) plane, Fig. 7(a), whose abscissa is u^A. This line intersects the c_1 simple wave curve BP at L and the c_3 simple wave curve BS at K. From (4.3),

$$v^A = v^0 + v_2^A \sin\theta^A + v_3^A \cos\theta^A.$$

Since v^0 and θ^A are unknowns, v^A can be anywhere on the line KL. If A is located above K, between KL or below L, we have wave pattern 2, 1 or 4, respectively. The wave curve BK in Fig. 7(a) corresponds to the wave curve BK in Fig. 7(b) when $v_3^A = 0$. Following the procedure explained earlier, we can obtain a circle through K shown by the solid line in Fig. 7(b) such that the wave curve BK in Fig. 7(a) corresponds to a family of solutions with (v_2^A, v_3^A) on this circle. Likewise, one can obtain a circle through L shown by another solid line in Fig. 7(b) such that the wave curve BL in Fig. 7(a) corresponds to a family of solutions with (v_2^A, v_3^A) on this circle. We then have the result that if (v_2^A, v_3^A) is located within the two circles, the solution belongs to wave pattern 1. If (v_2^A, v_3^A) is located outside (or inside) the circle passing through K (or L), we have wave pattern 2 (or wave pattern 4).

It should be pointed out that the two circles passing through K and L in Fig. 7(b) are for the fixed $u^A > 0$ shown in Fig. 7(a). For a different value of u^A, the circles would be different. For $u^A < 0$, a similar procedure can be employed to determine whether the solution for given (v_2^A, v_3^A) belongs to wave pattern 2, 3 or 4.

REFERENCES

[1] YONGCHI LI AND T. C. T. TING, *Plane waves in simple elastic solids and discontinuous dependence of solution on boundary conditions*, Int. J. Solids Structures, 19 (1983), pp. 989–1008.

[2] ZHIJING TANG AND T. C. T. TING, *Wave curves for the Riemann problem of plane waves in isotropic elastic solids*, Int. J. Eng. Sci., 25 (1987), pp. 1343–1381.

[3] T. C. T. TING, *The Riemann problem with umbilic lines for wave propagation in isotropic elastic solids*, in Notes in Numerical Fluid Mechanics, Nonlinear Hyperbolic Equations - Theory, Numerical Methods and Applications, ed. by Josef Ballmann and Rolf Jeltsch, 24, Vieweg, 1988, pp. 617–629.

[4] P. D. LAX, *Hyperbolic systems of conservation laws. II*, Comm. Pure Appl. Math., 10 (1957), pp. 537–566.

[5] A. JEFFREY, *Quasilinear Hyperbolic Systems and Waves*, Pitman, 1976.

[6] J. A. SMOLLER, *On the solution of the Riemann problem with general step data for an extended class of hyperbolic systems*, Mich. Math. J., 16 (1969), pp. 201–210.

[7] T. -P. LIU, *The Riemann problem for general systems of conservation laws*, J. Diff. Eqs., 18 (1975), pp. 218–234.

[8] C. M. DAFERMOS, *Hyperbolic systems of conservation laws*, Brown University Report, LCDS 83-5,(1983).

[9] D. G. SCHAEFFER AND M. SHEARER, *Riemann problem for nonstrictly hyperbolic 2 × 2 systems of conservation laws*, Trans Amer. Math. Soc., 304 (1987), pp. 267–306.

[10] C. TRUESDELL AND W. NOLL, *The Nonlinear Field Theories of Mechanics*, Handbuch der Physik, III/3, Springer, Berlin, 1965.

[11] D. G. SCHAEFFER AND M. SHEARER, *The classification of 2 × 2 systems of nonstrictly hyperbolic conservation laws, with application to oil recovery, Appendix with D. Marchesin and P. J. Paes-Leme*, Comm. Pure Appl. Math., 40 (1987), pp. 141–178.

[12] B. L. KEYFITZ AND H. C. KRANZER, *The Riemann problem for a class of conservation laws exhibiting a parabolic degeneracy*, J. Diff. Eqs., 47 (1983), pp. 35–65.

[13] E. ISAACSON AND J. B. TEMPLE, *Examples and classification of nonstrictly hyperbolic systems of conservation laws*, Abstracts of Papers Presented to AMS, 6 (1985), pp. 60.

[14] M. SHEARER, D. G. SCHAEFFER, D. MARCHESIN AND P. J. PAES-LEME, *Solution of the Riemann problem for a prototype 2 × 2 system of nonstrictly hyperbolic conservation laws*, Arch. Rat. Mech. Anal., 97 (1987), pp. 299-320.

[15] GUANGSHAN ZHU AND T. C. T. TING, *Classification of 2 × 2 non-strictly hyperbolic systems for plane waves in isotropic elastic solids*, Int. J. Eng. Science, 27 (1989), pp. 1621–1638.

[16] T. C. T. TING, *On wave propagation problems in which $c_f = c_s = c_2$ occurs*, Q. Appl. Math., 31 (1973), pp. 275–286.

[17] XABIER GARAIZAR, *Solution of a Riemann problem for elasticity*, Courant Institute of Mathematical Sciences Report (1989).